Norm

CW00921966

Structures Under Shock and Impact XI

WIT*PRESS*

WIT Press publishes leading books in Science and Technology.
Visit our website for the current list of titles.
www.witpress.com

WIT*eLibrary*

Home of the Transactions of the Wessex Institute.
Papers presented at SUSI XI are archived in the WIT elibrary in volume 113 of
WIT Transactions on The Built Environment (ISSN 1743-3509).
The WIT electronic-library provides the international scientific community with
immediate and permanent access to individual papers presented at WIT conferences.
http://library.witpress.com.

ELEVENTH INTERNATIONAL CONFERENCE ON STRUCTURES UNDER SHOCK AND IMPACT

SUSI XI

CONFERENCE CHAIRMEN

N. Jones
University of Liverpool, UK

C.A. Brebbia
Wessex Institute of Technology, UK

U. Mander
University of Tartu, Estonia

INTERNATIONAL SCIENTIFIC ADVISORY COMMITTEE

M. Avalle
E. Baker
X. Chen
P. Heffernan
M. Hoo Fatt
D. Makovicka
K.A. Marchand
M.R. Said
G.K. Schleyer
P. Verleysen
D. Weggel
E. Williamson

Organised by
Wessex Institute of Technology, UK

Sponsored by
WIT Transactions on the Built Environment

WIT Transactions

Transactions Editor

Carlos Brebbia
Wessex Institute of Technology
Ashurst Lodge, Ashurst
Southampton SO40 7AA, UK
Email: carlos@wessex.ac.uk

Editorial Board

B Abersek University of Maribor, Slovenia
Y N Abousleiman University of Oklahoma, USA
P L Aguilar University of Extremadura, Spain
K S Al Jabri Sultan Qaboos University, Oman
E Alarcon Universidad Politecnica de Madrid, Spain
A Aldama IMTA, Mexico
C Alessandri Universita di Ferrara, Italy
D Almorza Gomar University of Cadiz, Spain
B Alzahabi Kettering University, USA
J A C Ambrosio IDMEC, Portugal
A M Amer Cairo University, Egypt
S A Anagnostopoulos University of Patras, Greece
M Andretta Montecatini, Italy
E Angelino A.R.P.A. Lombardia, Italy
H Antes Technische Universitat Braunschweig, Germany
M A Atherton South Bank University, UK
A G Atkins University of Reading, UK
D Aubry Ecole Centrale de Paris, France
H Azegami Toyohashi University of Technology, Japan
A F M Azevedo University of Porto, Portugal
J Baish Bucknell University, USA
J M Baldasano Universitat Politecnica de Catalunya, Spain
J G Bartzis Institute of Nuclear Technology, Greece
A Bejan Duke University, USA
M P Bekakos Democritus University of Thrace, Greece
G Belingardi Politecnico di Torino, Italy
R Belmans Katholieke Universiteit Leuven, Belgium
C D Bertram The University of New South Wales, Australia
D E Beskos University of Patras, Greece
S K Bhattacharyya Indian Institute of Technology, India
E Blums Latvian Academy of Sciences, Latvia
J Boarder Cartref Consulting Systems, UK
B Bobee Institut National de la Recherche Scientifique, Canada
H Boileau ESIGEC, France
J J Bommer Imperial College London, UK
M Bonnet Ecole Polytechnique, France
C A Borrego University of Aveiro, Portugal
A R Bretones University of Granada, Spain
J A Bryant University of Exeter, UK
F-G Buchholz Universitat Gesanthochschule Paderborn, Germany
M B Bush The University of Western Australia, Australia
F Butera Politecnico di Milano, Italy
J Byrne University of Portsmouth, UK
W Cantwell Liverpool University, UK
D J Cartwright Bucknell University, USA
P G Carydis National Technical University of Athens, Greece
J J Casares Long Universidad de Santiago de Compostela, Spain
M A Celia Princeton University, USA
A Chakrabarti Indian Institute of Science, India
A H-D Cheng University of Mississippi, USA

J Chilton University of Lincoln, UK
C-L Chiu University of Pittsburgh, USA
H Choi Kangnung National University, Korea
A Cieslak Technical University of Lodz, Poland
S Clement Transport System Centre, Australia
M W Collins Brunel University, UK
J J Connor Massachusetts Institute of Technology, USA
M C Constantinou State University of New York at Buffalo, USA
D E Cormack University of Toronto, Canada
M Costantino Royal Bank of Scotland, UK
D F Cutler Royal Botanic Gardens, UK
W Czyczula Krakow University of Technology, Poland
M da Conceicao Cunha University of Coimbra, Portugal
L Dávid Károly Róbert College, Hungary
A Davies University of Hertfordshire, UK
M Davis Temple University, USA
A B de Almeida Instituto Superior Tecnico, Portugal
E R de Arantes e Oliveira Instituto Superior Tecnico, Portugal
L De Biase University of Milan, Italy
R de Borst Delft University of Technology, Netherlands
G De Mey University of Ghent, Belgium
A De Montis Universita di Cagliari, Italy
A De Naeyer Universiteit Ghent, Belgium
W P De Wilde Vrije Universiteit Brussel, Belgium
L Debnath University of Texas-Pan American, USA
N J Dedios Mimbela Universidad de Cordoba, Spain
G Degrande Katholieke Universiteit Leuven, Belgium
S del Giudice University of Udine, Italy
G Deplano Universita di Cagliari, Italy
I Doltsinis University of Stuttgart, Germany
M Domaszewski Universite de Technologie de Belfort-Montbeliard, France
J Dominguez University of Seville, Spain
K Dorow Pacific Northwest National Laboratory, USA
W Dover University College London, UK
C Dowlen South Bank University, UK
J P du Plessis University of Stellenbosch, South Africa
R Duffell University of Hertfordshire, UK
A Ebel University of Cologne, Germany
E E Edoutos Democritus University of Thrace, Greece
G K Egan Monash University, Australia
K M Elawadly Alexandria University, Egypt
K-H Elmer Universitat Hannover, Germany
D Elms University of Canterbury, New Zealand
M E M El-Sayed Kettering University, USA
D M Elsom Oxford Brookes University, UK
A El-Zafrany Cranfield University, UK
F Erdogan Lehigh University, USA
F P Escrig University of Seville, Spain
D J Evans Nottingham Trent University, UK
J W Everett Rowan University, USA
M Faghri University of Rhode Island, USA
R A Falconer Cardiff University, UK
M N Fardis University of Patras, Greece
P Fedelinski Silesian Technical University, Poland
H J S Fernando Arizona State University, USA
S Finger Carnegie Mellon University, USA
J I Frankel University of Tennessee, USA
D M Fraser University of Cape Town, South Africa
M J Fritzler University of Calgary, Canada
U Gabbert Otto-von-Guericke Universitat Magdeburg, Germany
G Gambolati Universita di Padova, Italy
C J Gantes National Technical University of Athens, Greece
L Gaul Universitat Stuttgart, Germany
A Genco University of Palermo, Italy
N Georgantzis Universitat Jaume I, Spain
P Giudici Universita di Pavia, Italy
F Gomez Universidad Politecnica de Valencia, Spain
R Gomez Martin University of Granada, Spain
D Goulias University of Maryland, USA
K G Goulias Pennsylvania State University, USA
F Grandori Politecnico di Milano, Italy
W E Grant Texas A & M University, USA

S Grilli University of Rhode Island, USA

R H J Grimshaw Loughborough University, UK

D Gross Technische Hochschule Darmstadt, Germany

R Grundmann Technische Universitat Dresden, Germany

A Gualtierotti IDHEAP, Switzerland

R C Gupta National University of Singapore, Singapore

J M Hale University of Newcastle, UK

K Hameyer Katholieke Universiteit Leuven, Belgium

C Hanke Danish Technical University, Denmark

K Hayami National Institute of Informatics, Japan

Y Hayashi Nagoya University, Japan

L Haydock Newage International Limited, UK

A H Hendrickx Free University of Brussels, Belgium

C Herman John Hopkins University, USA

S Heslop University of Bristol, UK

I Hideaki Nagoya University, Japan

D A Hills University of Oxford, UK

W F Huebner Southwest Research Institute, USA

J A C Humphrey Bucknell University, USA

M Y Hussaini Florida State University, USA

W Hutchinson Edith Cowan University, Australia

T H Hyde University of Nottingham, UK

M Iguchi Science University of Tokyo, Japan

D B Ingham University of Leeds, UK

L Int Panis VITO Expertisecentrum IMS, Belgium

N Ishikawa National Defence Academy, Japan

J Jaafar UiTm, Malaysia

W Jager Technical University of Dresden, Germany

Y Jaluria Rutgers University, USA

C M Jefferson University of the West of England, UK

P R Johnston Griffith University, Australia

D R H Jones University of Cambridge, UK

N Jones University of Liverpool, UK

D Kaliampakos National Technical University of Athens, Greece

N Kamiya Nagoya University, Japan

D L Karabalis University of Patras, Greece

M Karlsson Linkoping University, Sweden

T Katayama Doshisha University, Japan

K L Katsifarakis Aristotle University of Thessaloniki, Greece

J T Katsikadelis National Technical University of Athens, Greece

E Kausel Massachusetts Institute of Technology, USA

H Kawashima The University of Tokyo, Japan

B A Kazimee Washington State University, USA

S Kim University of Wisconsin-Madison, USA

D Kirkland Nicholas Grimshaw & Partners Ltd, UK

E Kita Nagoya University, Japan

A S Kobayashi University of Washington, USA

T Kobayashi University of Tokyo, Japan

D Koga Saga University, Japan

S Kotake University of Tokyo, Japan

A N Kounadis National Technical University of Athens, Greece

W B Kratzig Ruhr Universitat Bochum, Germany

T Krauthammer Penn State University, USA

C-H Lai University of Greenwich, UK

M Langseth Norwegian University of Science and Technology, Norway

B S Larsen Technical University of Denmark, Denmark

F Lattarulo Politecnico di Bari, Italy

A Lebedev Moscow State University, Russia

L J Leon University of Montreal, Canada

D Lewis Mississippi State University, USA

S lghobashi University of California Irvine, USA

K-C Lin University of New Brunswick, Canada

A A Liolios Democritus University of Thrace, Greece

S Lomov Katholieke Universiteit Leuven, Belgium

J W S Longhurst University of the West of England, UK

G Loo The University of Auckland, New Zealand

J Lourenco Universidade do Minho, Portugal

J E Luco University of California at San Diego, USA

H Lui State Seismological Bureau Harbin, China

C J Lumsden University of Toronto, Canada

L Lundqvist Division of Transport and Location Analysis, Sweden

T Lyons Murdoch University, Australia

Y-W Mai University of Sydney, Australia

M Majowiecki University of Bologna, Italy

D Malerba Università degli Studi di Bari, Italy

G Manara University of Pisa, Italy

B N Mandal Indian Statistical Institute, India

Ü Mander University of Tartu, Estonia

H A Mang Technische Universitat Wien, Austria

G D Manolis Aristotle University of Thessaloniki, Greece

W J Mansur COPPE/UFRJ, Brazil

N Marchettini University of Siena, Italy

J D M Marsh Griffith University, Australia

J F Martin-Duque Universidad Complutense, Spain

T Matsui Nagoya University, Japan

G Mattrisch DaimlerChrysler AG, Germany

F M Mazzolani University of Naples "Federico II", Italy

K McManis University of New Orleans, USA

A C Mendes Universidade de Beira Interior, Portugal

R A Meric Research Institute for Basic Sciences, Turkey

J Mikielewicz Polish Academy of Sciences, Poland

N Milic-Frayling Microsoft Research Ltd, UK

R A W Mines University of Liverpool, UK

C A Mitchell University of Sydney, Australia

K Miura Kajima Corporation, Japan

A Miyamoto Yamaguchi University, Japan

T Miyoshi Kobe University, Japan

G Molinari University of Genoa, Italy

T B Moodie University of Alberta, Canada

D B Murray Trinity College Dublin, Ireland

G Nakhaeizadeh DaimlerChrysler AG, Germany

M B Neace Mercer University, USA

D Necsulescu University of Ottawa, Canada

F Neumann University of Vienna, Austria

S-I Nishida Saga University, Japan

H Nisitani Kyushu Sangyo University, Japan

B Notaros University of Massachusetts, USA

P O'Donoghue University College Dublin, Ireland

R O O'Neill Oak Ridge National Laboratory, USA

M Ohkusu Kyushu University, Japan

G Oliveto Universitá di Catania, Italy

R Olsen Camp Dresser & McKee Inc., USA

E Oñate Universitat Politecnica de Catalunya, Spain

K Onishi Ibaraki University, Japan

P H Oosthuizen Queens University, Canada

E L Ortiz Imperial College London, UK

E Outa Waseda University, Japan

A S Papageorgiou Rensselaer Polytechnic Institute, USA

J Park Seoul National University, Korea

G Passerini Universita delle Marche, Italy

B C Patten University of Georgia, USA

G Pelosi University of Florence, Italy

G G Penelis Aristotle University of Thessaloniki, Greece

W Perrie Bedford Institute of Oceanography, Canada

R Pietrabissa Politecnico di Milano, Italy

H Pina Instituto Superior Tecnico, Portugal

M F Platzer Naval Postgraduate School, USA

D Poljak University of Split, Croatia

V Popov Wessex Institute of Technology, UK

H Power University of Nottingham, UK

D Prandle Proudman Oceanographic Laboratory, UK

M Predeleanu University Paris VI, France

M R I Purvis University of Portsmouth, UK

I S Putra Institute of Technology Bandung, Indonesia

Y A Pykh Russian Academy of Sciences, Russia

F Rachidi EMC Group, Switzerland

M Rahman Dalhousie University, Canada

K R Rajagopal Texas A & M University, USA

T Rang Tallinn Technical University, Estonia

J Rao Case Western Reserve University, USA

A M Reinhorn State University of New York at Buffalo, USA
A D Rey McGill University, Canada
D N Riahi University of Illinois at Urbana-Champaign, USA
B Ribas Spanish National Centre for Environmental Health, Spain
K Richter Graz University of Technology, Austria
S Rinaldi Politecnico di Milano, Italy
F Robuste Universitat Politecnica de Catalunya, Spain
J Roddick Flinders University, Australia
A C Rodrigues Universidade Nova de Lisboa, Portugal
F Rodrigues Poly Institute of Porto, Portugal
C W Roeder University of Washington, USA
J M Roesset Texas A & M University, USA
W Roetzel Universitaet der Bundeswehr Hamburg, Germany
V Roje University of Split, Croatia
R Rosset Laboratoire d'Aerologie, France
J L Rubio Centro de Investigaciones sobre Desertificacion, Spain
T J Rudolphi Iowa State University, USA
S Russenchuck Magnet Group, Switzerland
H Ryssel Fraunhofer Institut Integrierte Schaltungen, Germany
S G Saad American University in Cairo, Egypt
M Saiidi University of Nevada-Reno, USA
R San Jose Technical University of Madrid, Spain
F J Sanchez-Sesma Instituto Mexicano del Petroleo, Mexico
B Sarler Nova Gorica Polytechnic, Slovenia
S A Savidis Technische Universitat Berlin, Germany
A Savini Universita de Pavia, Italy
G Schmid Ruhr-Universitat Bochum, Germany
R Schmidt RWTH Aachen, Germany
B Scholtes Universitaet of Kassel, Germany
W Schreiber University of Alabama, USA
A P S Selvadurai McGill University, Canada
J J Sendra University of Seville, Spain
J J Sharp Memorial University of Newfoundland, Canada
Q Shen Massachusetts Institute of Technology, USA
X Shixiong Fudan University, China
G C Sih Lehigh University, USA
L C Simoes University of Coimbra, Portugal
A C Singhal Arizona State University, USA
P Skerget University of Maribor, Slovenia
J Sladek Slovak Academy of Sciences, Slovakia
V Sladek Slovak Academy of Sciences, Slovakia
A C M Sousa University of New Brunswick, Canada
H Sozer Illinois Institute of Technology, USA
D B Spalding CHAM, UK
P D Spanos Rice University, USA
T Speck Albert-Ludwigs-Universitaet Freiburg, Germany
C C Spyrakos National Technical University of Athens, Greece
I V Stangeeva St Petersburg University, Russia
J Stasiek Technical University of Gdansk, Poland
G E Swaters University of Alberta, Canada
S Syngellakis University of Southampton, UK
J Szmyd University of Mining and Metallurgy, Poland
S T Tadano Hokkaido University, Japan
H Takemiya Okayama University, Japan
I Takewaki Kyoto University, Japan
C-L Tan Carleton University, Canada
M Tanaka Shinshu University, Japan
E Taniguchi Kyoto University, Japan
S Tanimura Aichi University of Technology, Japan
J L Tassoulas University of Texas at Austin, USA
M A P Taylor University of South Australia, Australia
A Terranova Politecnico di Milano, Italy
E Tiezzi University of Siena, Italy
A G Tijhuis Technische Universiteit Eindhoven, Netherlands
T Tirabassi Institute FISBAT-CNR, Italy
S Tkachenko Otto-von-Guericke-University, Germany
N Tosaka Nihon University, Japan
T Tran-Cong University of Southern Queensland, Australia
R Tremblay Ecole Polytechnique, Canada

I Tsukrov University of New Hampshire, USA

R Turra CINECA Interuniversity Computing Centre, Italy

S G Tushinski Moscow State University, Russia

J-L Uso Universitat Jaume I, Spain

E Van den Bulck Katholieke Universiteit Leuven, Belgium

D Van den Poel Ghent University, Belgium

R van der Heijden Radboud University, Netherlands

R van Duin Delft University of Technology, Netherlands

P Vas University of Aberdeen, UK

W S Venturini University of Sao Paulo, Brazil

R Verhoeven Ghent University, Belgium

A Viguri Universitat Jaume I, Spain

Y Villacampa Esteve Universidad de Alicante, Spain

F F V Vincent University of Bath, UK

S Walker Imperial College, UK

G Walters University of Exeter, UK

B Weiss University of Vienna, Austria

H Westphal University of Magdeburg, Germany

J R Whiteman Brunel University, UK

Z-Y Yan Peking University, China

S Yanniotis Agricultural University of Athens, Greece

A Yeh University of Hong Kong, China

J Yoon Old Dominion University, USA

K Yoshizato Hiroshima University, Japan

T X Yu Hong Kong University of Science & Technology, Hong Kong

M Zador Technical University of Budapest, Hungary

K Zakrzewski Politechnika Lodzka, Poland

M Zamir University of Western Ontario, Canada

R Zarnic University of Ljubljana, Slovenia

G Zharkova Institute of Theoretical and Applied Mechanics, Russia

N Zhong Maebashi Institute of Technology, Japan

H G Zimmermann Siemens AG, Germany

Structures Under Shock and Impact XI

Editors

N. Jones
University of Liverpool, UK

C. A. Brebbia
Wessex Institute of Technology, UK

U. Mander
University of Tartu, Estonia

N. Jones
University of Liverpool, UK

C. A. Brebbia
Wessex Institute of Technology, UK

U. Mander
University of Tartu, Estonia

Published by

WIT Press
Ashurst Lodge, Ashurst, Southampton, SO40 7AA, UK
Tel: 44 (0) 238 029 3223; Fax: 44 (0) 238 029 2853
E-Mail: witpress@witpress.com
http://www.witpress.com

For USA, Canada and Mexico

Computational Mechanics Inc
25 Bridge Street, Billerica, MA 01821, USA
Tel: 978 667 5841; Fax: 978 667 7582
E-Mail: infousa@witpress.com
http://www.witpress.com

British Library Cataloguing-in-Publication Data

A Catalogue record for this book is available
from the British Library

ISBN: 978-1-84564-466-6
ISSN: 1746-4498 (print)
ISSN: 1743-3509 (online)

The texts of the papers in this volume were set individually by the authors or under their supervision. Only minor corrections to the text may have been carried out by the publisher.

No responsibility is assumed by the Publisher, the Editors and Authors for any injury and/or damage to persons or property as a matter of products liability, negligence or otherwise, or from any use or operation of any methods, products, instructions or ideas contained in the material herein. The Publisher does not necessarily endorse the ideas held, or views expressed by the Editors or Authors of the material contained in its publications.

© WIT Press 2010

Printed in Great Britain by MPG Books Group, Bodmin and King's Lynn.

All rights reserved. No part of this publication may be reproduced, stored in a retrieval system, or transmitted in any form or by any means, electronic, mechanical, photocopying, recording, or otherwise, without the prior written permission of the Publisher.

Preface

This book contains the papers presented at the eleventh International Conference on Structures under Shock and Impact held in Tallinn, Estonia, July 2010. The earlier meetings were held in Cambridge, Massachusetts, USA in 1989, Portsmouth, UK in 1992, Madrid, Spain in 1994, Udine, Italy in 1996, Thessaloniki, Greece in 1998, Cambridge, UK in 2000, Montreal, Canada in 2002, Crete, Greece in 2004, the Wessex Institute of Technology, UK in 2006 and in The Algarve, Portugal, May 2008. It was the objective of these meetings to bring together engineers and scientists from a wide-range of academic disciplines and industrial backgrounds who have an interest in the shock and impact response of structures and materials. In this way, the major developments in different areas can be brought to the attention of the entire community, thereby ensuring that industries benefit from the latest advances.

The shock and impact behaviour of structures is a challenging area, not only because of the obvious time-dependent aspects, but also because of the difficulties in specifying the external dynamic loading and connection characteristics for structural designs and hazard assessments, and in obtaining the dynamic properties of materials. Thus, it is important to recognise and utilise fully the contributions and understanding emerging from theoretical, numerical and experimental studies on structures, as well as investigations into the material properties under dynamic loading conditions.

The papers in this volume reflect the broad range of practical interest in the shock and impact response of structures. The topics include various aspects of protection, such as the protection of buildings from explosive loadings and the protection of the public through structural energy absorbing systems. Other articles examine the response of concrete, composite and metal structures under a variety of impact and blast loadings. Individual papers focus on seismic loadings, missile penetration, energy absorbing systems, structural crashworthiness and other important industrial problems, some of which are studied using numerical schemes. Several papers examine the dynamic properties of materials.

It is clear from the collection of papers in this volume that the shock and impact behaviour of structures is an active field and that the range of topics is ever expanding when viewed from the perspective of the eleven conferences. This situation bodes well for the future growth of the subject, particularly since 9/11 which brought about increased concern for protecting public buildings, nuclear plant and other important sites. In addition, there is a continuing interest in the influence of shock and impact loadings on the quest for optimised systems and the resistance of new materials. It is hoped that the contents of this book will encourage and motivate many research workers and designers to apply the methods presented to new practical problems and to contribute, in due course, to our better understanding of the shock and impact behaviour of structures to the benefit of the worldwide community.

The Editors are grateful to the members of the International Scientific Advisory Committee and other colleagues who have helped to review the papers included in this volume, as well to all the authors for their contributions.

The Editors
Tallinn, 2010

Contents

Section 6: Structural crashworthiness

Section 7: Seismic behaviour

Section 8: Behaviour of structures

Section 1
Energy absorbing issues

Composite crash absorber for aircraft fuselage applications

S. Heimbs[1], F. Strobl[1], P. Middendorf[1] & J. M. Guimard[2]
[1]*EADS Innovation Works, Munich, Germany*
[2]*EADS Innovation Works, Suresnes Cedex, France*

Abstract

A composite crash absorber element for potential use in z-struts of commercial aircraft fuselage structures was developed, which absorbs energy under crash loads by cutting the composite strut into stripes and crushing the material under bending. The design concept of this absorber element is described and the performance is evaluated experimentally in static, crash and fatigue test series on component and structural level under normal and oblique impact conditions. The physics of the energy absorption by high rate material fragmentation and delamination interactions are explained and numerical modelling methods in explicit finite element codes for the simulation of the crash absorber are assessed.
Keywords: composite crash absorber, z-strut, aircraft crashworthiness, energy absorption, crushing, fragmentation, delamination, finite element simulation.

1 Introduction

Modern commercial aircraft are designed for crashworthiness with the fuselage structure's crash behaviour typically being evaluated in vertical drop tests, as illustrated in Fig. 1 [1-3]. In case of metallic materials, the energy is normally absorbed by plastic deformation, while it is crushing and fracture for composite structures. Besides the deformation of the primary structure itself, additional energy absorbers can be incorporated to improve the crash behaviour, which can be based on different concepts. In the chain of energy absorption, the subfloor area of the lower fuselage is loaded first. A lot of research was conducted with respect to composite sine wave beams in the subfloor structure that are crushed under vertical crash loads [4-9]. Further concepts are based on foam [10] or honeycomb absorbers [11, 12] in the subfloor structure.

WIT Transactions on The Built Environment, Vol 113, © 2010 WIT Press
www.witpress.com, ISSN 1743-3509 (on-line)
doi:10.2495/SU100011

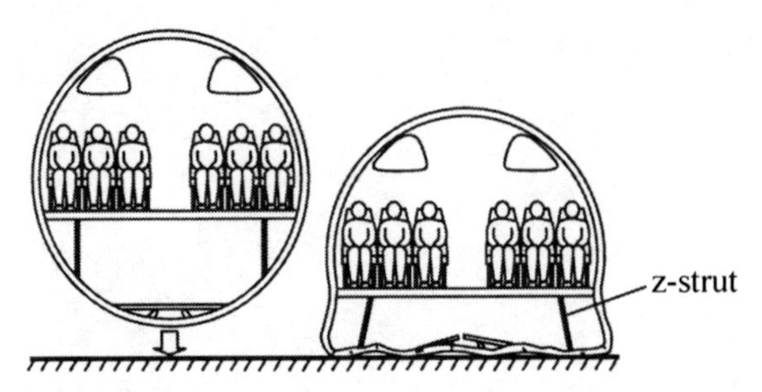

Figure 1: Illustration of aircraft fuselage drop test for crashworthiness evaluation.

A new approach, presented in this paper, includes energy absorbers in the z-struts of the fuselage (Fig. 1) [13]. Z-struts are the connection of passenger floor and lower frames, acting as the support in vertical (z-) direction. In the crash case, they are loaded in axial compression as soon as the lower fuselage part is flattened. In this study, a lightweight composite crash absorber element was developed, which absorbs energy under compression loads and meets at the same time design criteria like stiffness, buckling stability, trigger load or fatigue performance. Besides the design and experimental testing of this absorber device, this paper addresses the fundamentals of composite fragmentation and approaches for numerical modelling.

2 Crash absorber element

When it comes to the weight-specific energy absorption (SEA) of crash elements, that are also found in automotive or train applications, it is known that composite materials are superior compared to metallic absorbers. The fragmentation of fibre-reinforced composites happens under a nearly ideal constant crush load level, while the folding pattern of metallic crash boxes under compression typically leads to severe load amplitudes for each fold [14]. The characteristics of composite crash absorbers and the influence of various geometrical shapes, fibre architectures or trigger mechanisms have been investigated extensively in the past [15–19].

The idea behind the following study was to use the z-strut – made from a circular profile of composite material – as the crash element that is being crushed in its supporting device, allowing for a very long energy absorption length, basically the whole length of the z-strut (Fig. 2). The absorber element and its components are shown in detail in Fig. 3, their materials and functions are explained as follows:

Composite strut:

The composite strut is made from a carbon fibre/epoxy prepreg laminate with 50% fibres in 0° (axial) and ±45° direction, which was most suitable for current crush load requirements. Of course, the lay-up can be adjusted for other requirements. Although braided struts were also tested, prepreg material led to higher reproducibility. To avoid corrosion problems with the aluminium supports, an outside layer of glass fibres was used.

Figure 2: Composite z-strut with integrated energy absorber (before/after crash test).

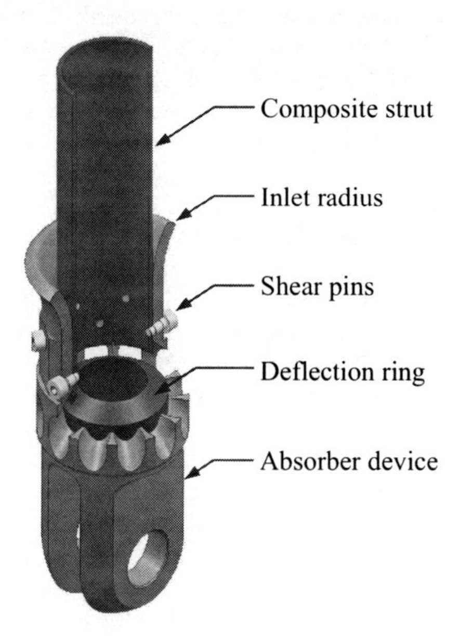

Figure 3: Energy absorber device and components.

Inlet radius:
The inlet radius at the upper end of the support device ensures the functionality under oblique impact conditions up to an angle of about 10°.

Shear pins:
All static loads are transferred from the composite strut to the metallic absorber device through shear pins (Fig. 4a). In case of the crash load, they are supposed to fail at a specific limit by shear failure, so that the crushing of the composite strut begins. Titanium pins were found to be superior compared to aluminium pins, as their yield and ultimate strength values are closer, leading to a more brittle failure without undesired nonlinearities. Composite CF/PEEK pins were also tested as an alternative but showed no improvement. Since solid pins led to bearing failure of the composite laminate, hollow titanium pins were used. The final geometry of the pins and their fillet radii were optimised with respect to fatigue demands.

Deflection ring:
The 7075-T651 aluminium inner ring inside the absorber element acts as a deflection device. Once the shear pins fail, the composite strut hits the inner ring and the composite material is deflected outwards. The angle of this deflection device primarily influences the crush load of the whole absorber system.
Absorber device:
The aluminium absorber device is basically the support for the z-strut and the connection to the fuselage structure. It consists of several circumferential holes, through which the composite z-strut is pushed after being deflected by the inner ring. Due to the sharp edges of the holes, the strut is cut into stripes. Through these holes the outflow of the material is ensured so that no blockage can occur, assuring the absorber functionality. Some trials were also performed with titanium absorber devices, leading to a slightly lower SEA, which may be attributed to the higher wear resistance and hence sharper cutting edges. The total weight of 765 g of this conceptual z-strut is lower than the weight of a state-of-the-art aluminium z-strut.

In summary, the energy is absorbed by cutting the z-strut into stripes and bending the material outward via the deflection device. During this process it delaminates and is crushed and fragmented to a large extent (Fig. 4b). For this application, this process turned out to be the optimum to meet the targeted crush load level, which is illustrated in Fig. 5. The first peak in the force-displacement diagram is the trigger load, when the shear pins fail and the absorber starts to work. It was specified to be 20% higher than the ultimate load of the static design and is therefore only reached in the crash case. After shear failure of the pins, the load level drops to zero, due to free displacement of the composite tube up to contact with the deflection ring, which is supposed to avoid the addition with the following peak load when the crushing begins. After this second peak a stable crush load plateau develops that lasts until all energy is absorbed.

For sure, the crush load level can be increased by crushing the whole tube instead of cutting it into stripes, but if the SEA increases the wall thickness of the tube would have to be reduced to meet the targeted load level, leading to buckling and bearing failure issues. Also the continuous material outflow would be more problematic and the initial peak load is much higher for the full tube crushing.

Figure 4: Shear pin failure (a) and failed material (b).

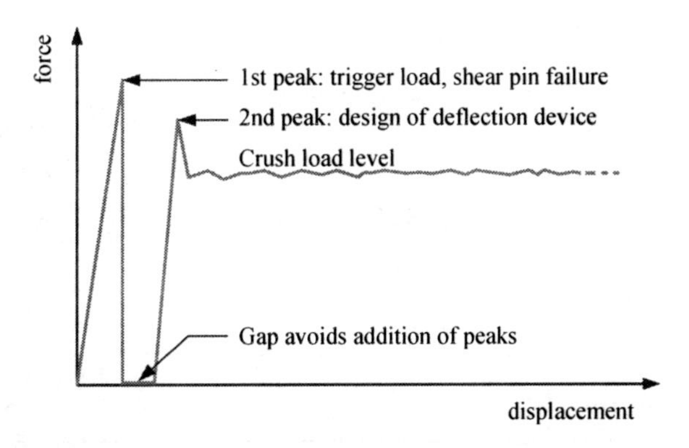

Figure 5: Absorber characteristics in force-displacement diagram.

The development of this crash absorber is based on a step by step testing pyramidal approach: starting with cylinder crush tests to identify the SEA of various laminates and materials, over static absorber tests to evaluate the shear pin failure, up to crash tests of absorber components and finally full scale crash tests of complete z-struts (Figs. 6 and 7). These latter tests were performed successfully both under normal and oblique impact conditions at the Institute of Composite Materials (IVW), Kaiserslautern. It is important to ensure that the absorber also works in the oblique configuration, as there is some rotation of the struts to be expected in the crash load case of the fuselage (see Fig. 1). The shear pin failure was proven not to be influenced by strain rate, which could be concluded from static and dynamic tests. The testing spectrum was finalised by fatigue tests under tension-compression loads. The main conclusion from this test spectrum was the robustness of the absorber design. While other absorber concepts often show their full performance only under a narrow range of ideal conditions, this system worked under various conditions and angles with an impressively high reproducibility.

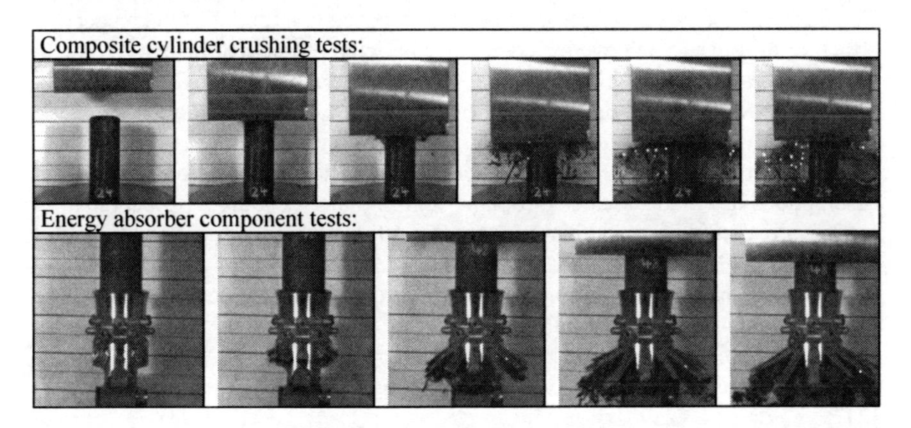

Figure 6: Extract of crash absorber test spectrum: component tests.

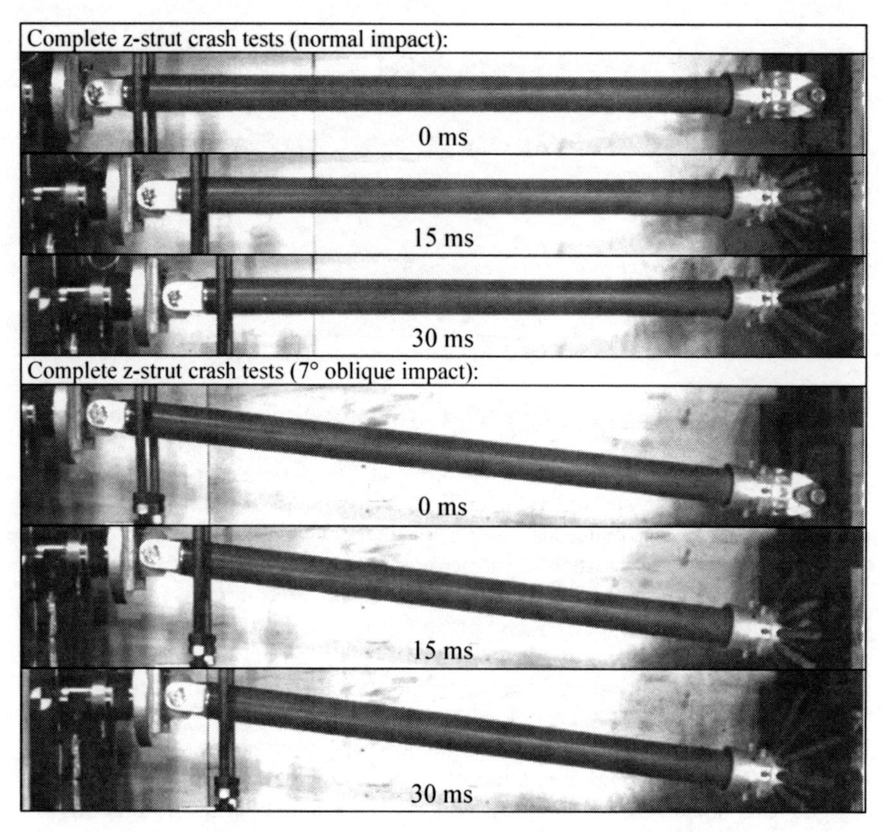

Figure 7: Extract of crash absorber test spectrum: complete z-strut tests.

3 Physics of fragmentation

The desire of today's engineer is to be able to generate a model of the respective structure in order to achieve efficient and predictive numerical design, reducing expensive testing efforts. For this reason, numerical analyses were also performed in the framework of the development of the crash absorber. However, since the failure behaviour is very complex, a fundamental understanding of the physical process of the composite crushing phenomena is mandatory before any model assumptions are made. Besides in-plane failure under bending, two degradation modes are dominating in this energy absorption process: fragmentation and delamination [20].

In this context, fragmentation can be seen as the last step of degradation, where the material is reduced to small particles. The initiation of fragmentation takes place at the microscale, where microscopic buckling of fibres occurs due to initial misalignments and fibre waviness, leading to kink band generation (Fig. 8, [21]).

The matrix material is supposed to support the fibres against microbuckling, but as soon as its yield strength is reached, fibre bending increases sharply and

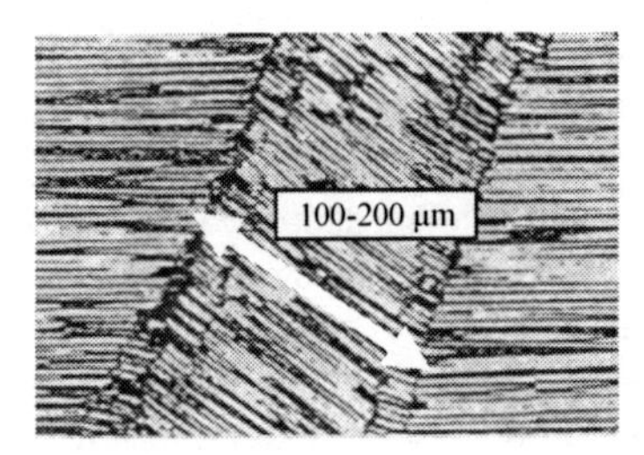

Figure 8: Microscopic observation of kink band [21].

stiffness drops with kink bands appearing that finally lead to catastrophic failure. An understanding of the energy dissipated in this process and influencing parameters was gained in [22] using a microscale modelling approach. To cover this energy absorption mechanism and also the fragmentation/delamination interaction in a meso- or macromodel for explicit finite element (FE) calculations typically used in industry today is still challenging and currently under investigations. Therefore, for first quantitative numerical evaluations, simplified phenomenological approaches were developed, e.g. either based on the definition of the mean crushing stress level of an element (CZone in ABAQUS [23]) or the reduction of the material strength in the vicinity of failed elements in order to generate stable crushing (crashfront algorithm in LS-DYNA [24, 25]). Indeed, the challenge remains to predict crush damage modes, its transitions and absorbed energy in mesoscale simulations based on regular material data.

Besides ply fragmentation, delamination as the interlaminar separation of two plies of the composite laminate is also a key mechanism in energy absorption. Since ply stiffness depends on the fibre orientation, adjacent layers with different fibre angles have different stiffnesses, leading to stress discontinuities and hence interlaminar shear stresses between the plies. In addition, the shearing crack propagation (mode II), which is most relevant for this absorber concept, was shown to be rate-dependent in [26]. Delamination modelling on the mesoscale typically involves interface models between separate plies or sublaminates represented with shell, continuum shell or solid elements. These interface models may e.g. be contact definitions or cohesive elements, with their failure behaviour classically being based on the cohesive zone model with a defined traction-separation law.

4 Modelling and simulation

On this basis, it was investigated if the features available in today's commercial explicit FE codes are able and accurate enough to predict the crush load level of the z-strut absorber. Explicit codes have to be used because of the highly nonlinear behaviour in combination with the very short duration of the problem. In some recent papers similar crushing and delamination phenomena were investigated numerically using ABAQUS/explicit, with composite shell elements separated by cohesive elements for delamination [27-29]. In this study, the three commercial codes ABAQUS/explicit, LS-DYNA and PAM-CRASH were used.

At first, the focus is on the ABAQUS/explicit model (Fig. 9). All aluminium parts, i.e. the absorber device, the inner ring and the support on the other side of the strut were modelled with C3D6 and C3D8R solid elements and an elastic-plastic material law with isotropic hardening. The composite strut was modelled with two layers of S4R shell elements and one interface layer of COH3D8 cohesive elements in-between for interlaminar separation (delamination). Intralaminar failure in the multi-layered composite shell elements is covered by the Hashin criteria [30] for damage initiation and a fracture energy-based formulation for damage evolution. The first row of elements in the strut was weakened to act as a trigger and initiate stable contact behaviour and crushing. It has to be mentioned that the cutting seams in the composite tube had to be predefined in the model like in [28] by lines of cohesive elements in order to achieve stable simulations and crack propagation. Although the shear pins were included in some first calculations on component level, they were excluded in the final crash simulation of the complete z-strut because the high loading rate in the crash test in combination with the limited sampling frequency and superimposed oscillations led to the fact that the shear pin failure could not be evaluated in the experimental force plots and therefore no comparison with the simulation was possible. A general contact definition was used to avoid penetration of the individual parts.

The boundary conditions were defined corresponding to the experimental crash tests with an impact velocity of 6.8 m/s and an initial kinetic energy of 1911 J.

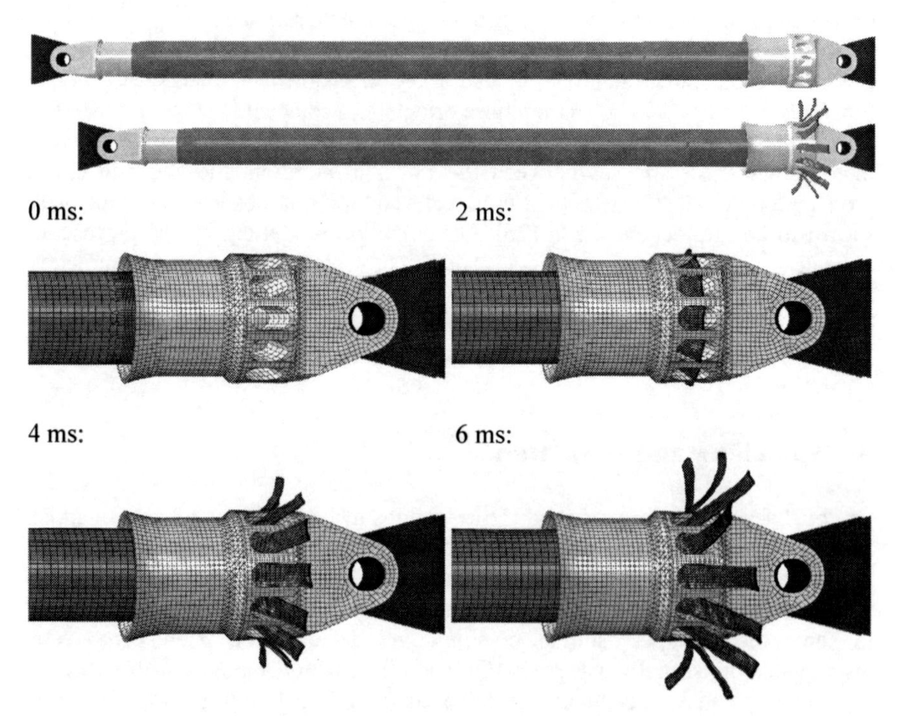

Figure 9: FE model for simulation of z-strut crushing.

The force-displacement diagrams of crash test and ABAQUS/explicit simulation are shown in Fig. 10, both being recorded with the same frequency and filtered with an SAE600 filter. The simulation was stopped earlier after the stable crushing load level was clearly identified to save computational time. It can be seen that the load peaks in the beginning cannot be covered by this model, as they are the result of the crack initiation. The initiation, however, is predefined in the model by the cohesive interface seams, the simulation only covers failure propagation. It can be seen that the experimental curve is slightly progressive in the crushing zone, which is a result of wear at the cutting edges. This effect is not covered by the model. Although stress concentrations are visible in Fig. 11, the yield stress of the elements is not reached. The metallic cutting edges would have had to be modelled with a very fine mesh for this purpose, making the explicit calculation inefficient. However, besides these drawbacks, the most important characteristic – the crush load level – can be predicted quite satisfactorily. This is due to the fact that the main contributors to the energy absorption are represented in this model, the cutting of the tube into stripes by the cohesive elements and the damage and delamination modes under bending. Again, it has to be recalled that the real physical process of fragmentation cannot be represented by this mesomodel, it is just approximated. The robustness of the z-strut absorber against oblique loading conditions could also be shown successfully with this model.

Similar models were developed in the codes LS-DYNA and PAM-CRASH, to a large extent based on the same modelling methods and material parameters (LS-DYNA: MAT24 for aluminium, MAT54 for composite laminate, cohesive zone tiebreak contact for delamination, spotweld beams for the cutting seams, surface-to-surface contact; PAM-CRASH: material type 1 for aluminium, bi-phase material type 131 for composite laminate, tied interface type 303 for delamination and cutting seams, contact types 33 and 34). The results, although not shown here in further detail, were very similar, both qualitatively and quantitatively.

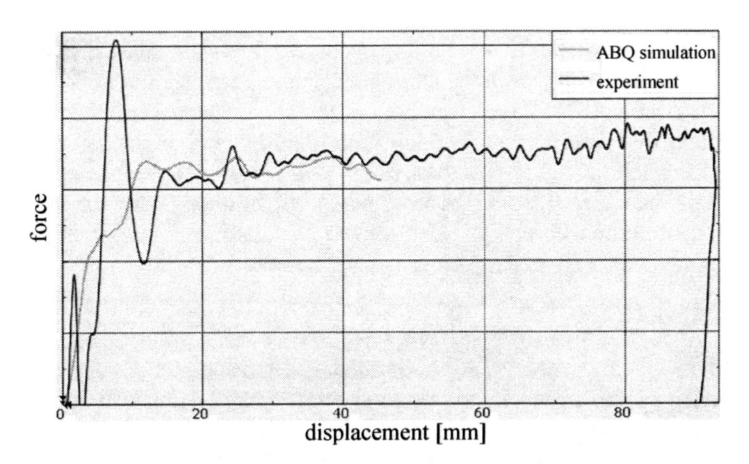

Figure 10: Crushing force vs. displacement diagram of complete z-strut, normal impact 1911 J.

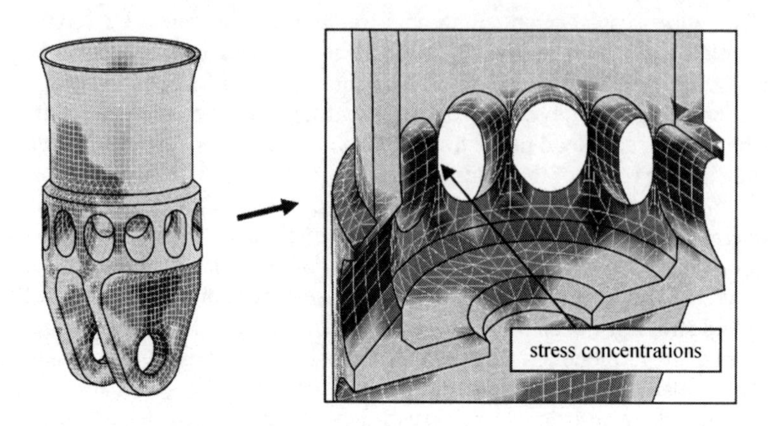

Figure 11: Von Mises stress concentrations at metallic absorber device cutting holes (no plasticity).

5 Conclusions

A lightweight composite crash absorber was developed, which can be used in the z-struts of a commercial aircraft to improve the crashworthiness behaviour. Component tests and complete z-strut crash tests under normal and oblique loading conditions provided consistent results within all requirements and showed a very high degree of robustness and reproducibility of the results. Although the absorber in this study was designed for specific load requirements, it can be adjusted to individual trigger and crush load levels by an appropriate choice of

- composite tube material, lay-up and thickness,
- shear pin material and design,
- angle of the deflection device,
- and number of cutting holes.

The use of explicit FE simulations with commercial codes for an overall prediction of the crush load level using mesomodels was shown to be successful to a certain extent. However, specific peaks in the load curve could not be represented, this would make a much more detailed and at the same time less efficient modelling approach necessary. The real physical fragmentation phenomena can also just be approximated, highlighting that the numerical prediction of composite energy absorption for industrial use cases is still a big challenge and currently under further investigations.

References

[1] Jackson, K.E.; Boitnott, R.L.; Fasanella, E.L.; Jones, L.E.; Lyle, K.H.: A history of full-scale aircraft and rotorcraft crash testing and simulation at NASA Langley Research Center. *4th Triennial International Fire & Cabin Safety Research Conference*, Lisbon, Portugal, 2004.

[2] Rassaian, M.; Byar, A.; Ko, J.: Numerical simulation of 737 fuselage section drop test. *NAFEMS World Congress*, Crete, Greece, 2009.

[3] Hashemi, S.M.R.; Walton, A.C.: A systematic approach to aircraft crashworthiness and impact surface material models. *Journal of Aerospace Engineering*, **214(5)**, pp. 265–280, 2000.
[4] Kindervater, C.M.; Georgi, H.: Composite strength and energy absorption as an aspect of structural crash resistance. In: *Structural Crashworthiness and Failure*, N. Jones, T. Wierzbicki (eds.), Elsevier, pp. 189–235, 1993.
[5] Kindervater, C.M.; Kohlgrüber, D.; Johnson, A.: Composite vehicle structural crashworthiness - A status of design methodology and numerical simulation techniques. *Int. J. of Crashworthiness*, **4(2)**, pp. 213–230, 1999.
[6] Mahé, M.; Ribet, H.; Le Page, F.: Composite fuselage crash FE modelling dedicated to enhance the design in correlation with full scale drop test. *Mécanique & Industries*, **2(1)**, pp. 5–17, 2001.
[7] Wiggenraad, J.F.M.; Michielsen, A.L.P.J.; Santoro, D.; Le Page, F.; Kindervater, C.; Beltran, F.; Al-Khalil, M.: Finite element methodologies development to simulate the behaviour of composite fuselage structure and correlation with drop test. *Air & Space Europe*, **3(3–4)**, pp. 228–233, 2001.
[8] Kindervater, C.M.: The crashworthiness of composite aerospace structures. *Workshop 'The Crashworthiness of Composite Transportation Structures'*, TRL, Crowthorne, UK, 2002.
[9] Arnaudeau, F.; Deletombe, E.; Mahé, M.; Le Page, F.: Crashworthiness of aircraft composites structures. *ASME International Mechanical Engineering Congress & Exposition (IMECE2002)*, New Orleans, LA, 2002.
[10] Fasanella, E.L.; Jackson, K.E.; Sparks, C.E.; Sareen, A.K.: Water impact test and simulation of a composite energy absorbing fuselage section. *American Helicopter Society 59th Annual Forum*, Phoenix, AZ, 2003.
[11] Meng, F.X.; Zhou, Q.; Yang, J.L.: Improvement of crashworthiness behaviour for simplified structural models of aircraft fuselage. *International Journal of Crashworthiness*, **14(1)**, pp. 83–97, 2009.
[12] Fasanella, E.L.; Jackson, K.E.; Kellas, S.: Soft soil impact testing and simulation of aerospace structures. *10th International LS-DYNA Users Conference*, Dearborn, MI, 2008.
[13] Filsinger, J.; Middendorf, P.; Gessler, A.: Crash energy absorber element, connecting element with a crash energy absorber element of said type, and aircraft. Patent PCT/EP2007/062825, 2007.
[14] Farley, G.L.: Energy absorption of composite materials. *Journal of Composite Materials*, **17(3)**, pp. 267–279, 1983.
[15] Farley, G.L.; Jones, R.M.: Crushing characteristics of continuous fiber-reinforced composite tubes. *Journal of Composite Materials*, **26(1)**, pp. 37–50, 1992.
[16] Brachos, V.; Douglas, C.D.: Energy absorption characteristics of hybrid composite structures. *27th International SAMPE Technical Conference*, Albuquerque, NM, pp. 421–435, 1995.
[17] Mamalis, A.G.; Robinson, M.; Manolakos, D.E.; Demosthenous, G.A.; Ioannidis, M.B.; Carruthers, J.J.: Crashworthy capability of composite material structures. *Composite Structures*, **37(2)**, pp. 109–134, 1997.

[18] Pein, M.; Krause, D.; Heimbs, S.; Middendorf, P.: Innovative energy-absorbing concept for aircraft cabin interior. *International Workshop on Aircraft System Technologies (AST2007)*, Hamburg, Germany, 375–384, 2007.

[19] Garner; D.M.; Adams, D.: Test methods for composites crashworthiness: A review. *Journal of Advanced Materials*, **40(4)**, pp. 5–26, 2008.

[20] Guimard, J.M.; Allix, O.; Pechnik, N.; Thevenet, P.: Energetic analysis of fragmentation mechanisms and dynamic delamination modelling in CFRP composites. *Computers & Structures*, **87(15–16)**, pp. 1022–1032, 2009.

[21] Effendi, R.R.; Barrau, J.; Guedra-Degeorges, D.: Failure mechanism analysis under compression loading of unidirectional carbon/epoxy composites using micromechanical modelling. *Composite Structures*, **31(2)**, pp. 87–98, 1995.

[22] Guimard, J.M.; Allix, O.; Pechnik, N.; Thevenet, P.: Statistical energy and failure analysis of CFRP compression behavior using a uniaxial microbuckling model. *Journal of Composite Materials*, **41(23)**, pp. 2807–2828, 2007.

[23] Indermuehle, K.; Barnes, G.; Nixon, S.; Schrank, M.: Simulating composites crush and crash events using ABAQUS. *50th AIAA/ASME/ASCE/AHS/ASC Structures, Structural Dynamics, and Materials Conf.*, Palm Springs, 2009.

[24] Rassaian, M.; Byar, A.; Bolukbasi, A.; Feraboli, P.; Deleo, F.: Crashworthiness of composite structures: Numerical and experimental guidelines. *NAFEMS World Congress*, Crete, Greece, 2009.

[25] Heimbs, S.; Strobl, F.; Middendorf, P.; Gardner, S.; Eddington, B.; Key, J.: Crash simulation of an F1 racing car front impact structure. *7th European LS-DYNA Users Conference*, Salzburg, Austria, 2009.

[26] Guimard, J.M.; Allix, O.; Pechnik, N.; Thevenet, P.: Chracterization and modelling of rate effects in the dynamic propagation of mode-II delamination in composite laminates. *Int. J. of Fracture*, **160(1)**, pp. 55–71, 2009.

[27] Guillon, D.; Rivallant, S.; Barrau, J.J.; Petiot, C.; Thevenet, P.; Malherbe, B.: Experimental and numerical study of the splaying mode crush of CFRP laminates. *17th Int. Conference on Composite Materials (ICCM-17)*, Edinburgh, UK, 2009.

[28] Palanivelu, S.; Van Paepegem, W.; Degrieck, J.; Kakogiannis, D.; Van Ackeren, J.; Van Hemelrijck, D.; Wastiels, J.: Numerical energy absorption study of composite tubes for axial impact loadings. *17th International Conference on Composite Materials (ICCM-17)*, Edinburgh, UK, 2009.

[29] Swaminathan, N.; Averill, R.C.: Contribution of failure mechanisms to crush energy absorption in a composite tube. *Mechanics of Advanced Materials and Structures*, **13(1)**, pp. 51–59, 2006.

[30] Hashin, Z.: Failure criteria for unidirectional fiber composites. *Journal of Applied Mechanics*, **47(2)**, pp. 329-334, 1980.

An experimental comparison of half-scale rockfall protection sandwich structures

A. Heymann[1, 3], S. Lambert[1], E. Haza-Rozier[2], G. Vinceslas[2] & P. Gotteland[3]
[1]*Cemagref, Grenoble, France*
[2]*CER, Rouen, France*
[3]*L3S-R, Grenoble, France*

Abstract

Protection against falling rocks often requires the building of civil engineering structures such as soil reinforced embankments. A recent development consists of building a sandwich cellular structure for this purpose. Cellular structures are efficient technological solutions widely used in civil engineering for various applications. These structures also appear to be well suited to resist rockfall and to act as protective structures against impacts. This paper investigates the behaviour of three sandwich structures based on half-scale experiments. The 1.5 m high cellular sandwich structures were leaned against a concrete wall with the facing made of geocells filled with a coarse granular material. Three different granular materials were used for the kernel part of the sandwich (between the facing and the wall). The experiments were carried out with dead load "pendular" impacts by a 260 kg spherical boulder with maximal impact energy of 10 kJ. The aim was to evaluate the ability of each kernel material for reducing the stress on the concrete wall.
Keywords: impact, gabion, scrapped tyres.

1 Introduction

Passive structural countermeasures against rockfall consist of structures placed in the vicinity of the elements at risks in order to intercept or deflect the rocks falling from slopes or cliffs. Among the possible structures, some are partly or totally constructed from natural granular materials as for instance galleries covered with cushion layers and embankments, the latter being appropriate for

WIT Transactions on The Built Environment, Vol 113, © 2010 WIT Press
www.witpress.com, ISSN 1743-3509 (on-line)
doi:10.2495/SU100021

medium to high energy impacts (2 to 50 MJ). Even though several experimental campaign and numerical studies were carried out [1–3], the design of rockfall protection embankments suffers from the lack of knowledge concerning the dynamic interaction between the rock and the structure. As a consequence, this design is most often based on empirical approaches. Therefore, research is needed to improve their efficiency and to propose their optimized structures, taking fully into account the dynamics.

For instance, sandwich structures seem to be a promising technical solution. Pioneered in this domain by Yoshida [1], this concept was recently explored using geocells to build the structure [4]. The geocells are metallic wire netting cages. Using different fill materials allows the building of vertical layers to constitute the sandwich. With such a sandwich structure, the aim is to reduce the stresses transmitted within the structure, increasing the diffusion of the stress, as well as the dissipation of the impact energy.

This study is part of a research program dealing with the concept of cellular sandwich protection structures (the Rempare project). This program couples experiments with numerical developments with investigations at the various scales from the constitutive materials to the real-scale structure [4–7].

This paper focuses on half-scale structures with special attention on the transmission of stress within the structure in the impact direction. Three different structures are exposed to dynamic loading with the aim of improving the effectiveness of sandwich structures exposed to rock impacts.

2 Materials and methods

2.1 Impacted structures

The structures consist in two-layered sandwiches, 1.5 m in height, 2.5 m in length and 1 m in thickness (Figs. 1 and 2).

The first layer, or front layer, is made of 15 gabion cages filled with a coarse granular material. These cages are cubic in shape, 500 mm in height and made up

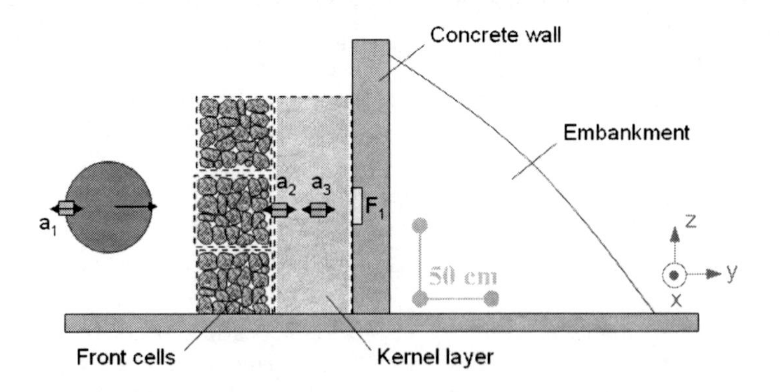

Figure 1: Impacted structure and measurement devices.

Figure 2: A structure before impact.

of a hexagonal wire mesh. The fill material is a crushed quarry limestone, 80 to 150 mm in grain size.

The three structures differ in their kernel fill material (Tab. 1). The sand is a well-graded sand with a 0.2 to 5 mm grain size distribution (Seine sand). Scrapped tyres are 20 to 150 mm in length. The ballast is 30 to 50 mm in grain size distribution. The kernel material is dumped behind the front layer and contained in a geotextile.

This structure is leaned on a reinforced concrete wall, with a ground compacted backfill. This latter is assumed to be rigid compared to the sandwich.

The sandwich structure aims at reducing the force applied to the wall.

Table 1: Kernel fill material.

Structure ref.	Kernel fill material
S1-sand	Seine sand, 0.2-5 mm in grain size distribution
S2-mixture	Mixture : 70% Sand - 30 % scrapped tyres (in mass)
S3-ballast	Ballast

2.2 Experiments

Experiments consisted of pendular impacts by a projectile on the structure. The projectile is made of a 54 cm in diameter steel spherical shell, filled with concrete and having a mass of 260 kg [4].

The pendulum system consists of two metallic beams, 7 m high, connected by a cross beam on which are fixed two metal chains that support the sphere (Fig. 3). The projectile can be lifted up to a maximal height of 4 m using a hand cable winch. The maximal impact energy that can be developed is 10 kJ.

The continuous measurements during these tests are:

- The acceleration of the projectile: a piezoresistive accelerometer is mounted on the shell opposite the impact point (sensor a_1, Fig. 1).

- The acceleration of different points within the structure. These points are mainly in the impact direction, at the interface between the front layer and the kernel layer and in the middle of the kernel (sensors a_2 and a_3 respectively). These piezoresistive uni-axial accelerometers are The force transmitted to the concrete wall at various points, and notably along the impact direction (sensor F_1 on Fig. 1). The sensitive surface of the sensor is 0.1 m². The stress transmitted to the wall, σ_{trans}, is deduced from this measurement.

Sensors a_2, a_3 and F_1 are placed along the impact axis, that is to say at the same height as the impact point (mid-height of the structure). The sample rate is 40 kHz. In order to minimize the noise due to high frequency phenomena, signals are submitted to a low-pass Butterworth filter with a cut frequency of 1 kHz.

All the signals are submitted to the same filter to avoid any time lag bias resulting from this treatment.

Curves plotted give the variation of the signal during the impact.

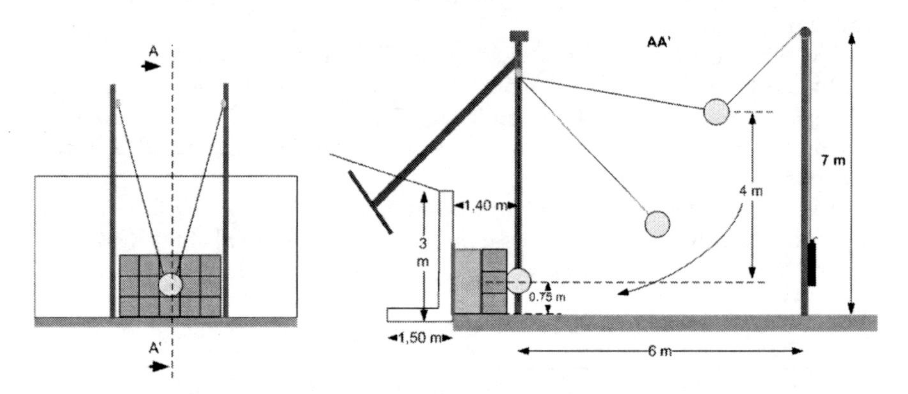

Figure 3: Sketch of the experimental device.

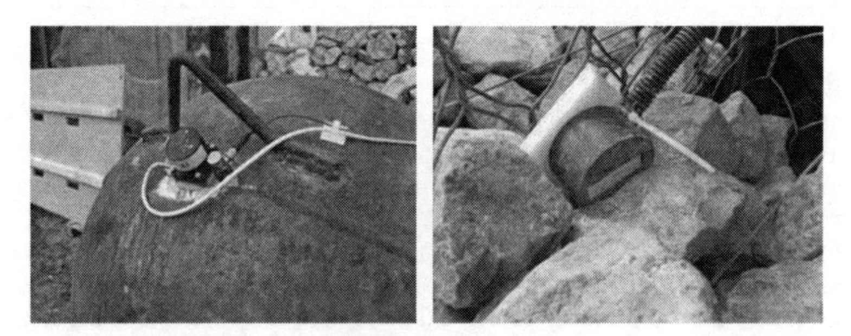

Figure 4: Accelerometer: a_1 (left) and a_2 (right).

Four successive tests with increasing energy (2, 4, 8 and 10 kJ) were carried out on the same structure, without repairing. The 10 kJ impact was repeated once.

3 Results

As the structure is expected to reduce the load on the concrete wall, the assessment of the response of the different structures is primarily based on the forces (or stress) transmitted to this wall. The other data are intended to characterize the impact and also to understand the phenomena explaining the responses of the three structures. It is expected that these data will give evidence of the influence of energy dissipation and of stress spreading on the concrete wall.

3.1 Stress on the concrete wall

The sensor along the impact axis (F_1) is considered first as it is presumed to be exposed to the higher load. Figure 5 illustrates the time evolution of the stress in the case of the 4 kJ impacts on the three structures. The curves are plotted so that the signal starts changing at t=0, without any consideration for the impact time as determined from accelerometer a_1.

The stress curve shape and stress amplitude are different from one structure to the other. The lower stress is obtained for structure S1-sand whose kernel is composed of sand. By contrast, the higher stress is obtained for structure S3-ballast, whose kernel material is ballast. The ratio between these extremes is of about 2. Concerning the shape of the curves, the main conclusion drawn is that the maximum is reached later in the case of structure S2-mixture. Moreover, for this structure the load increase rate before reaching the peak is lower than for the other structures.

Comparison based on the maximum value of the transmitted stress shows the same trends for all the impact energies (Fig. 6). Ballast as kernel material leads

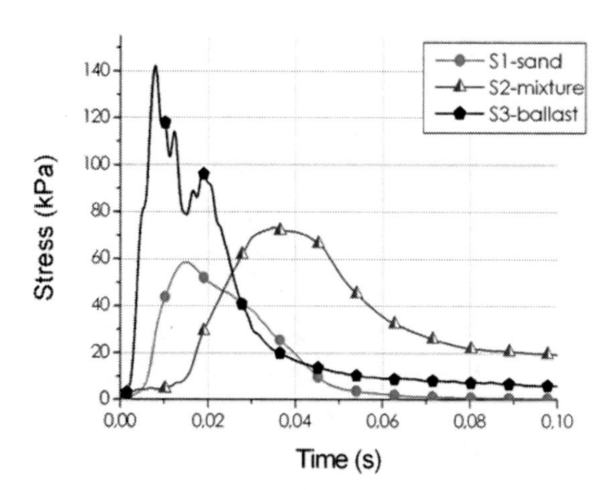

Figure 5: Time evolution of the stress on the wall during the 4 kJ impact (sensor F_1).

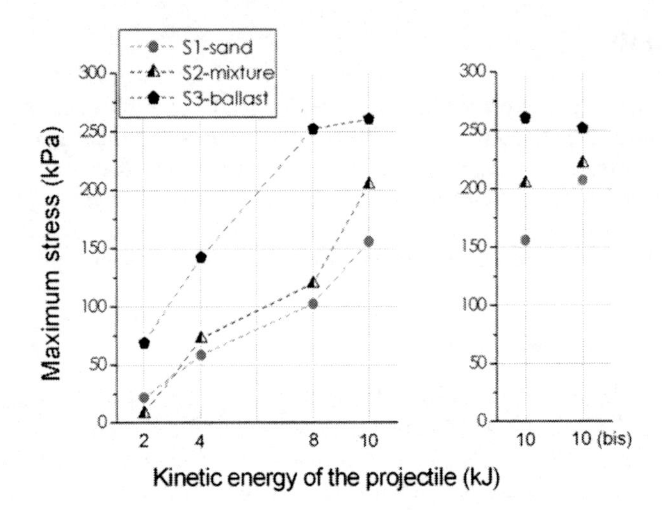

Figure 6: Stress on the concrete wall (sensor F_1) - all the tests.

to the higher stress while the lower stress is obtained for sand as kernel material. Values obtained with the sand-tyre mixture are slightly higher than those obtained with sand. In addition, in the case of ballast as kernel material, the maximum stress seems to reach a threshold value from the 8 kJ impact at a stress of about 250 kPa, even for the second 10 kJ impact. For the others kernel materials, the maximum stress increases almost linearly, without reaching this threshold value.

In order to explain these differences the other measurements will be analysed in the following. The first step consists in assessing the diffusion within the structure. Indeed, the diffusion within the structure may be affected by the characteristics of the different materials.

3.2 Diffusion

During the impact, the stress propagates with time from the contact area to the wall with a spreading. In soils, it is generally assumed that the stress diffuses

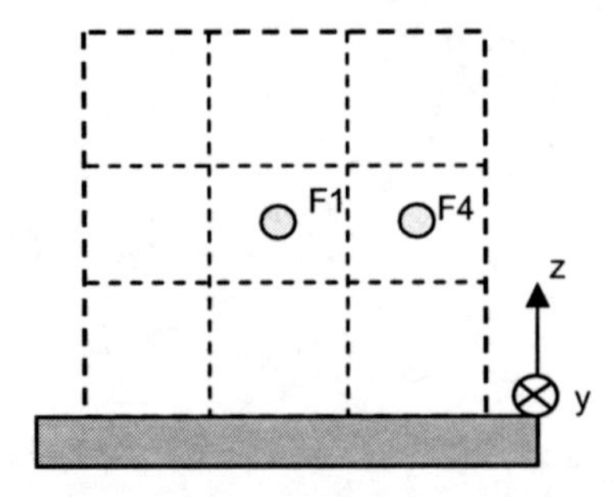

Figure 7: View of the position of stress sensors F_1 and F_4.

www.witpress.com, ISSN 1743-3509 (on-line)

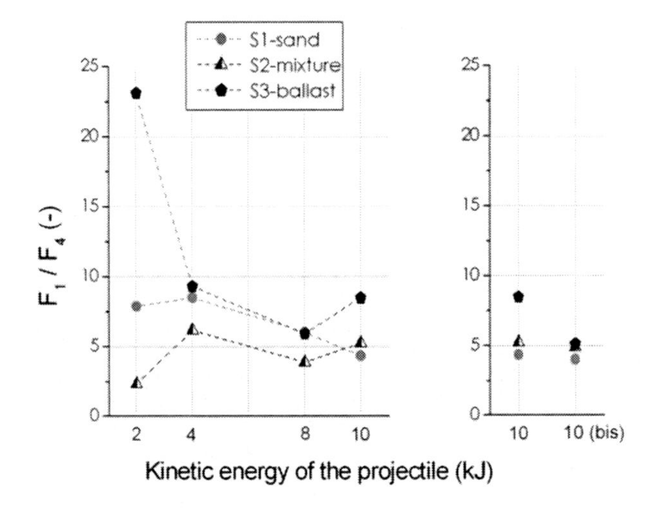

Figure 8: Illustration of the diffusion effects: ratio F_1 to F_4 - all the tests.

within a cone. As a consequence, the loading on the wall is not concentrated along the impact axis but also concerns points apart from this axis. To explore the consequence of the diffusion mechanisms a force sensor is placed on the concrete wall 50 cm aside the impact vertical plane on the same horizontal plan (F_4 on Figure 7).

The ratio of the maximum values of F_1 to those of F_4 is plotted in Figure 8. A high ratio value reveals a stress concentration in the impact axis. Figure 8 shows that for the first impact, there is a great variability among the three structures.

Ballast leads to an important load concentration in the impact axis. From the second impact the values rapidly converge on 5 for all the structures.

Based on this maximum stress criterion there is thus no significant difference in terms of diffusion. Diffusion mechanisms do not explain the differences in stress value in the impact axis from one structure to the other.

3.3 Projectile acceleration

Another way to analyse the structures response is to investigate the projectile acceleration (sensor a_1, Fig. 1). Figure 9 shows that the maximum projectile acceleration is four time higher with sand as kernel material than with ballast. The acceleration in the case of the sand-tyre mixture as kernel material is slightly higher than that with ballast. This is exactly the opposite of what is observed on the transmitted stress. While the ballast structure presents the higher transmitted stress, the projectile acceleration is the lower.

In addition, in both cases the maximum acceleration seems to reach a plateau from the third impact (8 kJ) while in the case of mixture as kernel material it increases, even for the second 10 kJ impact.

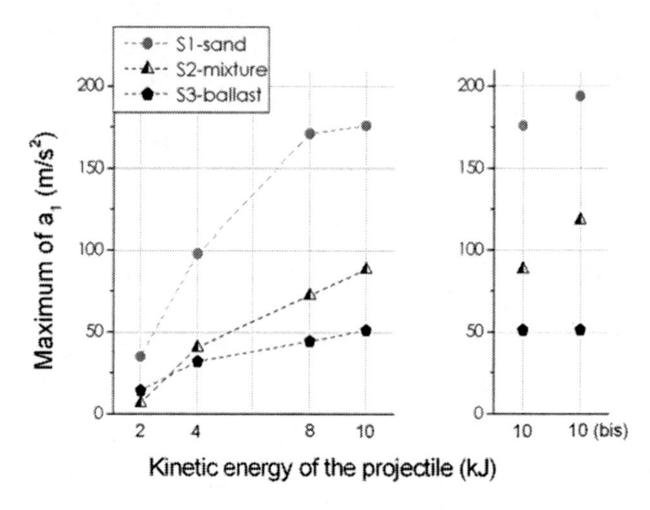

Figure 9: Projectile acceleration (a_1) - all the tests.

3.4 Acceleration within the structure

In order to understand the energy transfer inside the kernel layer, the acceleration measured by the accelerometer a_2 is analysed. The a_2 accelerometer is positioned at the interface between the front and the kernel layers (Fig. 1). The maximum acceleration value is presented in Figure 10. Previous analysis has shown that this peak was not affected by reflection of the compression wave on the concrete wall [7].

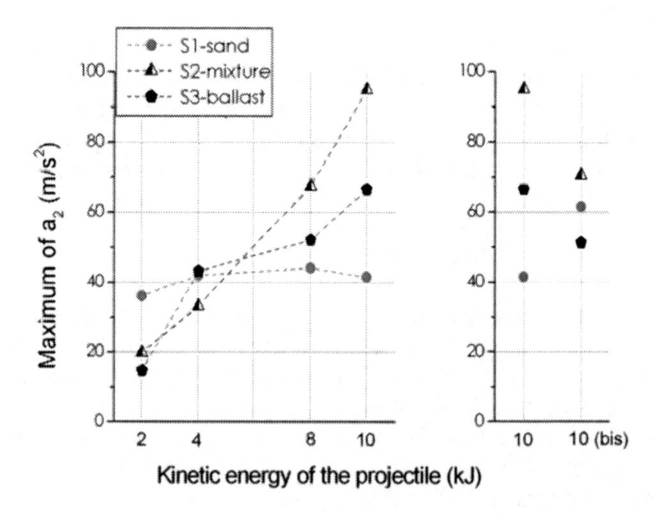

Figure 10: Acceleration at the interface between the front and kernel layers - all the tests.

Trends are different compared to results presented in Figure 9. In the case of sand-tyre mixture or ballast as kernel material, the acceleration increases almost linearly up to the 10 kJ impact. The increase is higher with the mixture. With sand, the acceleration is almost constant from the 2 kJ impact to the 10 kJ impact (40 +/- 5 m ^{2}/s). For the third impact (10 kJ) a ratio of two is obtained between the maximum acceleration, obtained with the mixture and the minimum obtained with sand. With the mixture or the ballast, the acceleration measured during the second 10 kJ impact is lower than during the first one.

4 Discussion

4.1 General comments

The response of the three structures appears to be very complex to understand (Tab. 2). No simple analysis allows interpreting the stress values based on the other measurements.

Table 2: Trends from the measurements along the impact direction.

Measurement	Ranking based on the maximum value
Projectile acceleration	Ballast < Mixture < Sand
Front/kernel acceleration	Mixture < Ballast < Sand
Stress on the wall	Sand ≤ Mixture < Ballast

Comparison of observed trends must account for the fact that the test consisted of successive impacts on two-layered structures involving granular materials. In this context, phenomenon such as compaction and particles crushing are expected, inducing changes in the mechanical and geometrical characteristics of both the kernel and the front layers, at least in the impacted area.

Moreover, as the compression wave propagates through the structure, a temporal analysis would be necessary for interpreting the data from the different sensors. For instance, in the case of sand in the kernel and a 8kJ impact, it takes 2.5 ms and 10 ms for the compression wave to reach accelerometer a_2 and sensor F_1 respectively [7]. These values decrease with the increase of number of impacts and similar values are obtained with the other kernel materials.

4.2 On the influence of the kernel material characteristics

Before the first impact, the kernel material is rather loose. In the case of sand and ballast the successive impacts leads to particle rearrangement and compaction. In the case of ballast the rapid change of the ratio F1/F4 is assumed to be due to stones rearrangement. In a loose particles assembly, diffusion is less important than in a dense one.

In the case of sand, the limited changes observed on the acceleration at the interface between the front and kernel layers are due to compaction. The kernel rigidity increases with successive impacts resulting in a decrease in the acceleration at this interface, compensating the boulder acceleration increase. This is confirmed by the increase in stress on the wall from the first to the second 10 kJ impact.

By contrast, the successive impacts have a different influence on the sand-tyre mixture. With this material, the maximum values of the different measurements along the impact axis increase almost linearly during the test series. This is to be associated with the elasticity of this material. The compression wave induces minor changes in the characteristics of this material.

With increasing impact energy, coarse particle crushing is expected to occur. Actually, this phenomenon has been observed in both the front layer and the kernel layer composed of ballast. With the other kernel materials, crushing concerns the only front layer. Crushing may explain the plateau observed on the stress curve in the case of ballast. Indeed, crushing tends to restrict the amplitude of the stress transmitted to the load [4]. This phenomenon is associated to the amplitude of the force transiting through force chains in the coarse material. The stress on the wall is not really appropriate for this purpose at it gives an average calculated on a large surface compared to the particles size.

By extrapolation the stress on the wall would still be 250 kPa for higher energy impacts, as long as there are particles to crush between the projectile and the wall.

4.3 Practical implications

The first practical conclusion is that sand is the most efficient as kernel material for reducing the load to the concrete wall resulting from the impact by the projectile. The difference with the sand-tyre mixture is little. This is consistent with previously published results concerning singles geo-cells [4].

The second practical implication concerns the design of rockfall protection embankments: neither the kinetic energy of the projectile nor its acceleration are appropriate for assessing the response of the structure. For a same kinetic energy very different structurc bchaviours arc obscrved and a high projectile deceleration does not lead to a high stress within the structure. This conclusion should be considered as the current design of classical rock fall protection embankments is generally based on one of these two data.

4.4 Perspectives

In order to better understand the behaviour of the three structures, complementary analyses are necessary.

For this purpose, the data from the other sensors placed in the structure will be analysed. Actually, three other force sensors and six other acceleration sensors were positioned outside the impact axis. Deformations of the front face of the structure are also measured after each impact. This analysis will account for temporal effects.

In addition this data will allow validate the numerical modelling tools that are being developed [6, 8]. In return these models will help providing simple physical models for understanding the response of these structures.

5 Conclusion

This paper presented the very first results and analysis from impact tests on two-layered half-scale rockfall protection structures.

The experiments have shown that the most efficient material to be used as kernel fill material to reduce the stress on the concrete wall was sand: its efficiency is up to twice that of ballast. The efficiency of the sand-tyre mixture is intermediate. Particle crushing and compaction appears to explain the difference of behaviours observed. These two phenomenons do not affect the three different fill materials in the same way.

This paper provided a large number of data related to the dynamical response of these structures. Nevertheless, these data do not allow interpretation directly on the observed responses. A comprehensive understanding and assessment of the mechanisms at work in the structure is however necessary for optimising the design of such structures. With this aim, the next step will consist of processing all of the measurements, taking into account the temporal effect, and using numerical tools.

Acknowledgements

This study was conducted in the framework of the research development project REMPARe (www.rempare.fr) supported by the French National Research Agency (ANR). The authors gratefully acknowledge all the partners of this project.

References

[1] Yoshida, H., Recent experimental studies on rockfall control in Japan. *Proc. of the Joint Japan-Swiss scientific seminar on impact by rock falls and design of protection structures,* pp. 69-78, 1999.

[2] Hearn, G., Barrett, R. & Henson, H., Development of effective rockfall barriers. *Journal of transportation engineering*, **121(6)**, pp. 507-516, 1995.

[3] Peila, D., Oggeri, C. & Castiglia, C., Ground reinforced embankments for rockfall protection: design and evaluation of full scale tests. *Landslides*, **4**, pp. 255-265, 2007.

[4] Lambert, S., Gotteland, P., & Nicot, F., Experimental study of the impact response of geocells as components of rockfall protection embankments. *Natural Hazards and Earth Systems Sciences,* **9**, pp. 459-467, 2009.

[5] Bertrand, D., Nicot, F., Gotteland, P., & Lambert, S., Modelling a geo-composite cell using discrete analysis. *Computers and Geotechnics*, **32**, pp. 564-577, 2006.

[6] Nicot, F., Gotteland, P., Bertrand, D., & Lambert, S., Multi-scale approach to geo-composite cellular structures subjected to impact, *International Journal for Numerical and Analytical Methods in Geomechanics*, **31**, p. 1477-1515, 2007.
[7] Heymann, A., Gotteland, P. & Lambert, S., *Impact load transmission within a half scale sandwich rockfall protection wall.* Proc. of AGS'10, Djerba, 6 p, 10-12 May, 2010.
[8] Bourrier, F., Gotteland, P., Nicot, F., Lambert, S., A model for rockfall protection structures based on a multi-scale approach. *Proc. of Geoflorida 2010*, West Palm Beach, Florida, pp. 2280-2290, 2010.

Damage simulation for the disaster merit of light substances

P. Brož
Professional Association for Science, Research and Consultancy, Czech Republic

Abstract

In general, dynamic failures of metal construction embrace a large variety of plastic strain rates. These influences must be taken via viscoplasticity, coupled with damage. For this problem two damage models, Gologanu's and Lemaitre's, are convenient; they are modified by the Hill potential to take account of material anisotropy. To stipulate the damage parameters of the model concerned, an inverse routine employing the optimizer to correlate the experimental and numerical responses of notched specimen tension tests is applied. This method was used to measure components being composed of the selected material and resulted in suitable values of damaged parameters in an appropriate time for numerical modelling. These quantities come from an optimization method that yields the best solution potential for a variety of possibilities. Both damage models by virtue of the damage mechanics are put on Gologanu's simulation, which concentrates on the progress of ellipsoidal microvoids in the course of plastic strain, and Lemaitre's, which applies a comprehensive damage variable that evolves the strain energy density release rate. When elaborating the damage parameter identification method, static and dynamic tests are utilized. The said damage system is used for the aluminium structural component.

Keywords: aluminium material, crash failure ductility, ellipsoidal microvoid, failure modelling, finite element codes, high strain rate, inverse technique.

1 Introduction

A great saving of weight, and disaster merit protection, is possible merely by the application of modern design intentions and use of lightweight substances, which have restricted ductility and a complicated failure. The failure possibility is thus

WIT Transactions on The Built Environment, Vol 113, © 2010 WIT Press
www.witpress.com, ISSN 1743-3509 (on-line)
doi:10.2495/SU100031

a reality that could sharply vary the progression of deformation effects. The current crashworthiness codes are unsuitable to predict failure in such materials, or jointing methods that have raised serious doubts over their incomes. In order to prevent going back to a "prototype based design" and to maintain the high level of safety reached in recent years, it is an urgent necessity to better the failure prediction potentials of model systems of rules. This is why it has been urgent to take on a European research project called IMPACT [1], in order to evolve simulations and methodologies for failure prediction. The issues given involve metallic materials that suffer from complex failures under dynamic loadings. Thus, definite damage simulations based on the framework of continuum damage mechanics are implied to characterize the development of damage under mechanical loadings, its advance up to the initiation of a macro-crack and lastly the propagation of this macro-crack up to failure of the constituent. Gologanu's damage model, which describes all of these phenomena by introducing an accurate characterization of the evolution of microvoids under dynamic loadings, is applied [2]. The form and orientation of the microstructural cavities are considered in the finite element model developed to make better prediction of damage and fracture phenomenon, and furthermore the growth, nucleation and coalescence stages are taken into account to describe the evolution of the microvoid volume fraction, in their majority, according to [3]. The law of porosity evolution owing to nucleation is adapted to consider damage evolution in the course of pure shear loading.

A macroscopic way of tackling this problem, based on Lemaitre's model, is applied to symbolize pertinently the damage evolution subject to dynamic loading. Extended Lemaitre's simulation for ductile damage was set up to involve anisotropy in the damage development. The damage law is applied either in monotonic loadings for the ductile fracture or in cyclic loadings for low cycle or high cycle loadings. This law depends on the strain energy density release rate, which is the main variable for governing the damage event. The evolution damage law and threshold parameters have to be identified in the large plastic strain region when an inverse method is applied. Concurrently, the identification of the damage parameters lies in correlating an experimental and numerical macroscopic measurement. Dynamic tensile tests on notched specimens are used to measure macroscopic responses. The test data obtained is the change of the inner radius of the specimens in compliance with their elongation and also the change of forces according to the stated elongation. The damage models given simulate specimens and components for metal materials. Their capacity to predict the stress softening, damage and failure path is explained.

2 Damage simulations

A micro-mechanical approach using solid Finite Elements is applied to stipulate a quite exact prediction of the 3D stress classification in the contraction failure location and around a propagating crack. If suitably fine meshing and appropriate material laws are selected then both failure initiation and crack propagation can be represented; clearly this approach leads to large FE models and is thus restricted to constituent failure studies.

The term 'micro-mechanical models' is used to arrange a simulating approach using detailed 3D solid elements which attempt to represent local necking, the formation of a crack and crack growth.

2.1 Gologanu damage model

Gologanu model takes into account the alterations in microvoid form which occur during deformation. In fact, the Gologanu model considers cavities of ellipsoidal form, whose shape and orientation can evolve. The plastic potential is a quadratic formulation which can also be used as a yield function in which σ_{eq} is the macroscopic equivalent stress, σ_M is the elasto-viscoplastic flow stress, q_1; α_1 and α_2 are material parameters introduced in order to converge the model with full numerical analyses of periodic arrays of voids; f^* is the Tvergaard and Needleman's coalescence function:

$$\phi_{evp} = C\frac{\left\|\sigma' + \eta\sigma_H X\right\|^2}{\sigma_M^2} - \varphi = 0 \tag{1}$$

where $\varphi = \left|1 + \left(q_1 f^*\right)^2 - 2q_1 f^* cosh(\nu)\right|, \quad \nu = \frac{\kappa\sigma_H}{\sigma_M}, \quad \sigma_H = \left(1 - 2\alpha_2\right)\sigma_{11} + \alpha_1\sigma_{22}$

(2)

The parameters κ, η, C and X depend on the geometry of the ellipsoid void, σ_H is the mean stress and the von Mises norm can be written down as

$$\left\|\sigma_{ij}\right\| = \sqrt{\frac{3}{2}\sigma_{ij}\sigma_{ij}} \tag{3}$$

In pure shear loading, there is no damage evolution and consequently no rupture. In order to consider damage due to shearing the damage evolution law is modified to comprise of the sum of the classical law and a new part due to shear loading yielding,

$$\dot{f} = \dot{f}_{growth} + \dot{f}_{nucleation} + \dot{f}_{shear} \tag{4}$$

In pure shear it is commonly accepted that the voids experience a rotation without any change in growth and it appears that nucleation can be generated. Consequently, the damage evolution law due to shearing takes the form of a statistical law, similar to the nucleation evolution law, but taking into consideration the shearing strain and the shearing strain rate. It is given by,

$$\dot{f}_{shear} = \frac{f_s}{S_s\sqrt{2\pi}}\exp\left(-\frac{1}{2}\left(\frac{\varepsilon_{xy} - \varepsilon_s}{S_S}\right)^2 \dot{\varepsilon}_{xy}\right) \tag{5}$$

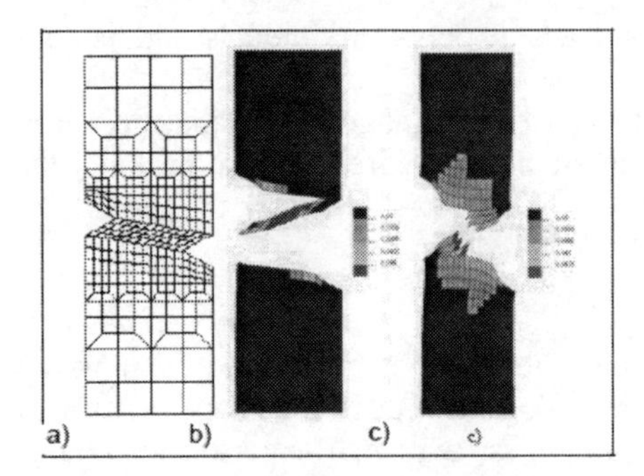

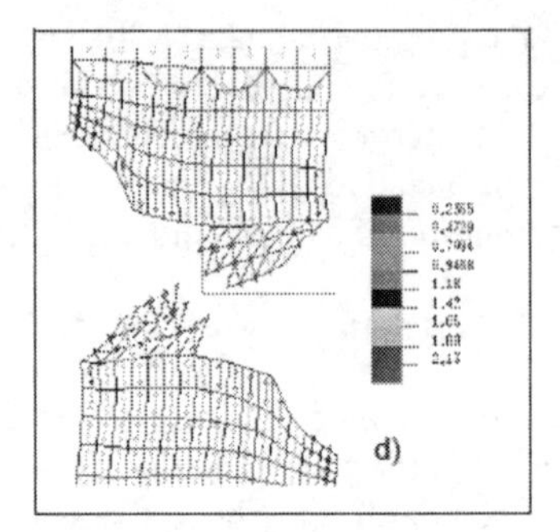

Figure 1: Comparison of rupture mechanisms for Gurson and Gologanu models (taken from [3]).

The identification of the damage parameters controlled by the shearing strains is carried out by the inverse method using an Arcan type test. Fig. 1 shows the tensile simulation of a non-axisymmetric double V-notched specimen; a) shows the initial finite element modelling; b) and c) are the damage distribution at the end of the process using Gologanu model. Fig. 1d gives the orientation and shape evolution S at the end of the rupture process for an initial prolate void shape using Gologanu model. From this figure it can be seen that including changes in the void shape does allow anisotropic damage to be represented and more correctly captures the failure process. It should be noted, that this implementation was only for shell element which restricts precision as triaxial stresses, especially in the necking zone, are ignored. A solid element would end in better results but would be CPU time consuming; an additional progression could be to apply an energy criterion to dissipate energy because each failed element is got rid of.

2.2 Lemaitre damage model

The framework of continuum damage mechanics is used to describe the development of damage under mechanical loadings, its progression up to the initiation of a macro-crack and finally the growth of this macro-crack during failure of the component. The original model for ductile fracture was established for isotropic damage conditions and later extended to include anisotropic damage development [11, 12]. The basic ingredient of the model is the damage law used either in monotonic loadings for ductile fracture, or in cyclic loadings for low cycle or high cycle loadings. This damage law depends on the damage variable D and the strain energy density release rate which is the principal variable governing the phenomenon of damage and is expressed by,

$$Y = \frac{W_e}{1-D} \tag{6}$$

The strain energy rate W_e is split into its shear and hydrostatic parts, and leads to the following expression,

$$W_e = \frac{(1+\nu)}{2E(1-D)}\langle\sigma_{ij}\rangle\langle\sigma_{ij}\rangle - \frac{\nu}{2E(1-D)}\langle\sigma_{ij}\rangle^2 \tag{7}$$

in which $\langle\sigma_{ij}\rangle = \sigma_{ij}$ if $\sigma_{ij} > 0$ and $\langle\sigma_{ij}\rangle = 0$ if $\sigma_{ij} < 0$. Lastly, the damage evolution during plastic straining is formulated by the expression as follows,

$$\dot{D} = \left(\frac{Y}{S}\right)^S \dot{\varepsilon}^p \quad \text{if} \quad \varepsilon^p > \varepsilon_D \tag{8}$$

where S and s are material coefficients, ε^p is the effective plastic strain and ε_D is the plastic strain at damage threshold.

Since the damage evolution is localized in the large plastic strain zone, the evolution damage law and the threshold parameters have to be consequently identified. A direct identification approach or local approach is used essentially on uniaxial monotonic and cyclic tests, Fig. 2. For a better identification experiments in the largest possible domains of stresses, strains, time and number of cycles are needed. An identification approach using inverse techniques is used to find the damage parameters by correlating experimental and numerical macroscopic measurement strongly dependent on the parameters. Tensile tests on thin notched specimens are used as mechanical tests to measure macroscopic responses and variations of inner radius and force with respect to elongation of the specimen employed to correlate experiment and numerical simulations.

In order to better the characteristics of an aluminium alloy, an anisotropic potential can be applied; in the stated case, a Hill 48 potential is considered. The above damage model for anisotropic materials was implemented in 2D shell and 3D solid elements. The damage parameters were identified as treated above and applied to a validation example consisting of the three points bending of an

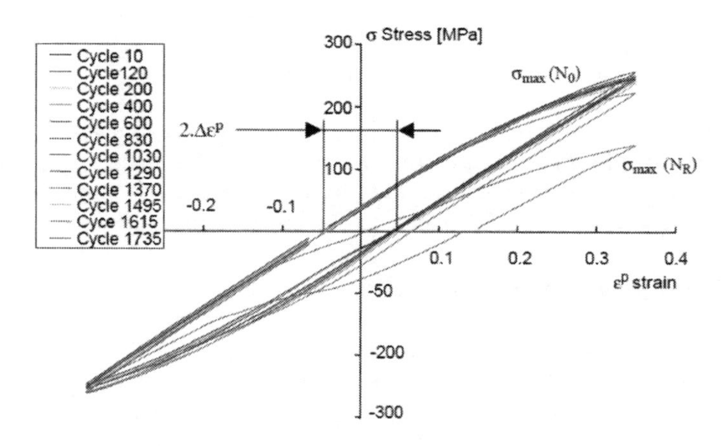

Figure 2: Monotonic cyclic test to determine Lemaitre damage parameters.

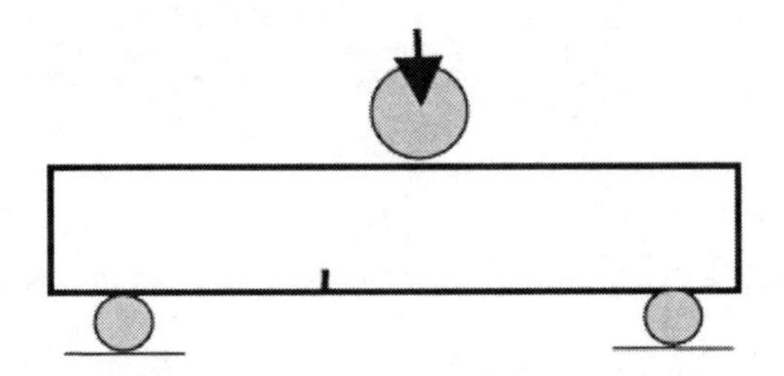

Figure 3: FE model for the 3 point bending of an aluminium section with 'starter crack'.

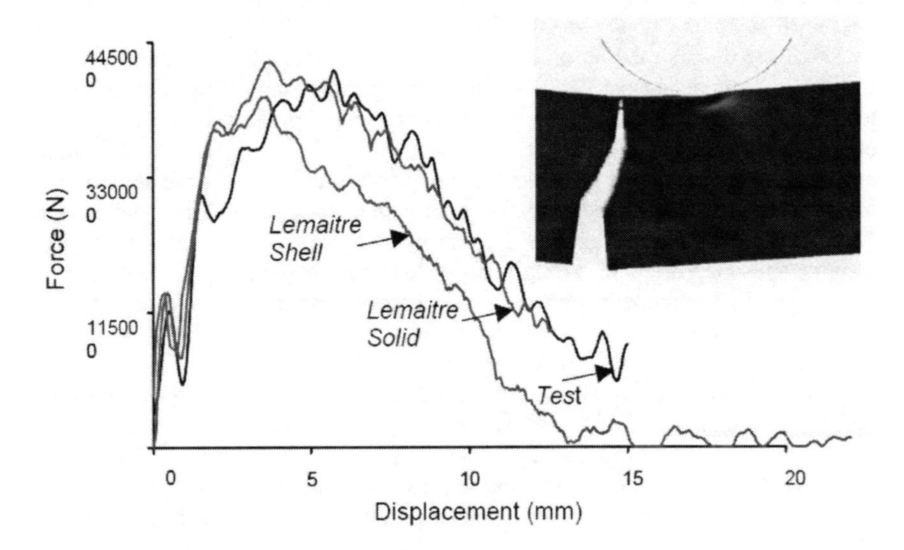

Figure 4: Experimental and simulation load versus displacement curves for shells and solids (Lemaitre model).

aluminium extruded section with an initial narrow opening (initial crack) on the lower face to localize failure initiation, Fig. 3. This test was performed dynamically and the numerical and experimental results are found to be in good correspondence as regards failure path and energy rank, Fig. 4.

3 Damage parameters identification

Founded on an inverse technique, an identification method is used. This procedure lies in the identification of the damage parameters by correlating, with an optimization process, an experimental and numerical macroscopic measurement considerably dependent on the parameters [4, 5]. In tensile tests, the notched flat specimens 2, 3 and 4 mm radii are applied. The variation of both in the bottom of the notch and the variation of the axial force in function of the

elongation of the notched specimen are employed as macroscopic measurements. The finite element simulations of the tests are performed with the exact boundary conditions of the experimental set up. For strain rate sensitive materials like steel, the tensile tests are performed for various plastic strain rates, covering the range of crash tests (0 to 500 s^{-1}) by employing a quasi-static test machine, servohydraulic device and Hopkinson bar [6]. A high speed camera, laser extensometer and representation analyzer are used. In the case of dynamic loading, to measure the macroscopic response, an imaging correlation program was evolved to obtain the experimental information. Next, the measurements before defined are calculated following the evolution of the form of the specimen by dint of a speed camera and home-made software. For Hopkinson bars merely the strength against elongation of the specimen is employed owing to the pure number of images gained at high speed loadings.

Lastly, the correlation is obtained by minimizing a cost function that is explained by the least square approximation being:

$$Q(\alpha) = \sum_{i=1}^{nb\_point} \frac{\left[Z_i^{sim}(\alpha) - Z_i^{\exp}\right]^2}{\left[Z_i^{\exp}\right]^2} \tag{9}$$

where α are material parameters to identify, Z^{sim} and Z^{exp} are simulated and experimental macroscopic responses *nb*_point is the number of experimental points of the response.

The scalar material parameters are $\alpha = \{q_1, f, f_N, S_N, \varepsilon_N, S, f_c, f_F\}$ and $\alpha = \{\varepsilon_D, D_c, S, s\}$ in the case of Gologanu's model and Lemaitre's model. A first initial curve named "numerical non optimized" in Fig. 5 is obtained with the initial parameters and following some iterations comes to be the "numerical optimized" using the optimized parameters.

The damage parameters of 6014 T7 aluminium alloy are recognized using the identification routine indicated (Table 1). This aluminium material has a high ductility with a weak anisotropy and damage owing to the extrusion process. As a consequence, the critical damage D_c is low for Lemaitre's model and the critical microvoid volume fraction at coalescence beginning f_c is low, too, in view of the initial damage f_0 for Gologanu's model – for this simulation the nucleation belongs especially to a very low equivalent plastic strain that fast increases the damage value.

Table 1: Damage parameters of 6014 T7 aluminium alloy.

Gologanu's model	q_1(-)	S_0(-)	f_N(-)	S_N(-)	ε_N(-)	f_c(-)	f_F(-)	f_0(-)
	1.52	0.001	0.040	0.1	0.19	0.06	0.08	0.001
Lemaitre's model	ε_D(-)	s(-)	S(MPa)	D_c(-)				
	0.05	2	1.22	0.34				

4 Extruded pipe

A notched extruded pipe was dynamically tested on an impact bending at 3 meters /second with 50 mm radius impactor of 275 kg weight. The model consists of 15 614 shell elements. For the solid model, the mesh comprises three elements over the thickness. The failure is modelled by element elimination for both considered damage models.

For the 6014 T7 which is an orthotropic substance, material parameters are identified using 0°, 45° and 90° tensile experiments from the rolling direction (Table 2). The law of characteristics in 0° direction is indicated in Fig. 5. Experimental and numerical results are in good conformity as regards the failure moment, failure path and energy dissipation primarily in the case of Lemaitre's model. Gologanu's model using shell elements is exact till the first element elimination adding after that, the strain control of damage results in an underestimation of the energy dissipation along the failure. For all that, when using shell element, finite element models lead to an underestimation of the dissipated energy which is defined by the fact that shell elements cannot consider the through thickness stresses even for seven integration points through the thickness. In the case of solid elements, the outcomes are extra auspicious.

5 Effect of stress triaxility and strain rate on the characteristics of a constructive material

Constitutive relations for metallic materials at high strain rates change from empirical to theoretical relationships, based on the micro-mechanical processes

Table 2: HILL 48 material parameters.

F	G	L	M	N
1.049	1.3	3	3	3.13

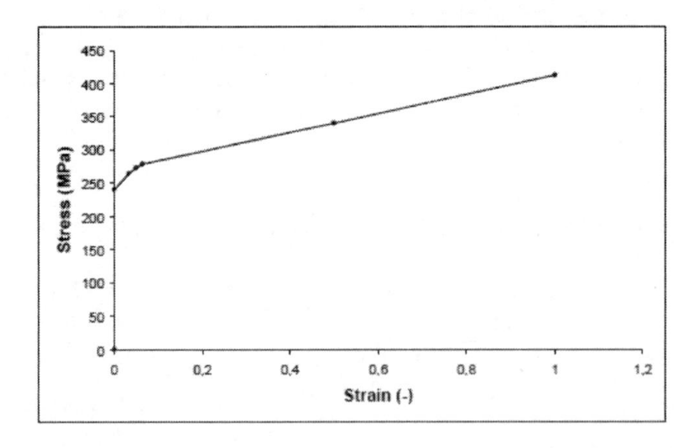

Figure 5: Behaviour law in the 0° direction for 6014 T7 aluminium alloy [3].

supposed to determine the macroscopic characteristics of the material. Normally, the constitutive relation defines the equivalent stress σ in terms of equivalent (or accumulated) plastic strain ε_p, equivalent plastic strain rate ε_p and temperature T. In principle, the material parameters in the constitutive equations are identified founded on test data brought from tension or torsion tests for considerable extends of strain rate and temperature.

Although the strength of metallic materials depends on strain, strain rate and temperature, much experimental checking indicates that the ductility depends substantially on the triaxiality of the stress state. The stress triaxiality is characterized by the dimensionless parameter σ_m/σ, where σ_m stands for the mean stress. Computational fracture simulations were presented that include the effect of stress triaxility and the influences of strain rate and temperature. The influence of the stress triaxility in these models is based on the void growth equation. It is familiar to fit the material parameters of the computational fracture models to date obtained from tensile tests with notched axisymmetric specimens. Various stress triaxility levels are obtained by changing the notch radius of the specimens.

The goal is to investigate the effect of stress triaxility and strain rate on the characteristics of a structural steel and to get experimental data that may be used to legalize constitutive relations and fracture criteria. At last, notched axisymmetric specimens of S 460 steel with three different notch radii were tested in tension at two different strain rates. The tests were carried out in a Split Hopkinson Tension Bar, using a digital high-speed camera system to record the local deformation in the notch. Next, computer models are performed to validate the constitutive relation and fracture model, and to detailed analysis of the material characteristics in the course of high-rate tension tests of smooth and notched samples.

5.1 Experimental program

The specimens were made from a 12 mm thick plate of 460 E steel that is a thermo – mechanically rolled ferritic fine grained structural steel proposed high strength combined with a high degree of ductility. All specimens were taken parallel to the rolling direction of the steel plate. A slight normal anisotropy was achieved.

The geometry and dimensions of the notched specimens are given in Fig. 6. Tests were carried out for three different notch root radii R_0 equal to 0.4 mm, 0.8 mm and 2 mm. Two levels of loading rate were applied, determined by a pre-loading force N_0 of the modified Split Hopkinson bar apparatus equal to 20 kN and 40 kN. As a result, the total number of tests was 25. A test identification system is chosen here where, e.g. R0.8 – 20 – 4 has the following meaning: R0.8 – notch radius R_0 = 0.8 mm; 20 – pre-loading force N_0 = 20 kN; 4 – parallel test No. 4.

The tensile tests of the notched specimens were carried out in a Split Hopkinson Tension Bar at the European Commission Joint Research Centre in Ispra, Italy. A drawing of the Split Hopkinson Tension Bar is demonstrated in Fig. 7.

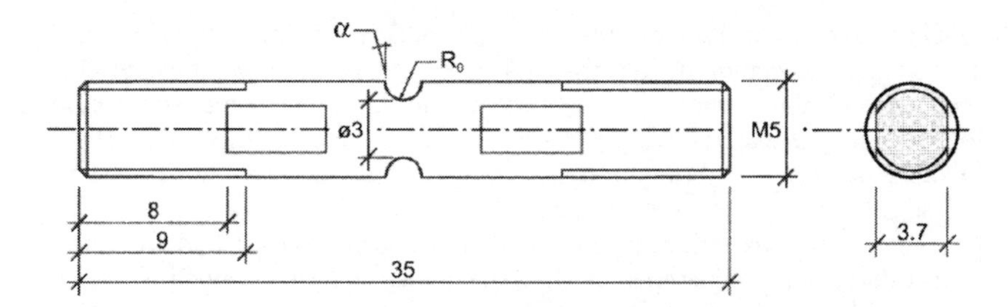

Figure 6: Geometry and dimensions (in mm) of notched test specimens for high rate experiments in the Split Hopkinson Tension Bar (taken from [6]).

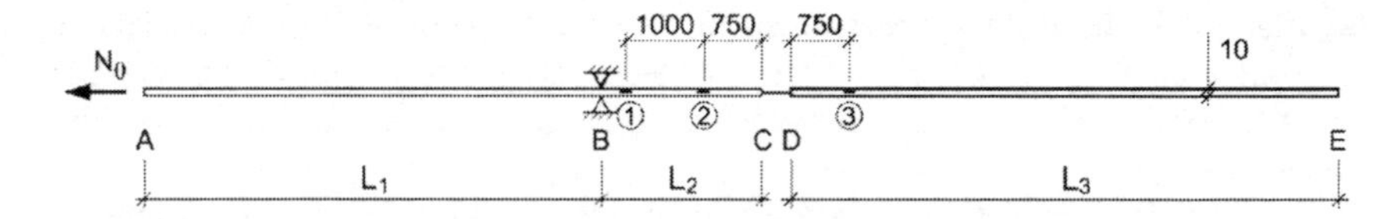

Figure 7: Sketch of the Split Hopkinson Tension Bar (in mm), where L_1 = 5870 mm, L_2 = 2055 mm and L_3 = 5800 mm. The diameter of the input bar A – B – C and the output bar D – E is 10 mm.

The test apparatus consists of two half – bars named the input bar A – B – C and the output bar D – E, and the specimen C – D is inserted between these two bars. The pre-loading force N_0 is applied to the left further point of the input bar A, while the bar is fixed in the longitudinal direction with a brittle intermediate portion at B. By rupturing the brittle intermediate piece be, a tension wave with a rise-time of about 50 μs is transmitted along the input bar and load the specimen to fracture. The strain rate is varied by changing the pre-loading force N_0 of the loading device and so the stored elastic energy.

According to the theory of one-dimensional elastic wave propagation in circular rods, the elastic stress wave propagates along the input bar with a constant velocity and form. When the incident pulse achieves the specimen, it is partly reflected at the interface between the input bar and the specimen, and partly transmitted throughout the specimen and into the output bar. The relative amplitudes of the event, reflected and transmitted pulses depend on the mechanical properties of the specimen. In is supposed that the specimen, which is short relative to the input and output bars, reaches a force equilibrium state after several reflections of elasto-plastic waves in it.

6 Conclusion

Gologanu and Lemaitre damage models were adapted by bringing in Hill potential to consider anisotropy of the material. An inverse routine applying to correlate experimental and numerical responses of notched specimen tensile tests is used to stipulate the damage parameters of both damage simulations. This

technique was applied to an aluminium component and resulted in relevant values of damage parameters in a suitable time for numerical modelling. These values originate from an optimization method which yields the best solution.

An impact bending test on a notched aluminium extruded tube to locate the failure was carried out. Generally, the predicted energy dissipation and the failure moment and growth were determined to be in good harmony with the experimental steps. Anyhow, because both identification techniques are mesh dependent, a regularization procedure should be applied to prevent this dependency. Both damage models can be employed for crash simulations. Gologanu damage simulation is very precious and the use of ellipsoidal microvoids permits to possess a damage anisotropic evolution. Indeed, the complexity of the model is a liability for industrial applications. Lemaitre damage model is easier to employ and identify owing to the small number of parameters. The stress check of the damage evolution is definitely an asset in the simulation.

Acknowledgement

The author gratefully acknowledges the financial support of the presented research by the Grant Agency of the Czech Republic (project No. 103/09/1580).

References

[1] Framework V project. Improved failure prediction for advanced crashworthiness of transportation vehicles. G3RD-CT-2000-00186, 2000-2003.

[2] Gologanu, M., Leblond, J.B. & Devaux, J., Numerical and theoretical study of coalescence of cavities in periodically voided solids. *Computational Material Modelling*, **42**, pp. 223-244, 1994.

[3] Bennani, B. & Lauro, F., damage models and identification procedures for crashworthiness of automotive light materials. *Latin American Journal of Solids and Structures*, **3**, pp. 75-87, 2006.

[4] Lauro, F., Bennani, B., Croix, P. & Oudin, J., Identification of the damage parameters for anisotropic materials by inverse technique: application to an aluminium. *Journal of Materials Processing Technology*, **118**, pp. 472-477, 2001.

[5] Brož, P. & Dobiáš, D., Determining material parameters of damage. *Acta Technica*, AS CR, Prague, **54(4),** pp. 415-436, 2009.

[6] Hopperstad, O.S., Børvik, T., Langseth, M., Labibes, K. & Albertini, C., On the influence of stress triaxiality and strain rate on the behaviour of a structural steel. Part I. Experiments, *European Journal of Mechanics A/Solids*, **22**, pp. 1-13, 2003.

Section 2
Hazard mitigation and assessment

Comparative analysis of bus rollover protection under existing standards

C. C. Liang & L. G. Nam
Department of Mechanical and Automation Engineering,
Da-Yeh University, Taiwan, ROC

Abstract

Both Europe and the United States (US) have enforced the legislations for bus rollover protection: Regulation number 66 of the Economic Commission for Europe (ECE R66) and standard number 220 of the American Federal Motor Vehicle Safety Standards (FMVSS 220) in order to prevent catastrophic rollover accidents. Therefore, this paper studied the legislation for bus rollover protection including both ECE R66 and FMVSS 220. Satisfying the rollover requirements by buses is obligatory by law. However, the scope of those two regulations does overlap for some group of vehicles. Thus, this study firstly presents a physical meaning comparative analysis of the ECE R66 with the FMVSS 220. The LS-DYNA 971/MPP was used for numerical analysis. The analysis models were constructed by the eta/FEMB that is a preprocessing module integrated in the LS-DYNA 971 package. The validation was turned from experimental data of body knots extracted from the real vehicle. This investigation performed the comparative analysis following ECE R66 and FMVSS 220 assessments, then moved to demonstrate the distortion configuration of the vehicle superstructure through the absorbed energy and its distribution in the vehicle and the vehicle frame sections, as well as the violation of the passenger compartment under the rollover testing conditions of both ECE R66 and FMVSS 220. Great differences were found between the rollover strength of bus superstructures depending on which regulations are followed. The results also demonstrate that the passenger compartment and residual space are more violated and more dangerous under the lateral rollover testing condition of the ECE R66 than the other. Above findings could be used for the automobile manufacturers in a new design of bus

WIT Transactions on The Built Environment, Vol 113, © 2010 WIT Press
www.witpress.com, ISSN 1743-3509 (on-line)
doi:10.2495/SU100041

superstructure, incorporating the rollover safety legislation and lightweight designs.
Keywords: bus rollover, ECE R66, FMVSS 220, LS-DYNA.

1 Introduction

Today, transit buses are an integral part of each nation's transportation system. Although buses are one of the safest means of transportation, occupant injuries and fatalities in bus crashes do occur.

Thus, it was observed that rollover seriously threatens the lives of coach passengers. Rollovers are complex, chaotic, and unpredictable events involving the interaction of the driver, road, vehicle, and environmental factors. A rollover is a crash in which a vehicle revolves at least one-quarter turn (which would be on its side), regardless of whether the vehicle ends up laying on its side or roof, or even returning upright on all four wheels [1]. Thus, rollover strength has become an important issue for bus and coach manufacturers. For this problem, both Europe and the United State of America have enforced bus rollover safety regulation and standard to prevent catastrophic rollover accidents. Economic Commission for Europe had enforced Regulation No.66 (ECE R66) for the Bus Strength of Superstructure since 1987 in order to provide protection to the bus and coach occupants during rollover accidents through the maintenance of a survival space [2, 3]. Department of Transportation, the United States of America had enforced the FMVSS 220 standard for the school bus rollover protection since 1977 which included transit buses and vans, having the length less than 35 feet [4, 5].

In the social of globalize economic, Bus and Coach Manufacturers want to bring their products to Europe or the US or both of the markets. While the same problem of bus rollover safety for large bus, in Europe is controlled by ECE R66, and in the USA is controlled by FMVSS 220, although the scope of these regulations does overlap for some groups of vehicles. In recent years, automotive industries are concentrating more on vehicle rollover. There were many researchers to study the structure strength of buses and the injury analysis of passengers in accordance with tests of the ECE R66 [2, 3]. Although many studies have been done on bus structure strength, most of them are following in or based on ECE R66 to carrying out their researches. However, the comparative analysis between ECE R66 and FMVSS 220 is still limited.

Nowadays, with the advances in both computer technology and structural analysis via finite element method, the capacity of computer and FE software are confirmed in predictive analysis and computing assistances of bus structure. That is also new point in ECE R66 version 2006 in which the computer simulation with full scale model is officially used as an assessing method for the bus rollover protection requirements [3]. As a result, rollover strength of bus superstructure has been a topic of interest over the years, and a number of numerical studies of bus structure have been established. While a number of researches have been performed, a thorough development of rollover strength of bus superstructure under production costs and fuel economy has not yet received the same level of attention. In this study, both ECE R66 and FMVSS 220 were analyzed systematically and specified by the numerical study. In addition, distortion configurations of bus frame structure based on ECE R66 and FMVSS 220 tests were also be implemented. Finally, a guideline of bus rollover safety regulation then is recommended for the studies of bus body strengthening and optimal design of bus superstructure as follows.

2 Legislation for bus rollover protection

2.1 ECE R66 regulation

ECE R66 regulation was issued from on Jan 30th, 1987, and enforced by Economic Commission for Europe because of serious status of rollover accidents. It applies to single-decked vehicles constructed for carrying more than 22 passengers, whether seated or standing, in addition to the driver and crew. "Superstructure" refers to the parts of a vehicle structure that contribute to the strength of the vehicle in the event of a rollover accident. The purpose of this regulation is to ensure that the vehicle superstructure has sufficient strength so that the residual space during and after the rollover test on the complete vehicle is unharmed. This means that no part of the vehicle that is outside the residual space at the start of the rollover, like luggage, is intruding into the residual space and no part of the residual space projects outside the deformed structure. The envelope of the vehicle's residual space is defined by creating a vertical transverse plane within the vehicle which has the periphery described in Fig. 1. The SR points are located on the seat-back of each forward or rearward facing seat, 500 mm above the floor under the seat, 150 mm from the inside surface of the sidewall of the vehicle [3].

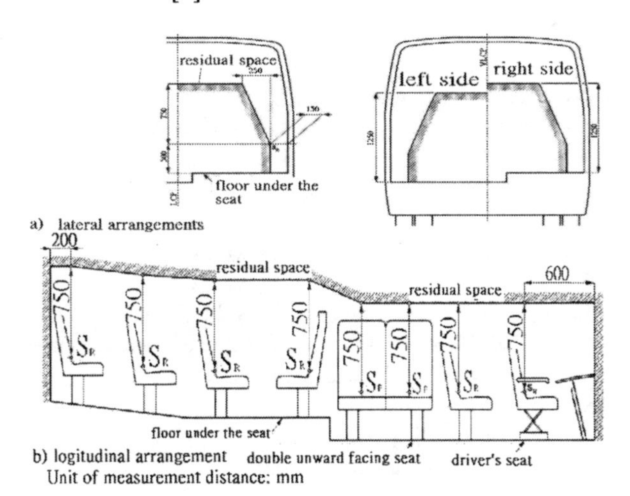

Figure 1: The residual space of a bus [3].

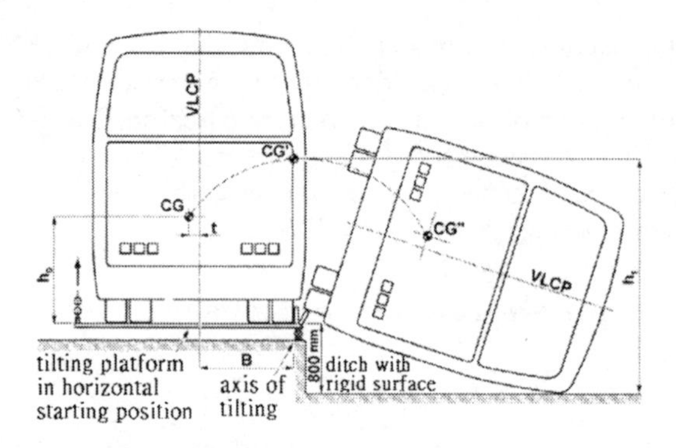

Figure 2: Specification of the rollover test [3].

This regulation is continuously updated based on actual requirements. And it is using as an international bus rollover regulation. The current version was issued on Feb 22nd, 2006. The rollover test is a lateral tilting test (see Fig. 2). The complete vehicle is standing on the tilting platform, with blocked suspension and is tilted slowly on its unstable equilibrium position. If the vehicle type is not fitted with occupant restraints it will be tested at unladen curb mass. If the vehicle is fitted with occupant restraints, it will be tested at total effective vehicle mass.

The rollover test starts in this unstable vehicle position with zero angular velocity, and the axis of rotation passes through the wheel-ground contact points. The vehicle rollover into a ditch, having a horizontal, dry, and smooth concrete ground surface with a nominal depth of 800 mm [3].

The rollover test shall be carried out on the side of the vehicle that is more dangerous with respect to the residual space. This decision was made by the technical service on the basis of the manufacturer's proposal, considering at least the following:

– The lateral eccentricity of the center of gravity and its effect on the reference energy in the unstable starting position of the vehicle.

– The asymmetry of the residual space.

– The different asymmetrical construction features of the two sides of the vehicle, and

the support given by the partition or inner boxes (e.g. wardrobe, toilet, and kitchenette).

The side with less support shall be chosen as the direction of the rollover test. The latest version of ECE R66, version 2006, with above requirements, describes a test to be chosen among five different methods:

(1). Complete Vehicle Rollover Test.

(2). Body Section Rollover Test.

(3). Body section test with Quasi-static load.

(4). Component testing base on Quasi-static calculation.

(5). Complete vehicle rollover test base on computer simulation.

Method (1) was accepted as the standard method. Others are equivalent methods. In which, method (3) and (4) are new methods in ECE R66, version

2006. (1), (2) and (3) are experimental methods base on real test. The Body Section Pendulum Impact Test of previous version had been deleted from Regulation No. 66 due to experimental tests and numerical studies found that it was not equivalent [4, 5]. The method (5) is officially accepted with full scale computer simulation [3]. In this study, method (5) was used to perform a numerical analysis because of the LS-DYNA ability and a powerful computation of the computer.

2.2 FMVSS 220 standard

The FMVSS 220 was effected from on Jan 4th, 1977, enforced by Department of Transportation, the United States of America. The FMVSS 220 is the school bus rollover protection regulation, specifies performance requirements for school bus rollover protection. This standard increases the structural resistance of school buses in rollover-type accidents. It only applies to school buses and covers all styles of school bus and transits buses, vans having less than 35 feet length. The requirements of this regulation are displacements of the application plate shall not exceed 5-1/8 inch (130.175 mm) and capable of being opening of emergency exits during the full application of the force and after release the force, with a force equal to 1.5 times the unloaded vehicle weight (UVW) shall be applied to the roof of vehicle's body structure through a flat, rigid, rectangular force application plate at any rate not more than 12.7 mm (1/2 inch) per second, as in Fig. 3 [6]. The FMVSS 220 test is real test however, this research makes use of full advantage of computer simulation power tries to perform by numerical simulation meet package of FMVSS 220 requirements.

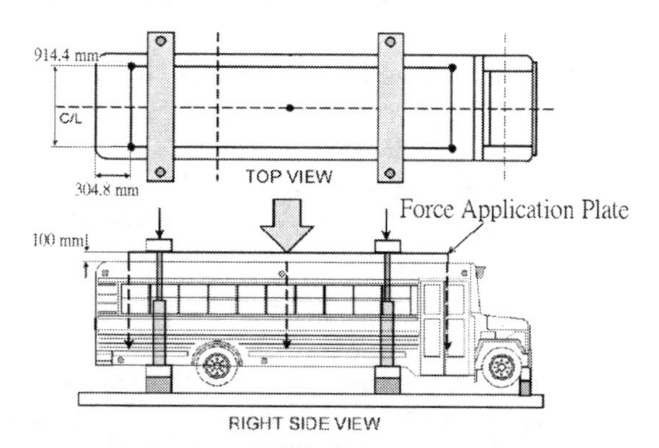

Figure 3: The FMVSS 220 standard test [6].

Comparing with the ECE R66, the FMVSS 220 only supports one testing method that is a quasi-static test of vertical compressing of vehicle roof without concern of CG position. Whilst the ECE R66 support five methods for testing, where the lateral rollover test concerns the impact of CG position on the evaluation results.

3 LS-DYNA introduction

LS-DYNA was developed by LSTC (Livermore Software Technology Cooperation). It is a multifunctional applicable explicit and implicit Finite-Element program to simulate and analyze highly nonlinear physical phenomena obtained in real world problems. Usually such phenomena manifest large deformations within short time durations, e.g. crashworthiness simulations. The significant features of LS-DYNA are the fully automatic definitions of contact areas, the large library of constitutive models, the large library of element types and the special implementations for the automobile industry [7, 8].

This study uses the FE software to carry out the bus rollover and the bus roof compressing simulation. The behavior of the bus rollover simulation belongs to the area of transient, dynamic, nonlinear, large deformed problems. And the bus roof compressing simulation belongs to the area of transient, quasi-static, nonlinear, large deformed problems. The Finite Element Analysis code, LS-DYNA, is a favorite tool for both of these two problems which often include contact and impact. The main solution is based on explicit time integration. The package LS-DYNA software contains a pre-processing finite element model builder (FEMB), an LS-DYNA solver and a post-processing LS-PREPOST. With LS-DYNA, the standard input such as geometry, mesh density, materials, element properties, boundary conditions, and contact modes can be used. The LS-DYNA solver will produce solutions. The output results such as stress and strain on elements, displacement, velocity and acceleration of nodes and energy distribution etc. can be shown clearly through the user interface (Hallquist [7]; LSTC [8]). The main solution is based on explicit time integration.

4 Numerical analysis procedures for bus rollover protection

Computational models of vehicles are convenient for the comparison of respective performance between the ECE R66 and FMVSS 220 regulations. This study prepares and considers two models; one is an original model, and one is a strengthened model of vehicle.

4.1 Original model – model I

This FE vehicle model used for simulation is based on a full scale bus model developed at Da-Yeh University, Taiwan for rollover crashworthiness investigation and evaluation of reinforcement structures [9–11]. It includes 68132 elements. These consist of 67084 quadrilateral elements, 914 triangular elements, 35 hexagons and 99 mass elements. The description is based on the shell elements and their materials as in the Table 1. All deformable parts are modelled with the 4-noded Belytschko-Tsay shell elements with three integration points through the shell thickness. The shell element formulation is based on Belytschko-Lin-Tsay formulation with reduced integration available in LS-DYNA [7, 8]. This element is generally considered as computationally efficient

Table 1: Material for simulating bus superstructure, chassis, tyres, axis and tilting plate.

	Bus Superstructure	Chassis of Bus	Axis and tilting plate	Tires
Density (ton/mm3)	7.83E-09	2.783E-08	7.83E-09	2.85E-09
Young's Modulus (N/mm2)	210,000	205,000	200,000	11,000
Poisson's Ratio	0.28	0.3	0.3	0.3
Yield Stress (N/mm2)	282	270		
Plastic Fracture Pressure (N/mm2)	3.76E08	1.0E08		

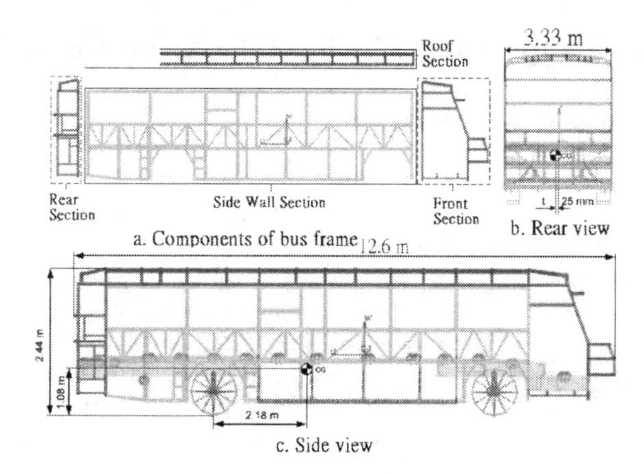

Figure 4: Full-scale FE bus model.

and accurate [12]. The shell element that has been, and still remains, the basis of all crashworthiness simulation is the 4-noded Belytschko-Tsay shell.

The CG (Center of Gravity) of the vehicle was measured using a test platform at the ARTC (Automotive Research & Testing Center, Taiwan, R.O.C). The measured values were in good agreement with the ones coming from the FEA model. To exactly match the measures and calculated CG, the CG of engine, gearbox and the axles were fine tuned in the FEA model. The unloaded vehicle weight is 7716.47 kg (7.71647 ton), and its capacity is 49 passengers. The vehicle size and its position of CG are shown in Fig. 4.

4.2 Strengthened model – model II

The strengthened model was developed from the original model following the design criteria of Roca et al. [13]. Each structural connection used the reinforcement obtained by Chiu [9], and bus frame thickness considered followed Liang and Le [14]. This model II has the same design style and vehicle size as model I. The unloaded vehicle weight of model II is 7916 [kg], and its height of CG is 1.10 [m].

4.3 Survivor space definition of a bus

For estimation of ECE R66 requirements, the survivor space was specified in the FEMB in line with the statement in the ECE R66. Throughout the whole vehicle, the SR points are located on the seat-back of each outer forward or rearward facing seat, 500 [mm] above the floor under the seat, and 150 [mm] from the inside surfaces of the side walls of the vehicle. The model of the survivor space consists of a rigid shell frame in each section along the vehicle interior (Fig. 1), rigidly mounted in the stiff region under the floor. There is no stiff connection between these rigid shell frames because these shell elements are modeled with "NUL_MATERIAL" for visualization only.

5 Numerical experiments for bus rollover protection

5.1 ECE R66 numerical simulation

The testing model is established by full scale bus model as in Fig. 4 and tilting platform model as in Fig. 5. According to ECE R66, the initial condition is that the angular velocity of the tilted platform shall not exceed 5 degrees / sec (0.087 radians / sec) [3]. To reduce the computing time, the testing model can be already rotated to reach its just before unstable position. The boundary condition is the vehicle model shall be tilted without rocking and without dynamic effects until it reaches unstable equilibrium and commences its rollover. The solving algorithm for that contact and this ECE R66 simulation is based on the explicit LS-DYNA solver. The problem time is 7 sec and a fixed time step of 0.9 microseconds is used. The simulation is carried out by the LS-DYNA version 971 with the 4-CPU Workstation. The CPU time for each of these two ECE R66 simulations of the original model and strengthened model is about 12 hours. The simulation process is shown in Fig. 6.

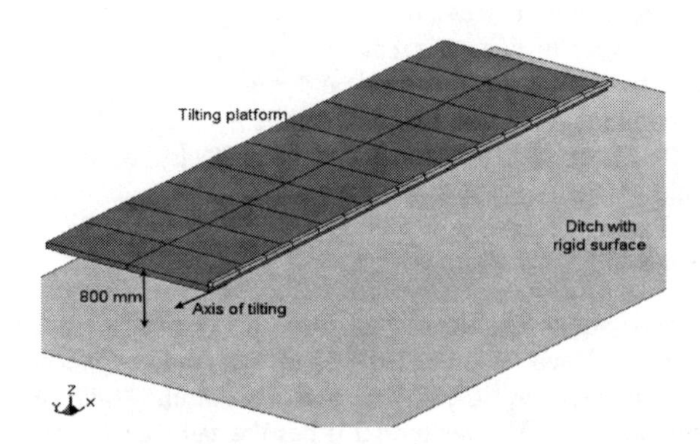

Figure 5: Tilting platform.

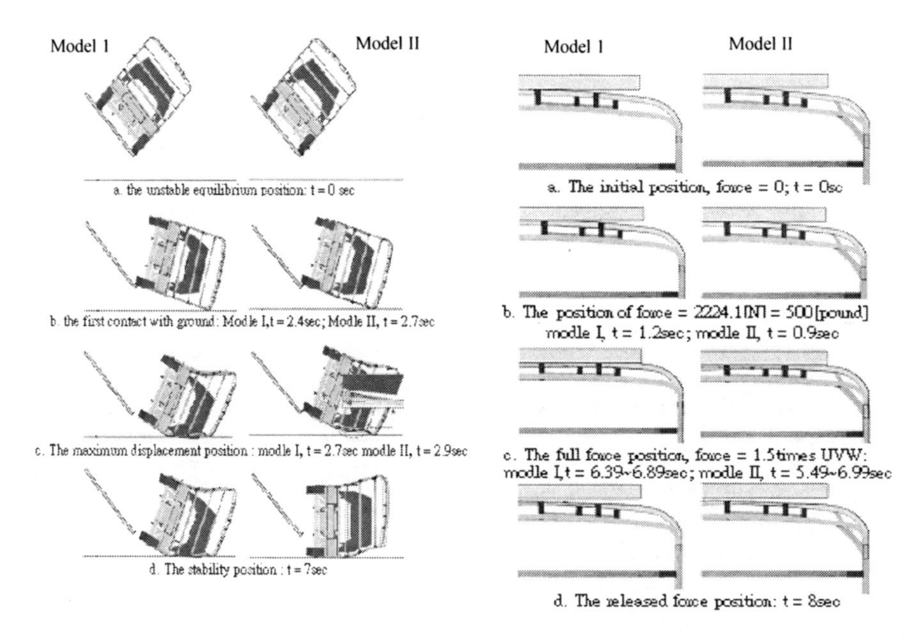

Figure 6: Rollover process versus time for model I and II.

Figure 7: FMVSS 220 simulation process versus time for model I and II.

5.2 FMVSS 220 numerical simulation

The testing model is established by the full scale bus model as in Fig. 4 and a force application plate model as in Fig. 3. The force application plate is determined with respect to longitudinal and lateral centerline. It is 304.8 mm (12 inches) shorter than the roof and 914.4 mm (36 inches) wide. This place weight 1528.289 kg. According to FMVSS 220, the initial condition is the direction of the force to application plate at continuous rate of not more than 12.7 mm (½ inch) per second until the force is equal to 1.5 times the UVW (unloaded vehicle weight). The boundary condition is evenly distributed vertical force, and the record on the distance versus time plots the deflection where the downward force is 500 lbs in order to eliminate slack from the system, and the solving algorithm for that contact and this FEVSS 220 simulation is based on the explicit LS-DYNA solver. The problem time is 8 sec and a fixed time step of 0.9 microseconds was used. These simulations are carried out by the LS-DYNA version 971 with the 4-CPU Workstation. The CPU time for each of these two FMVSS 220 simulations is about 13 hours. The simulation process is shown in Fig. 7.

6 Results

By FE analysis using the LS-DYNA 971solver, two rollover problems were simulated, one following ECE R66, one following FMVSS 220. Each simulation was performed with two vehicle models: model I is the original vehicle model, and model II is a strengthened model following Roca et al. [13], Chiu [9] and Liang and Le [14].

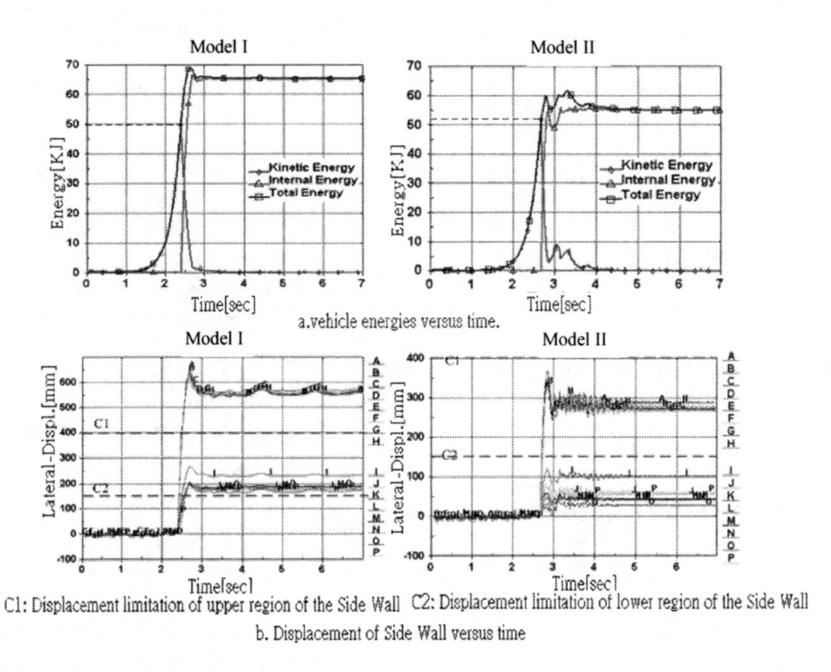

Figure 8: Vehicle energies and displacements of the side wall versus time for models I and II.

Fig. 8 displays the vehicle energies and deformations of vehicle frame versus the survivor space of both models following ECE R66 during and after rollover. The energy balance, Fig. 8a, is one of the rules for verification of simulation by itself. Where, the energies may be observed. The kinetic energy drops and transform into internal energy. When the kinetic energy is gone to zero, the total energy is the internal energy (a summary of plastic and elastic strain energy), which is one of the indications for correct analysis results. The deformations at considered points are shown in Fig. 9 of outside parts versus the survivor space which is shown in Fig. 8b. These figures display clearly the status of the two vehicle models and that the vehicle model I does not satisfy the ECE R66 and model II satisfies the ECE R66.

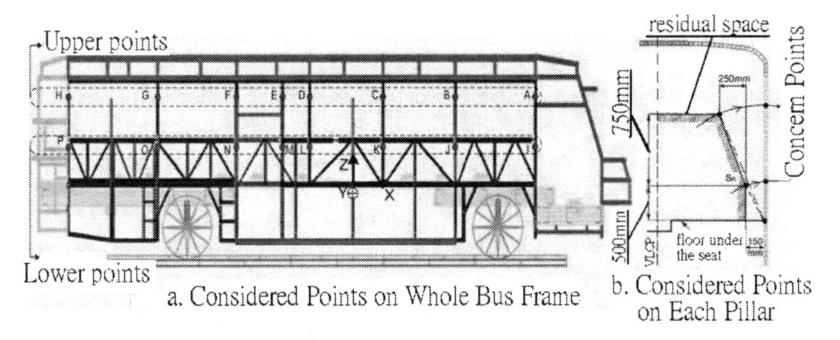

Figure 9: Concerned points on the bus frame for ECE R66 test.

Fig. 10 displays the application forces applied to the vehicle roofs and the displacements of the force application plates of two vehicle models. Loading curve with three stages of the roof compressing, including quasi-static increasing force up to 1.5 times UVW, maintaining the maximum value of force, and releasing the force that are shown in Fig. 10a for model I and model II. These are the FMVSS 220 requirements for the application force. Displacements of the force application plate (Fig. 10b), express the status of those two vehicle models following the FMVSS 220. Maximum displacements of the force application plate of both vehicle models do not exceed 130.175 mm.

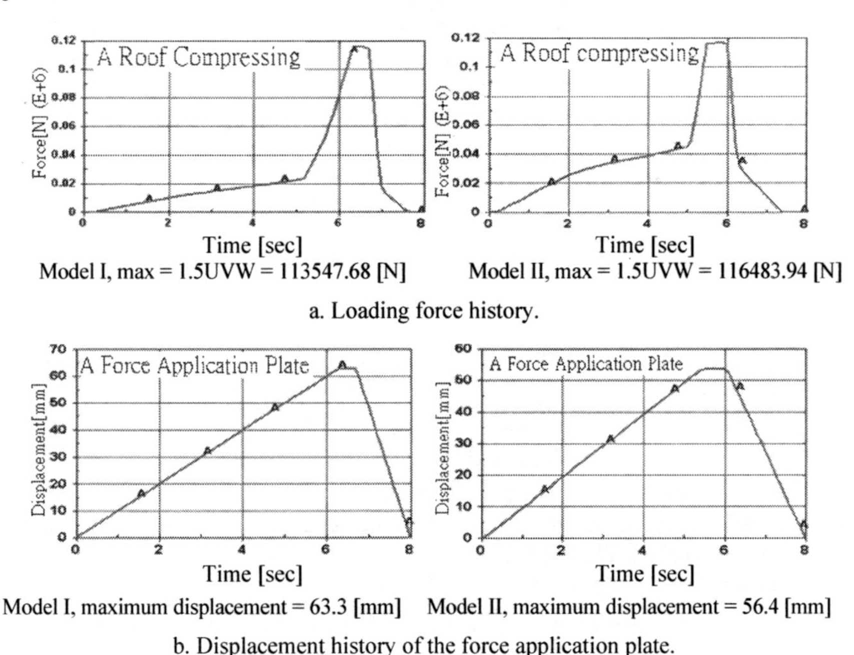

Figure 10: Model I and model II, loading force and displacement histories for the force application plate.

7 Conclusions

The objectives of this research are vehicle deformation configuration analysis in accordance with energy absorption and large bus rollover protection analysis following the legislation for standard. The following conclusions can be drawn:

(1) This study established two numerical simulation processes of the bus rollover event, one following ECE R66, and one following EMVSS 220. Thus, the strength of vehicle superstructure as well as the survivor space status and deformation modes of vehicles can be estimated.

(2) The absorbed energy of the vehicle superstructure and its distribution (Table 2) expressed the distortion configuration of body structure according to ECE R66 and FMVSS 220. A significant difference between the requirements of ECE R66 and FMVSS 220 for the body structure strength is discovered. With 57.16% total absorbed energy, the side wall section is the highest requirement following ECE R66. However, with 50.01% of the total absorbed energy, the roof section is the highest requirement following FMVSS 220.

(3) The structural behaviors of a coach were different when submitted to the two different assessment procedures prescribed by the ECE R66 and FMVSS 220. However, the investigation of distortion configuration of two testing models

(4) showed that the passenger compartment and residual space were more violated and more dangerous under the lateral rollover testing condition of ECE R66. The physical meanings of the bus rollover safety regulations (ECE R66 and FMVSS 220) are considered followed the regulations' requirements as well as the distortion configuration of vehicle superstructure under rollover conditions. In this study shock resistant capability of the ECE R66 and the FMVSS 220 was compared: one bus superstructure passing the FMVSS 220 assessment may not pass the ECE R66 assessment.

References

[1] NHTSA, Initiatives to Address the Mitigation of Vehicle Rollover, 2003.

[2] JASIC, ECE Regulation No.66 S1 - Strength of Super Structure, Economic Commission for Europe, pp1-19, 1998.

[3] JASIC, ECE Regulation No.66 01 - Strength of Superstructure, Economic Commission for Europe, pp 1-49, 2006.

[4] UNECE, GRSG-93-04 http://www.unece.org/trans/main/wp29/wp29wgs-/wp29grsg/grsginf93.html, 2006.

[5] Schwartz, W. Simulations with LS-DYNA for Registration Approval of a Coach according to ECE R66 Regulation. International Congress Center Dresden, Germany, 2004.

[6] NHTSA, FMVSS 220 standard – School Bus Rollover Protection, TP-220-02, Department of Transportation, Washington, DC 20590, 1991.

[7] Hallquist, J.O. LS-DYNA Theoretical Manual, Livermore Software Technology Corporation, 2006.
[8] LSTC, LS-DYNA Users Manual, Livermore Software Technology Corporation, 2007.
[9] Chiu, X.-T. The design and Evaluation of Reinforcement Structure for Bus Frame. Master thesis, Da-Yeh University, 2007.
[10] Chai, Y.-X. Numerical Simulation and Analysis of Bus Rollover, Master thesis, Da-Yeh University, 2005.
[11] Chang, R.-H. A Study on Increasing Structural Strength of Bus, Master thesis, Da-Yeh University, 2006.
[12] Guler, M.A. Elitok, K. Bayram, B. Stelzmann, U. The influence of seat structure and passenger weight on the rollover crashworthiness of an intercity coach. International journal of Crashworthiness, 12/6, pp.567-580, 2007.
[13] Roca, T. Arbiol, J. and Ruiz, S. Development of Rollover Resistant Bus Structures. Society of Automotive Engineers 970581, 1997.
[14] Liang C.-C. and Le, G.N. A Developing Method of the Bus Rollover Strength by Investigating and Re-distributing the Ability of Energy Absorption. The13th National Conference on Vehicle Engineering, Taiwan, 2008.

Damage detection of strengthened RC frame model with FRP sheets under lateral loads

A. Vimuttasoongviriya, N. Kwatra & M. Kumar
Department of Civil Engineering, Thapar University, India

Abstract

For the purpose of assuring seismic safety in north India, it is necessary to monitor the damaged state of existing reinforced concrete (RC) buildings. Non-ductile structures often result in the need of strengthening to increase the lateral load carrying capacity. This paper intends to investigate the effect of lateral load on damage indexes of RC frame model strengthened with fiber reinforce polymer (FRP) sheets. Park and Ang damage theory and stiffness damage index method were used. These damage indexes are expressed as a formula based on deformation, energy dissipation and change in stiffness. The damage indexes based on the change in dynamic characteristics namely modal plastic softening index and modal flexibility damage index were used with the help of impact hammer excitation test. Results of this study show that the use of FRP wrapped for structural strengthening provides significant lateral load capacity increases when compared to original specimen. Damage indexes based on deformation and change in stiffness show a much acceptable accuracy correlation with modal parameters indexes in general.
Keywords: strengthened RC frame model, FRP sheets, damage indexes, displacement ductility, non-linear FE analysis.

1 Introduction

For the purpose of assuring seismic safety, it is necessary to monitor the damaged state of structures. Many existing buildings in north India are non-ductile RC structures. These buildings often result in the need of strengthening or retrofitting to increase lateral load carrying capacity. Un-strengthening building may potentially lead to more seismic damage under future earthquake. The strengthening technique using FRP sheets has been successfully used to prevent

WIT Transactions on The Built Environment, Vol 113, © 2010 WIT Press
www.witpress.com, ISSN 1743-3509 (on-line)
doi:10.2495/SU100051

its brittle shear failure and significantly improved the displacement ductility and energy dissipation capacity [8]. This process is also an effective method for upgrading deficient RC connections [9]. FRP wrapped is one popular strengthened method because FRP with epoxy resin have received considerable attention due to its high strength, light weight, quick and easy manageability on-site and high resistance against corrosion [3].

It is necessary to monitor occurrence, location and extent of damage status of structures. The well known damage index is proposed by Park and Ang [11]. This index is calculated as a linear combination of maximum displacement response and total hysteretic energy dissipation. Biddah et al. [1] and Kanwar et al. [5] suggested the method of stiffness damage assessment that explains damage indicator based on changes in structural stiffness. This method is represented by a decrease in stiffness. Damage detection by calculating the change ratio of modal frequency has been widely applied in damage alarming in health monitoring systems of highway bridges [6, 7]. The modal flexibility damage index is the most well known one [10]. The principle of this method is on the basis of the comparison of the flexibility matrices obtained from two sets of mode shapes. Another advantage index is structural ductility. It is used to defined the ability of structure undergo inelastic deformation with stiffness and strength reduction. The collapse of brittle system always occurs suddenly beyond the maximum resistance.

Response of whole RC frame model subjected to external loads is a problem which is important to understand and there is little point in performing analysis without testing since the accuracy of the analysis cannot be verified. This paper intends to investigate both damage detection and effects on strengthened RC frame model under quasi-static load. Finite element (FE) method was used for modelling, nonlinear analysis and results processing of the specimen.

2 Damage detection methods

Traditionally, damage indexes have been used to monitor damage status of structures. It expresses performance in terms of a value between 0 (undamaged) and 1 (collapse or ultimate state). The damage detection methods can be calculated as follows.

2.1 Park and Ang damage index

Park and Ang [11] have formulated a damage index to estimate the level of damage in RC structures subjected to cyclic loading:

$$DI_{Park} = \frac{\delta_{\max}}{\delta_u} + \frac{\beta}{\delta_u P_y} \int dE_h \tag{1}$$

where $\delta_{\max}$ is the maximum experienced deformation, δ_u is the ultimate deformation of the element, P_y is the yield strength of the element, $\int dE_h$ is the hysteretic energy absorbed by the element and β is a model constant parameter. DI_{Park} is combined between the change in deformation ratio and damage due to the energy dissipation. The deformation ratio between ultimate and yield point calls displacement ductility. It is an advantage index to defined structural ability undergoes inelastic deformation with stiffness and strength reduction. Relation between DI_{Park} and various damage states is presented in table 1.

Table 1: Relation between *DI* and various damage states [11].

Damage State	Damage Index, *DI*	State of Building
No Damage	0.0	No Damage
Slight Damage	0.0-0.1	No Damage
Minor Damage	0.1-0.25	Minor Damage
Moderate Damage	0.25-0.4	Repairable
Severe Damage	0.4-1.0	Beyond Repair
Collapse	> 1.0	Loss of Building

2.2 Stiffness damage index method

Biddah et al. [1] proposed stiffness damage index method that uses an indicator based on the relationship between the material stiffness properties of the undamaged and the damaged member of the structure. According to this method severity of damage is expressed as the fractional change in stiffness of an element [5, 12]:

$$DI_k = \frac{k_j - k_j^*}{k_j} = 1 - \frac{1}{\nu_j} \tag{2}$$

where ν_j is stiffness ratio, k_j and k_j^* are the initial stiffness and damage stiffness of the jth member. The asterisk (*) denotes the damage state.

2.3 Dipasquale and Cakmak damage index

Dipasquale and Cakmak [2] defined the modal plastic softening index for the one-dimensional case, where the fundamental eigen frequency is considered. This damage index is given by

$$DI_{Dip} = 1 - \frac{\omega^{*2}}{\omega^2} \tag{3}$$

where ω and ω^* are the fundamental eigen frequency and damage frequency, respectively.

2.4 Modal flexibility damage index method

The principle of modal flexibility damage index method is based on the comparison of flexibility matrices obtained from two sets of experimental fundamental frequency and mode shape [6, 13]. The method is applicable if the mode shapes are mass normalized to unity. The damage index for the lth story using modal flexibility is defined as

$$DI_{MFDI} = 1 - \frac{F_l}{F_l^*} = 1 - \frac{\sum_{i=1}^{N} \phi_{li}^2 / \omega_i^2}{\sum_{i=1}^{N} \phi_{li}^{*2} / \omega_i^{*2}} \tag{4}$$

where F_l is the static displacement due to a unit static load applied at the lth degree of freedom which was used directly as a damage indicator. ϕ_{li} is the ith mode shape and ω_i is modal frequency.

3 Experimental setup

3.1 Specimen detail and testing procedure

A three story non-ductile RC moment resisting frame model was manufactured. Beam-column joints were built without transverse reinforcement. Each story was identical in most of the geometrical aspects. The frame model consisted of three slabs 2000 mm x 2000 mm x 50 mm. Each column was equally sized rectangular of the cross section 100 mm x 100 mm (four 8 mm diameter bars) with height of 950 mm floor-to-floor. All the beams were equally sized rectangular of 100 mm x 150 mm (two 10 mm diameter bars at the tension and compression faces). All columns and beams were provided with 6 mm diameter stirrups. Each column cast integrally with 150 mm x 200 mm x 400 mm stub foundation. The stub was in turn bolted firmly on strong floor. The material details, schematic drawing and test set-up of the frame model are presented in table 2 and fig. 1. Each floor was equipped with one displacement dial gauge and one accelerometer of 5 kHz frequency in the horizontal direction. A hydraulic jack of 200 kN capacity was horizontally installed along the desired direction at top floor. The frame model tested under quasi-static loads as shown in fig. 2 to simulate a change in structural damage. After applied each load step, an impact hammer of sensitivity 0.25 mV/N was used to excite the structure. Before initiating the monitoring calibration of impact hammer and accelerometers was carried out, according to which the sensitivity was assigned to eight channels of the Fast Fourier Transforms (FFT) spectrum analyzer for recording dynamic characteristic data.

Table 2: The details of the materials.

Compressive strength of concrete			20 MPa	
Tensile strength of steel bars (MPa)		Diameter	Yield	Ultimate
		10 mm	475.68 MPa	586.60 MPa
		8 mm	516.65 MPa	628.91 MPa
Mechanical properties of FRP laminate				
Fibre	Thickness	Density	Tensile	*E*
GFRP	0.34 mm	2.6 g/cm^3	3.4 GPa	63 GPa

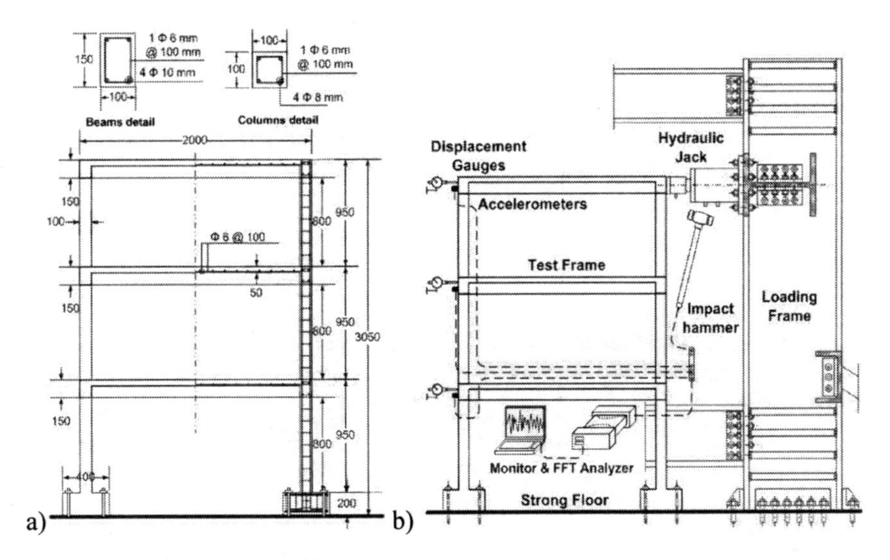

Figure 1: a) Schematic drawing of control frame model and b) experimental test set-up. All dimensions in millimetres.

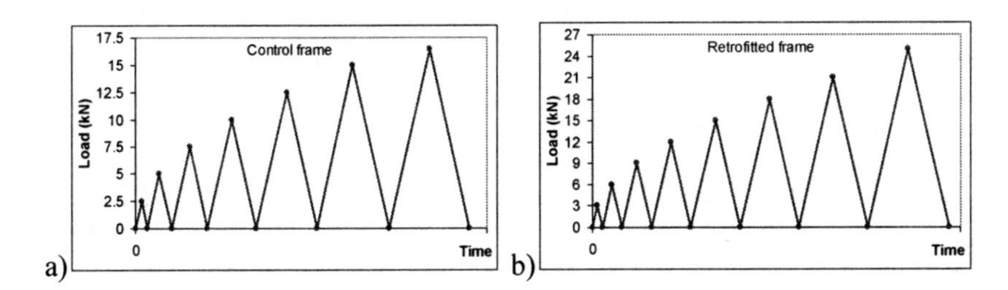

Figure 2: Applied load history: a) control frame and b) strengthened frame.

3.2 Strengthening scheme

As earlier explained, the horizontal load was applied to the top floor of the model till the desired damage state was reached. The damaged control frame was then moved back to its initial state. Loose concrete was removed and the surfaces were cleaned of dirt. All the corners of damaged elements were bevelled and rounded to a radius of 10 mm. The small cracks were filled with adhesive epoxy. The concrete surface was applied by adhesive layer of MBrace primer and it was smoothed by MBrace concessive layer. Glass fibre reinforce polymer (GFRP) was used in this paper. Application of FRP wrap provided in two layers on the damaged elements as shown in fig. 3. The first layer was provided with fibre oriented along the beam or column axes, to increase their flexural strength capacity. The columns and beams were confined at each edge zones by wrapping the other layer in the transverse direction as well.

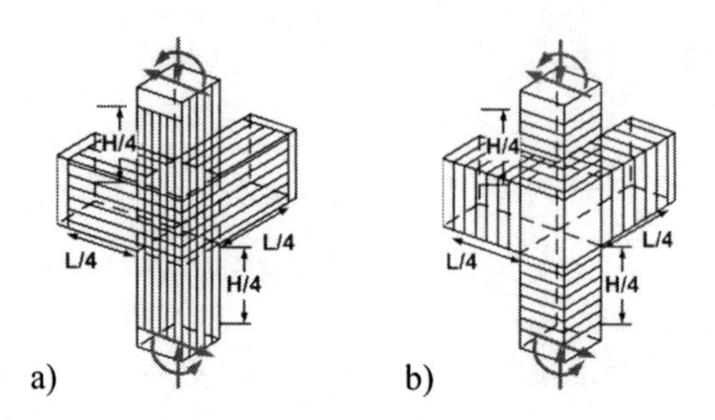

Figure 3: Application of FRP wrap: a) flexural layers and b) confinement layers.

4 Finite element modelling of retrofitted RC structure

4.1 Element types and material properties

The ATENA nonlinear finite element program (ATENA-3D v4) was used in this study to simulate the behaviour of the RC frame. An eight-node brick element was used to model the concrete. The element is capable of plastic deformation, cracking in three orthogonal directions and crushing. The nonlinear behaviour of compression is following Committee Euro International du Beton model (CEB-FIP Code 90). In tension, the stress-strain curve is approximately linearly elastic up to the maximum tensile strength. After this point, the concrete cracks and the strength decrease gradually to zero. The equivalent uniaxial stress-strain diagram for concrete is shown in fig. 4. A polyline element was used to model the steel bars. Two nodes are required for this element. Each node has three degrees of freedom. The steel for the FE models was assumed to be an elastic-perfectly plastic material, Poisson's ratio of 0.3.

A shell element with 20 nodes, quadratic 3D brick element, was used to model the FRP composite. This element allows for different material layers with different orientations. FRP composite is that consist of two constituents. The

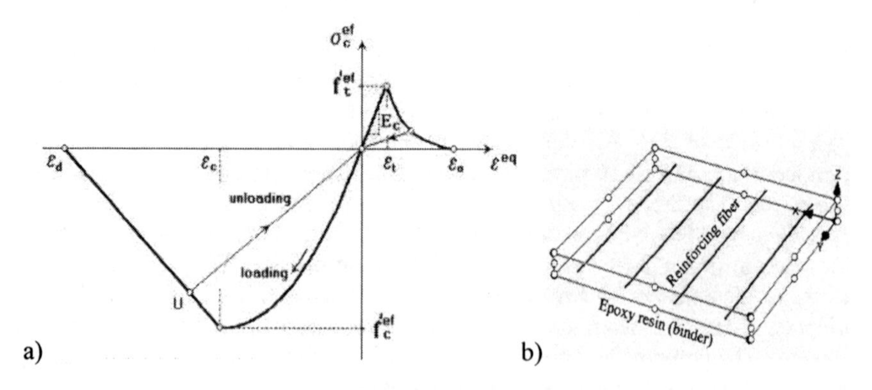

Figure 4: a) Uniaxial stress-strain law for concrete and b) Schematic of FRPs.

constituents are combined at a macroscopic level and are not soluble in each other. One constituent is the reinforcement, which is embedded in the second constituent of epoxy [4]. The reinforcing material is in the form of anisotropic materials of glass fibers, which are typically stiffer and stronger than the epoxy. Nodes of FRP layered shell elements were connected to those of adjacent concrete solid element in order to satisfy the perfect bond assumption as shown in fig.4. An eight-node brick element was used to model steel plates at the supports.

4.2 Loading, boundary conditions and nonlinear solution

By taking advantage of the symmetry of the frame, a symmetrical half of the full frame was used for modeling. The typical steel reinforcement location and strengthened frame for half of the entire model are shown in fig. 5. A one inch thick steel plate was added at the supporting and loading location in order to avoid stress concentration problems. At a plane of symmetry, the displacement in the direction perpendicular to the plane was held at zero. In nonlinear analysis, the loads applied to a finite element model are divided into a series of load increments called load history. The ATENA program uses Newton-Raphson equilibrium iterations for updating the model stiffness. The monitoring points were measured at the same location as for the experimental frame at the middle of each floor, where the largest horizontal displacements can be expected.

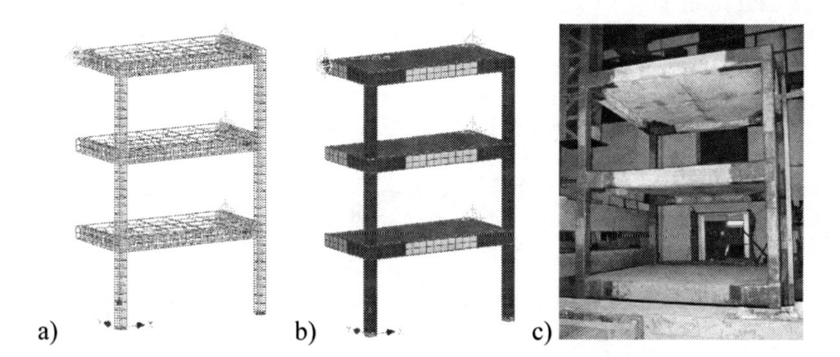

Figure 5: a) The typical steel reinforcement location model, b) strengthened haft frame model and c) experimental strengthened frame model.

5 Results and discussion

5.1 Structural behaviour and damage state of control frame

The control frame model was constructed. It was loaded in lateral direction at the top floor. Structural stiffness degradation can be observed from the load-displacement plot. After applied each load steps, the impact hammer was used to excite the testing frame model. The dynamic characteristics gave the records in FFT analyzer based on linear analysis setup. These records include trigger

hammer plot, time history plot which give damping, frequency response function (FRF) plot which give the amplitude of vibration along with frequency, respectively. The analyzer was set up to make a free zoom measurement with a frequency range of 0 to 50 Hz. In this frequency range there covered all three majority modes of this frame model. At undamaged state the natural frequency at first, second and third mode were 6.5 Hz, 19 Hz and 31.5 Hz respectively. The initial diagonal cracks occurred on the beam-column joints of the top floor at a load of 10 kN, DI_{Park} of 0.32, DI_k of 0.47, DI_{Dip} of 0.12 and DI_{MFDI} of 0.16, indicating that the elements of top floor are the most stressed, moderate damage state. At a load of 12.5 kN, large cracks started to open and small cracks occurred on connection joins of second floor. It indicated that the yield point was visible at a load 12.5 kN, displacement of 31 mm from initial state, DI_{Park} of 0.49, DI_k of 0.57, DI_{Dip} of 0.16 and DI_{MFDI} of 0.21, severe damage state. Ultimate damage state began at the load of 16.5 kN and with a displacement of 69 mm, leading to DI_{Park} of 0.95, DI_k of 0.76, DI_{Dip} of 0.34 and DI_{MFDI} of 0.34, respectively. The frequencies at ultimate state of first, second and third mode were 5.2 Hz, 15.5 Hz and 25.9 Hz respectively. The displacement ductility was nearly 2.22 and total energy dissipation was 567.57 kN-mm. Diagonal cracks occurred on connection zone of each floor and horizontal cracks occurred on columns near stub foundations. Failure mode and load-displacement plot of system are shown in fig. 6 and 7 respectively. Under quasi-static loads of non-ductile beam-column joints of this specimen, beams adjoining were subjected to shears and moments. Under these shears and moments, the top and bottom steel bars moved in the opposite direction. These forces were balanced by bond stress developed between concrete and steel bars. In such circumstances, the plastic hinges were formed and connection joints lost their capacity to carry load.

The relation between applied loads and damage indexes of control frame are shown in figs. 8 and 9 and summary in table 3. From these plots, it is worth mentioning here that the health of the non-ductile RC structure is said to be of reduces to more than 50% of initial state, and when modal parameters damage

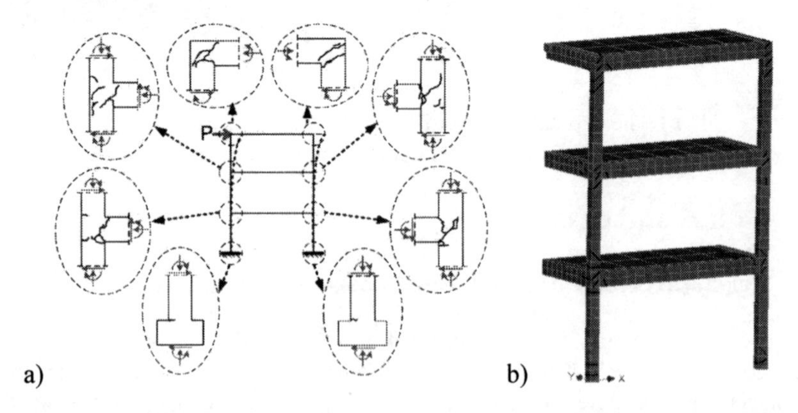

Figure 6: Failure mode of control frame: a) experimental frame, b) FE analysis.

a) 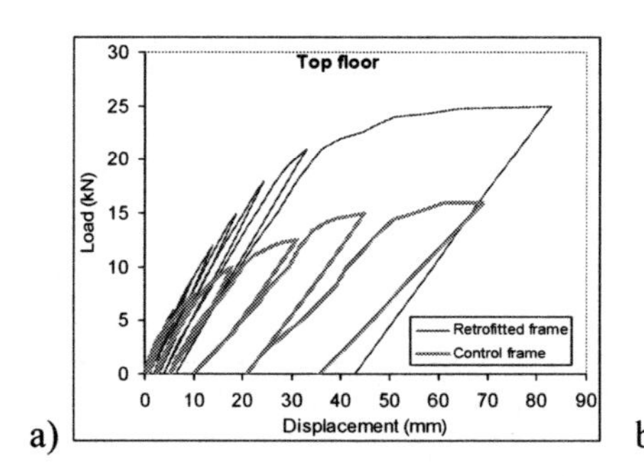

b) 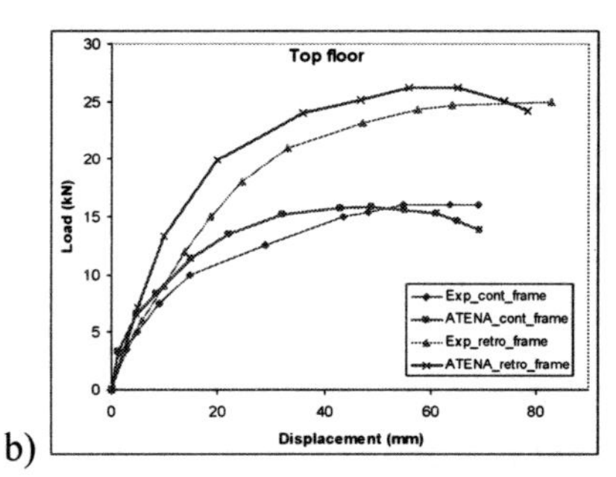

Figure 7: a) Load-displacement plot at top floor and b) comparison between backbone load-displacement plot of FE models and experimental results.

index increases to larger than 0.20. The vibrational damage detection methods seemed to increase slowly with increase in load. The average changed of corresponding natural frequencies decreased approximately 18.73%. In general, the effects of damaged structure on the changes in natural frequencies are found to be negligible.

Table 3: Damage index of structural system and appearance of control frame.

P (kN)	P/P_{max}	DI_k	DI_{Park}	DI_{Dip}	DI_{MFDI}	Appearance
0	0	0	0	0	0	Un-deformed
5.0	0.30	0.18	0.09	0.03	0.04	Un-cracked
7.5	0.45	0.25	0.18	0.06	0.09	Minor cracking
10.0	0.61	0.47	0.32	0.12	0.16	Moderate cracking
12.5	0.76	0.57	0.49	0.16	0.21	Severe cracking
15.0	0.91	0.65	0.61	0.25	0.26	Spalling of concrete cover
16.5	1	0.76	0.95	0.34	0.34	Loss of shear capacity

a) 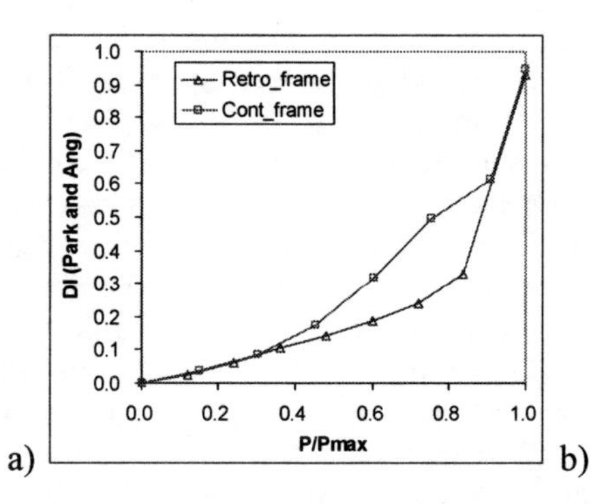

b) 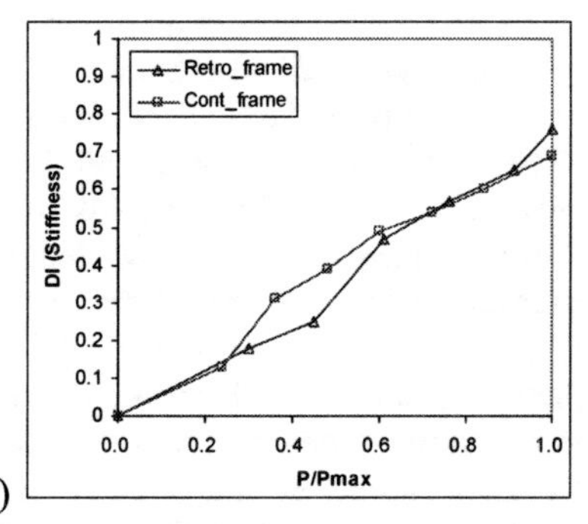

Figure 8: a) Park and Ang damage index and b) Stiffness damage index.

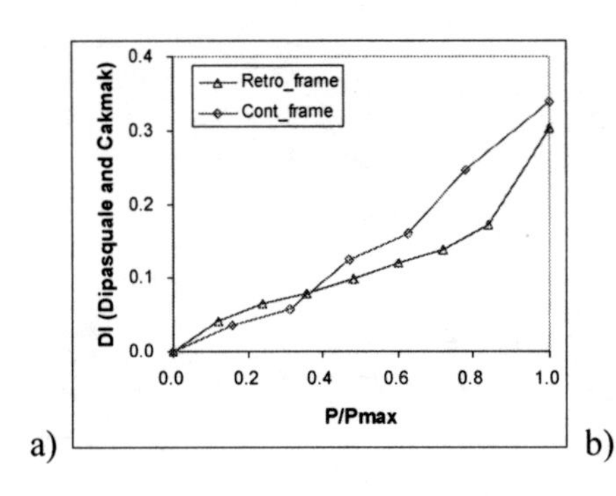

a)

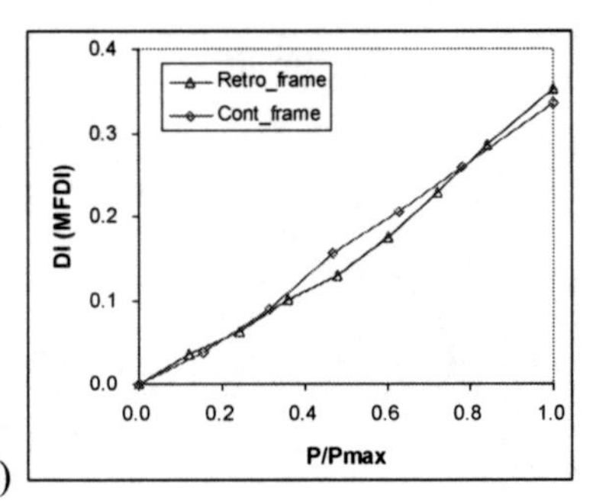

b)

Figure 9: a) Dipasquale damage index and b) Modal flexibility damage index.

5.2 Structural behaviour and damage state of strengthened frame

Test for the strengthened frame was performed in the similar manner as that for the control frame. At the final stage of the damaged control frame was grouting cracks by adhesive epoxy and wrapped it with FRP sheets. At initial state the natural frequency at first, second and third mode were 6.7 Hz, 20.7 Hz and 34.0 Hz, respectively. The yield damage state of the strengthened frame was visible at a load 18 kN, displacement of 24.8 mm, DI_{Park} of 0.24, DI_k of 0.54, DI_{Dip} of 0.14 and DI_{MFDI} of 0.23, severe damage state. Moving sound of FRP layers started from this load. Ultimate damage state was at the load of 25 kN at displacement of 83 mm from initial state and total energy dissipation was 1039.68 kN-mm. The frequencies at first, second and third mode were 5.5 Hz, 17.2 Hz and 29 Hz respectively. The damage index of DI_{Park} was 0.93, DI_k of 0.69, DI_{Dip} of 0.30 and DI_{MFDI} of 0.35. There were breaking sounds of fibre and epoxy layer from connection joints at ultimate state. Failure mode of strengthened frame was fibre layers slip as shown in fig. 10. The average changed of corresponding natural frequencies decreased approximately 16.51%. The load versus displacement behaviour is shown in fig. 7 along with the behaviour for the control frame. From this figure, the load displacement relation can be roughly considered to be linear when the load is smaller than or equal to 18 kN. After remove FRP layers, it was observed that in addition to old cracks which opened up, new flexural cracks also appeared at the connection joints and columns.

The relation between load and the damage indexes of strengthened frame are shown in figs. 8 and 9 and details are presented in table 4. Damage index curves of strengthened frame below the curves of control frame indicating better performance as compare to the control frame. The displacement ductility for strengthened frame was nearly 3.34. It shown that the FRP wrapped around the structural elements in this manner are intended to provide external confinement and crushing of the concrete cover at larger lateral displacements. Experimental results also approved that the use of FRP wrapped for structural strengthening provides significant lateral load capacity increases approximately 151.5% as compared to control frame. The ductile behaviour of the strengthened frame is largely restored after the FRP composite sheets are engaged.

Table 4: Damage index of structural system and appearance of retrofitted frame.

P (kN)	P/P_{max}	DI_k	DI_{Park}	DI_{Dip}	DI_{MFDI}	Appearance
0	0	0	0	0	0	Un-deformed
3.0	0.12	0.13	0.02	0.04	0.04	Un-cracked
9.0	0.36	0.39	0.10	0.08	0.10	Un-cracked
15.0	0.60	0.49	0.19	0.12	0.18	Noise of fibre moving
18.0	0.72	0.54	0.24	0.14	0.23	Severe damage
21.0	0.84	0.60	0.33	0.17	0.29	Breaking noise of fibre
25.0	1	0.69	0.93	0.30	0.35	Loss of shear capacity

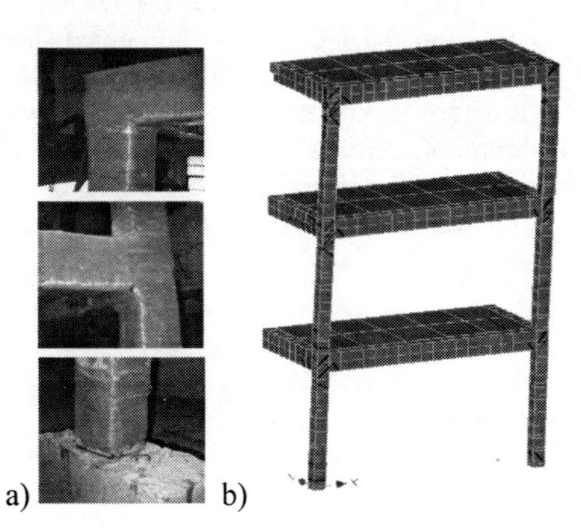

Figure 10: Failure mode of strengthened frame: a) actual frame, b) FE analysis.

5.3 Results of FE method and comparative analysis

Developed analytical models were validated by comparing the load-displacement results with existing experimental data. Fig. 7 shows that the backbone load-displacement plots from the FE analysis agree well with the experimental results. The model plots were stiffer than that from experimental results. The yield load for the FE analysis of control frame was 13 kN, which was higher than the load of 12.5 kN from the experimental results by 4%. Lastly, the ultimate load of 16 kN from model was lower than the ultimate load of 16.5 kN from the experimental data by 3%. Similar to the control frame, the yield load for the model of strengthened frame was 20 kN, which was higher than the actual frame of 18 kN by 11.11%. The ultimate load for the model was 26 kN, which was higher than the ultimate load of 25 kN for the actual frame by 4%. Figures 6 and 10 show the crack patterns at final state from FE models corresponded well with the observed failure mode of experimental control frame and strengthened frame after remove FRP sheets.

6 Conclusion

Based on the performed research investigation, the following main conclusions can be drawn: 1) Experimental results approve that the use of FRP wrapped for structural strengthening provides increased significant lateral load capacity. 2) The ductile behaviour of the strengthened frame is largely restored after the FRP composite sheets are engaged. 3) The health of RC structures of both original and strengthened specimen seemed to become unsafe when DI_{Park} and DI_k increases larger than 0.5 or dynamical damage indexes increase larger than 0.2. 4) The damage indexes of strengthened frame reduce indicating better performance as compare to the control frame. 5) Damage indexes based on deformation and change in stiffness show a much acceptable accuracy correlation with dynamical damage indexes in general. 6) Although the stiffness of the damaged RC structure is regained significantly by wrapping FRP jacket but it is not able to bridge the cracks fully. 7) The failure of the strengthened frame is due to breaking up of the bond between FRP layer and concrete. 8) The analytical results and crack patterns from FE method agree well with the experimental results.

References

[1] Biddah, A., Heidebrecht, A.C. & Naumoski, N., Use of push over test to evaluate damage of reinforced concrete frame structures subjected to strong seismic ground motions. *Proc. of the 7th Canadian Conf. On Earthquake Engineer,* Montreal, 1995.

[2] Dipasquale, E.l. & Cakmak, A.S., On the relation between local and global damage indices. *National Center for Earthquake Engineering Research*, Technical Report NCEER-89-0034, State University of New York at Buffalo, 1989.

[3] Ghobarah, A. & Elmandoohgalal, K., Seismic rehabilitation of short rectangular RC columns. *Earthquake Engineering*, **8(1)**, pp. 45-68, 2004.

[4] Kachlakev, D. & Miller, T., *Finite element modelling of reinforced concrete structures strengthened with FRP laminates.* Final report SPE 316: Oregon and Washington, 2001.

[5] Kanwar, v., Kwatra, N. & Aggarwal, P., Damage detection for framed RCC buildings using ANN modelling. *Damage mechanics.* **16(4)**, pp. 457-472, 2007.

[6] Ko, J.M., Sun, Z.G. & Ni, Y.Q., Multi-stage identification scheme for detecting damage in cable stayed Kap Shui Mun bridge. *Engineering Structures*, **24**, pp. 857-868, 2002.

[7] Lieven, N.A.J., & Ewins, D.J., Spatial correlation of mode shapes, the coordinate modal assurance criterion. *Proc. of the 6th Int. Conf. on Modal Analysis*, pp. 690-695, 1988.

[8] Memon, M.S. & Sheikh, S.A., Seismic resistance of square concrete columns retrofitted with glass-fiber reinforced polymer. *ACI Structural Journal*, **102(5)**, pp. 774-783, 2005.

[9] Mukherjee, A. & Joshi, M., FRPC reinforced concrete beam-column joints under cyclic excitation. *Composite Structures*, **70**, pp. 185-199, 2005.
[10] Pandey, A.K. & Biswas, M., Damage detection in structures using changes in flexibility. *Sound and Vibration*, **169(1)**, pp. 3-17, 1994.
[11] Park, Y.J. & Ang A.H.S., Seismic damage analysis of RC buildings, *Structural Engineering, ASCE*, **111**, pp. 740-757, 1985.
[12] Stubbs, N., Kim, J.T. & Tapole, K., An efficient and robust algorithm for damage localization in offshore platforms. *Proc. of '92 ASCE 10th Structures Congress*, Texas, 1992.
[13] Wang, J.F., Lin, C.C. & Yen, S.M., Story damage index of seismically-excited buildings based on modal parameters. *Proc. of the 18th Int. Conf. on Structural Mechanics in Reactor Technology*, pp. 3278-3289, Beijing, 2005.

Section 3
Impact and blast loading characteristics

Impact loading of ductile rectangular plates

N. Jones
Impact Research Centre, Department of Engineering,
University of Liverpool, UK

Abstract

In many industries, rigid-plastic methods of analysis are a useful design aid for safety calculations, hazard assessments and forensic investigations of ductile structures, which are subjected to large dynamic loads producing an inelastic rsponse. This paper examines the behaviour of a rectangular plate struck at the centre by a rigid mass impact loading. A theoretical method has been developed previously which retains the influence of finite transverse displacements, or geometry changes. It is used in this paper to predict the maximum permanent transverse displacements of plates having boundary conditions characterised by a resisting moment mM_0 around the entire boundary, where m = 0 and m = 1 give the two extreme cases of simply supported and fully clamped supports, respectively.

The theoretical predictions are compared with some experimental data recorded on fully clamped metal rectangular plates having a range of aspect ratios and struck by masses travelling with low impact velocities up to nearly 7m/s and which produce large ductile deformations without failure. The theoretical analysis gives reasonable agreement with the corresponding experimental data for masses having blunt, conical and hemisperical impact faces.

For sufficiently large initial impact energies, the projectile would perforate a plate and, for completeness, a useful design equation is presented which predicts perforation energies larger than all of the test data, as expected.

Keywords: rectangular plate, square plate, mass impact, rigid-plastic analysis, large permanent transverse displacements, metal plate, simply supported, fully clamped, experiments, perforation.

WIT Transactions on The Built Environment, Vol 113, © 2010 WIT Press
www.witpress.com, ISSN 1743-3509 (on-line)
doi:10.2495/SU100061

1 Introduction

Simple and reliable theoretical methods are still valuable for design purposes, particularly for preliminary design and hazard assessments, and for forensic investigations after accidents. A theoretical rigid-plastic method was developed in [1] which retained the influence of large transverse displacements (i. e., geometry changes, or membrane effects) and which has been used to predict the maximum permanent transverse displacements, or damage, of rectangular plates when subjected to a pressure pulse causing plastic strains. It was shown how this method can be simplified with an approximate yield condition which allows useful design equations to be obtained. This method was also used to examine the impulsive, or blast, loading of rectangular plates, and good agreement was found with experimental results recorded on ductile metal plates having various aspect ratios [2-4]. The method was extended to obtain the behaviour of circular plates [4-6] and square plates [6] struck by a solid mass at the centre, and again good agreement was reported with the maximum permanent transverse displacements observed in experimental tests on ductile metal plates.

This paper extends the above theoretical method to obtain the maximum permanent transverse displacements, or ductile damage, for a rectangular plate struck by a solid mass at the centre. It turns out that a relatively simple equation was obtained which gives good agreement with experimental data recorded on ductile metal rectangular plates having a range of aspect ratios from 0.4 to 1 and reported in [7].

Thus, design equations are now available for predicting the maximum permanent transverse displacements, or damage, for circular plates and rectangular plates (including square plates and beams) subjected to pressure pulses (including the limiting case of an impulsive loading) or central solid mass impacts. Moreover, these equations have been tested against experiments on ductile metal plates and are therefore suitable for design purposes, safety calculations and hazard assessments.

The next section of this paper provides a brief outline of the theoretical method which is used in section 3 to examine the behaviour of a rectangular plate subjected to a mass impact loading. Section 4 discusses briefly the experimental details of the data obtained on mild steel rectangular plates struck by a mass at the plate centre which produces large ductile deformations without any failure. Sections 5 and 6 contain a discussion and conclusions, respectively.

2 Theoretical method for dynamic loading of plates

A theoretical procedure was developed in [1] to provide the permanently deformed profile of an arbitrary shaped ductile plate, when subjected to large static or dynamic loads which produce plastic strains. The material is assumed to be rigid, perfectly plastic with a yield stress σ_o and the plate has a uniform thickness H and the governing equations can be simplified and written in the form

$$
\begin{aligned}
&-G\ddot{W}\dot{W} - \int_A \mu \ddot{w}\dot{w} dA = \\
&\int_A \left\{ \left(M_r + wN_r\right)\dot{\kappa}_r + \left(M_\theta + wN_\theta\right)\dot{\kappa}_\theta \right\} dA \\
&+\sum_{m=1}^{n} \int_{C_m} \left(M_r + wN_r\right)\left(\partial \dot{w} / \partial r\right)_m dC_m \\
&+\sum_{u=1}^{v} \int_{C_u} Q_r \left(\dot{w}\right) \; dC_u
\end{aligned} \tag{1}
$$

where G is an impact mass, and μ is the mass per unit surface area of a plate. The transverse displacement of a plate is w, while $\dot{w}$ and $\ddot{w}$ are the associated velocity and acceleration. W is the transverse displacement of the plate immediately underneath a striking mass.

The terms on the left hand side of equation (1) are the work rate due to the inertia forces, where A is the surface area of a plate. The first term on the right hand side of equation (1) is the energy dissipated in any continuous deformation fields. The second term gives the energy dissipated in n plastic bending hinges, each having an angular velocity $\left(\partial \dot{w} / \partial r\right)_m$ across a hinge of length C_m. The final term is the plastic energy absorption in v transverse shear hinges, each having a velocity discontinuity $\left(\dot{w}\right)_u$ and a length C_u. Equation (1) ensures that the external work rate equals the internal energy dissipation.

This general equation has been used to study the dynamic plastic response of beams, and of circular, square and rectangular plates subjected to dynamic pressure pulses and also blast loadings [1–4], and for beams, circular and square plates struck at the mid-span by a rigid mass [4–6]. It is used in this paper to examine a rectangular plate struck by a rigid mass at the centre, and, since large ductile deformations are studied without any material failure, or perforation, the transverse shear term in equation (1) is not considered further. Thus, the yield condition consists of four generalised stresses which can be related by the limited interaction surface shown in Figure 2 of [6]. However, if a deformation profile consists only of rigid regions separated by plastic hinges, then the exact yield condition in Figure 1 governs plastic flow along the hinge lines. A square yield condition circumscribes the exact yield condition (maximum normal stress yield criterion), while another one which is 0.618 times as large would inscribe it, thus providing upper and lower bounds on the exact yield condition according to a corollary of the limit theorems [4].

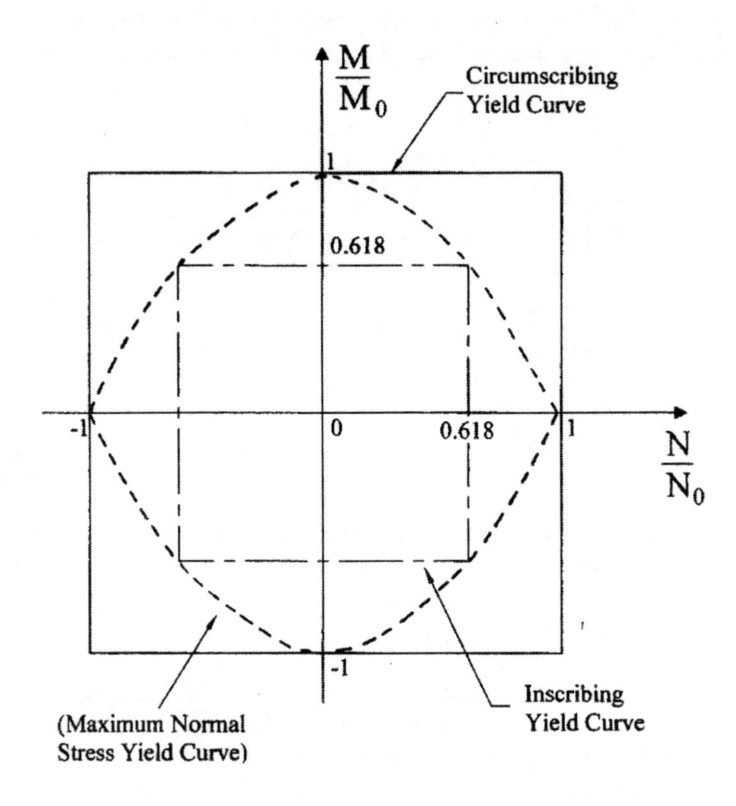

Figure 1: Yield conditions along the plastic hinge lines (including the supports) which develop within the rectangular plate in Figure 2.

3 Mass impact loading of a rectangular plate

The kinematically admissible transverse velocity field in Figure 2 is assumed to govern the response of the rectangular plate ($2Lx2B$) in Figure 3, which has a moment resistance mM_o around the supports (m = 0 and m = 1 give the simply supported and the fully clamped cases, respectively, but intermediate cases ($0 < m < 1$) are included in the analysis). It is assumed that the mass strikes the centre of a rectangular plate, has a negligible cross-section when compared with the plate dimensions, B and L, and produces a response characterised by the transverse velocity profile in Figure 2.

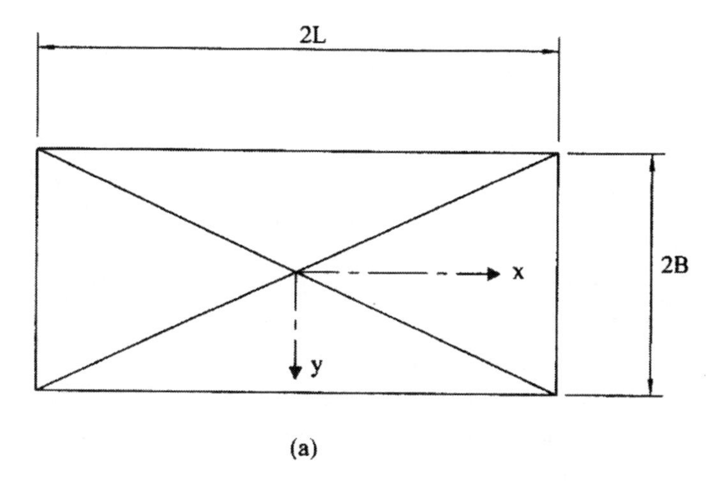

(a)

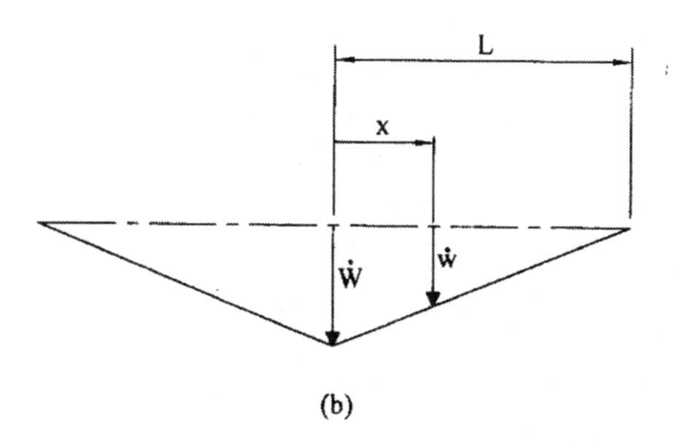

(b)

Figure 2: Pyramidal-shaped transverse velocity field for a rigid, perfectly plastic rectangular plate struck at the centre as shown in Figure 3. (a) plan view of plastic hinge lines which develop across diagonals and at boundaries (except for the simply supported case when $m = 0$ on the four boundaries). (b) side view (in y direction) of pyramidal-shaped transverse velocity field at time t.

Equation (1) can be written

$$-G\ddot{W}\dot{W} - \int_A \mu\ddot{w}\dot{w}dA = \sum_{m=1}^{r}\int_{l_m}(M + Nw)\dot{\theta}_m dl_m \tag{2}$$

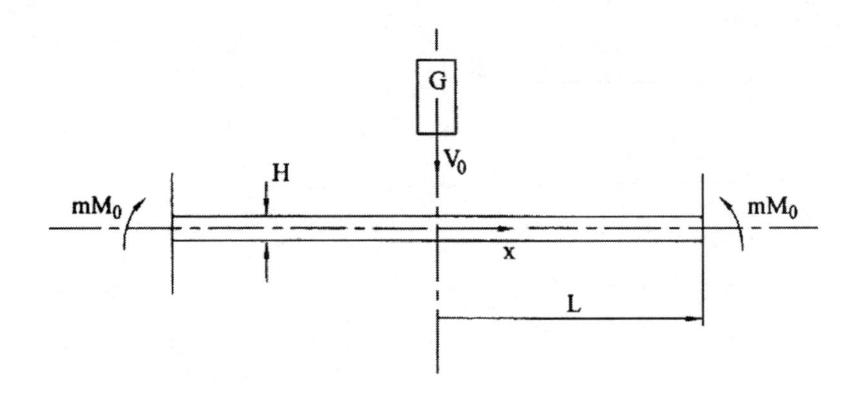

Figure 3: Rectangular plate struck at the centre by a rigid mass G travelling with an impact velocity V_0. The bending resistance around the four supports is mM_0, where $0 \le m \le 1$ and M_0 is the plastic limit moment per unit length of the plate cross-section.

for an initially flat, rigid, perfectly plastic plate which deforms into a number of rigid regions separated by r stationary straight line plastic hinges, each of length l_m. $\dot{\theta}_m$ is the relative angular velocity across a straight line hinge, w is the transverse displacement along a hinge, and N and M are the membrane force and bending moment, respectively, which act on a plane which passes through a hinge and is transverse to the mid-surface of a plate.

Equation (2) for the impact problem in Figure 3, when using the transverse velocity field in Figure 2 and the square yield condition in Figure 1, yields the governing equation

$$\ddot{W} + \alpha^2 W = -(1+m)H\alpha^2/2 \tag{3}$$

where $$\alpha^2 = \frac{12M_0(1+\beta^2)}{\mu HL^2(1+6\gamma)\beta^2}, \quad \beta = \frac{B}{L} \text{ and } \gamma = \frac{G}{4\mu BL} \tag{4a-c}$$

Solving equation (3) for the initial condition $w = 0$ at $t = 0$ and satisfying the conservation of linear momentum at $t = 0$ (i.e., $\dot{W} = V_0/\{1+1/3\gamma\}$, where V_0 is the initial velocity of the striking mass), then, for the fully clamped case with $m = 1$, the dimensionless maximum permanent transverse displacement is

$$\frac{W_f}{H} = \sqrt{\left\{1+\frac{2\beta\Omega}{(1+\beta^2)}\right\}} - 1, \tag{5}$$

where $$\Omega = \frac{GV_o^2}{4\sigma_0 H^3}. \tag{6}$$

Equation (5) for a square plate with β = 1 reduces to the corresponding equation developed in [6]. A more complete version of this analysis, which predicts the response duration and other features, will be published in due course.

4 Experimental details

The plates in the theoretical analysis reported in section 3 undergo wholly ductile deformations without any material failure or perforation. Some of the data for the perforation study reported in [7] on rectangular plates were obtained for lower impact velocities than those required for perforation and gave rise to wholly ductile behaviour of a rectangular plate without failure. This data is reported in Table 1.

Table 1: Experimental data for rectangular plates struck at mid-span. H = 4mm, σ_0 = 262 MPa, G = 11.8 kg for blunt-tipped projectiles and G = 19 kg for hemispherically-tipped and conically-tipped projectiles [7].

shape	β	$2L$x$2B$ (mm^2)	V_0 (m/s)	Ω	W_f / H
blunt	0.4	100x250	6.49	7.42	2.10*
blunt	0.5	100x200	6.49	7.42	2.03
blunt	0.66	100x150	6.26	6.90	2.20
blunt	0.66	100x150	6.42	7.25	1.83
blunt	1	100x100	6.42	7.25	1.84
blunt	1	200x200	6.57	7.60	2.12
hemisph'l	0.4	100x250	6.86	13.34	2.74*
hemisph'l	0.5	100x200	6.72	12.79	2.66
hemisph'l	0.66	100x150	6.57	12.23	2.51
hemisph'l	1	100x100	6.42	11.67	2.36
hemisph'l	1	200x200	6.86	13.32	2.91
conical	0.5	100x200	6.57	12.23	2.51*
conical	0.66	100x150	6.64	12.51	2.40*
conical	1	100x100	5.60	8.89	2.01
conical	1	200x200	6.64	12.51	2.76*

*Indicates that the plating has cracked underneath the projectile impact location.

The uniaxial tensile characteristics of the H = 4 mm thick mild steel plate are σ_0 = 262 MPa, σ_u = 359 MPa and ε_f = 37.5%. The projectiles have a 10.16 mm diameter (d) body, and the impact faces are blunt, hemispherical or conical (90° included angle). A drop hammer rig was used for the tests and the impact mass (G) was 11.8 kg for the blunt-tipped projectiles and 19 kg for the hemispherically-tipped and conically-tipped projectiles. The initial impact

velocity (V_0) was recorded using a laser Doppler velocimeter [8] and the maximum permanent transverse displacements were measured at the plate centre. Further details of the test arrangement is reported in [7, 9, 10].

5 Discussion

It is evident from equation (5) that the theoretical predictions for W_f / H collapse on to a single curve irrespective of the plate aspect ratio β when $2\beta\Omega/(1+\beta^2)$ is used in the abscissa, as shown in Figure 4. A yield stress $0.618\sigma_0$ inscribes the exact yield condition in Figure 1 and gives rise to the "upper bound" predictions according to equation (5) in Figure 4. The experimental data in Table 1 is also plotted in Figure 4 for values of $0.4 \leq \beta \leq 1$ and for three shapes of the projectile impact faces. First of all, it is evident that most of the

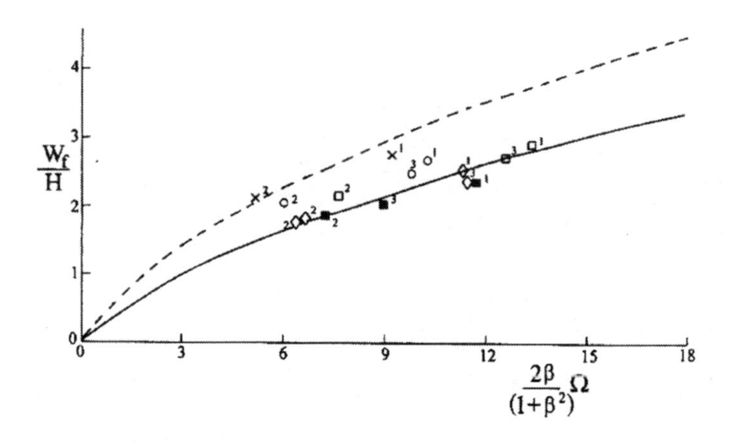

————: equation (5), circumscribing yield condition
– – – – – –: equation (5), inscribing yield condition
experimental data from Table 1:
X, ○, ◊; $\beta = 0.4, 0.5, 0.66$, respectively.
□, ■; $\beta = 1$ for 200mm x 200mm and 100 x 100mm plates, respectively.
superscripts 1, 2 and 3 denote projectiles with hemispherical, blunt and conical (90° included angle) impact faces, respectively.

Figure 4: Maximum dimensionless permanent transverse displacements at the centre of fully clamped rectangular plates versus the dimensionless initial kinetic energy.

experimental data is bounded by the "upper" and "lower" bound predictions of equation (5). Secondly, a trend is discerned which indicates that the experimental data migrates from the "lower bound" curve to the "upper bound" curve as β decreases for both the hemispherically-tipped and blunt-tipped projectiles. However, it should be noted that both of the $\beta = 0.4$ plates had some cracking underneath the impact site. The experimental results for the conical projectiles are not distinct from those for the two other projectile nose shapes and remained fairly close to the "lower bound" predictions of equation (5); all of these plates, except one, had cracked underneath the projectile impact location.

The rectangular plates in [7] are made from mild steel which usually is a strain rate sensitive material, although the dynamic material characteristics are not available for this particular plate material. The dynamic stress state in a rectangular plate is complex and is not uniaxial. In this regard, it is interesting to observe in Figure 11 of [11] that the maximum permanent transverse deflections of impulsively loaded rectangular plates made from aluminium alloy 6061T6, which is an essentially strain rate insensitive material, and mild steel, differ by about 10-15%, whereas the difference would be nearer 30-40% for a similar comparison on beams [3, 12] which have a less complex stress state. Nevertheless, if strain rate effects were incorporated into equation (5), as shown in [4] for blast loaded plates, then the theoretical curves in Figure 4 would be lower, although the strain rate effects for the low impact velocities in Table 1 are not expected to be large as those in [11] for blast loadings. It should be remarked that large plastic strains cause a reduction in the material strain rate effect [13], when compared with relatively small strain problems.

The experimental study reported in [7] has examined the threshold conditions for the perforation of plates which occurs at impact energies larger than those associated with the data for the wholly ductile deformations in Table 1. A perforation equation

$$\Omega_p = (\pi / 4)(d / H) + (d / H)^{1.53} (S / d)^{0.21} \tag{7}$$

was presented at the 1992 SUSI meeting [14] for circular plates, struck by cylindrical projectiles having blunt impact faces and travelling at low velocities. It turns out that this equation gave good estimates for the dimensionless perforation energy, Ω_p, of rectangular mild steel plates when the span, *S*, is taken as the plate breadth $2B$.

Equation (7) for $2B$ = 100mm predicts $\Omega_p = 8.72$, while $2B$ = 200mm gives $\Omega_p = 9.77$. These values are larger than the experimental results for the square plates struck by blunt-tipped projectiles in Table 1 and Figure 4 (identified with a superscript 2) since $2\beta\Omega_p / (1+\beta^2) = \Omega_p$ for $\beta = 1$. The factor $2\beta\Omega / (1+\beta^2)$ in equation (5), which is used for the abscissa in Figure 4, is 0.69Ω, 0.80Ω and 0.92Ω for β = 0.4, 0.5 and 0.66, respectively, and yields the corresponding values of $2\beta\Omega_p / (1+\beta^2) = 6.01$, 6.97 and 8.01. As anticipated, these values lie above all of the corresponding rectangular plate data struck by blunt-tipped projectiles, which are identified with a superscript 2 in Figure 4. Moreover, the experimental results in [7, 9] reveal that the threshold values of the dimensionless perforation energy, Ω_p , are higher for projectiles having hemispherical and conical impact faces in this series of experimental tests.

6 Conclusions

A theoretical rigid, perfectly plastic method, which has been developed previously for initially flat plates having an arbitrary shaped boundary and subjected to either static or dynamic loadings which produce large plastic strains and finite transverse displacements, has been used to study the behaviour of a rectangular plate struck by a mass at the centre. A relatively simple equation has been derived for the maximum permanent transverse displacements which can be used for preliminary design purposes and accident investigations. The theoretical predictions have been developed for a plate having a resisting moment mM_0, $0 < m < 1$, on the entire boundary. In the particular case of $m = 1$ for the fully clamped case, the theoretical predictions were compared with some experimental data recorded on fully clamped rectangular plates having aspect ratios between 0.4 and 1 (square) and struck transversely by masses with cylindrical bodies travelling up to about 7m/s. The theoretical predictions provide a reasonable estimate of the corresponding experimental data.

For completeness, an empirical equation has been used to predict the initial impact energy required to perforate a plate by a blunt-faced missile. This equation predicts perforation energies which lie above all of the corresponding experimental data for large-mass and low-velocity impacts on steel plates.

Acknowledgements

The author is grateful to Mrs J. Jones for her secretarial assistance and to Mrs I. Arnot for her assistance with the figures.

References

[1] Jones, N., A Theoretical Study of the Dynamic Plastic Behavior of Beams and Plates with Finite-Deflections, *International Journal of Solids and Structures*, **7**, pp. 1007-1029, 1971.

[2] Jones, N., Baeder, R. A., An Experimental Study of the Dynamic Plastic Behavior of Rectangular Plates, *Symposium on Plastic Analysis of Structures*, **1**, pp. 476-497, pub. by Ministry of Education, Polytechnic Institute of Jassy, Civil Eng. Faculty, 1972.

[3] Jones, N., A Literature Review of the Dynamic Plastic Response of Structures, *The Shock and Vibration Digest*, **7(8)**, pp. 89-105, 1975.

[4] Jones, N., *Structural Impact*, Cambridge University Press, pp. 575, 1989, Paperback edition, 1997. Chinese edition translated by Ping Jiang and Lili Wang, Sichuan Education Press, Chengdu, 1994

[5] Jones, N., Kim, S-B., Li, Q. M., Response and Failure Analysis of Ductile Circular Plates Struck by a Mass, *Trans. ASME, Journal of Pressure Vessel Technology*, **119(3)**, pp. 332-342, 1997.

[6] Jones, N., On the Mass Impact Loading of Ductile Plates, *Defence Science Journal*, Defence Research and Development Organisation, India, **53(1)**, pp. 15-24, 2003.

[7] Jones, N., Birch, R. S., Duan, R., Low Velocity Perforation of Mild Steel Rectangular Plates with Projectiles having Different Shaped Impact Faces, ASME, *Journal of Pressure Vessel Technology*, **130(3)**, pp. 031206-1 to 031206-8, August 2008.

[8] Birch, R. S., Jones, N., Measurement of Impact Loads Using a Laser Doppler Velocimeter, *Proc. I. Mech. E.,* **204(C1)**, pp. 1-8, 1990.

[9] Jones, N., Birch, R. S., Low Velocity Perforation of Mild Steel Circular Plates with Projectiles having Different Shaped Impact Faces, ASME, *Journal of Pressure Vessel Technology*, **130(3)**, pp 031205-1 to 031205-11, August 2008.

[10] Birch, R. S., Jones, N., Jouri, W. S., Performance Assessment of an Impact Rig, *proc. I. Mech. E.,* **202 (C4)**, pp. 275-285, 1988. Corrigenda **204 (C1)**, p.8, 1990.

[11] Jones, N., Uran, T. O., Tekin, S. A., The Dynamic Plastic Behavior of Fully Clamped Rectangular Plates, *International Journal of Solids and Structures*, **6**, pp. 1499-1512, 1970.

[12] Jones, N., Griffin, R. N., Van Duzer, R. E., An Experimental Study Into the Dynamic Plastic Behavior of Wide Beams and Rectangular Plates, *International Journal of Mechanical Sciences*, **13(8)**, pp. 721-735, 1971.

[13] Jones, N., Some Comments on the Modelling of Material Properties for Dynamic Structural Plasticity, *International Conference on the Mechanical Properties of Materials at High Rates of Strain,* Oxford. Ed. J. Harding, Institute of Physics Conference Series No. 102, pp.435-445, 1989.

[14] Wen, H. M., Jones, N., Semi-Empirical Equations for the Perforation of Plates Struck by a Mass, *2nd International Conference on Structures Under Shock and Impact*, Ed. P. S. Bulson, Computational Mechanics Publications, Southampton and Boston and Thomas Telford, London, pp. 369-380, 1992.

WIT Transactions on The Built Environment, Vol 113, © 2010 WIT Press
www.witpress.com, ISSN 1743-3509 (on-line)

Numerical determination of reflected blast pressure distribution on round columns

Y. Qasrawi[1], P. J. Heffernan[2] & A. Fam[1]
[1]*Queen's University, Canada*
[2]*Royal Military College of Canada, Canada*

Abstract

Blast load parameters are reasonably easily determined for rectangular columns and can be derived from either the literature or numerous utility programs. Little compiled information is available with respect to exposed round columns. A series of numerical simulations were carried out to investigate the design pressure imparted to a round column by an explosion. The column diameter and charge weight were varied and pressure-time histories recorded at regular radial intervals along the face of the column from the closest point of first contact (front) to the extreme point at the side. A numerical model was created in AUTODYN, which simulated a blast wave diffracting around a rigid round cross section. The results indicate that as the diameter of the section increases, the peak reflected pressure at the point of first contact rapidly approaches that of a flat wall. However, the pressure varies sinusoidally between this peak at the point of first contact to a minimum equal to approximately the incident pressure at the furthest point at the side. The results support the obvious advantages when designing against blast to be realized by the use of round vs. rectangular columns, particularly when in the near field.
Keywords: AUTODYN, numerical modelling, round column, reflected pressure, pressure distribution.

1 Introduction

Fujikura et al. [1] have shown experimentally that the reflected over pressure experienced by a round column is substantially lower than the indicated design values. This fact and the intuitive expectation that round columns would deflect

WIT Transactions on The Built Environment, Vol 113, © 2010 WIT Press
www.witpress.com, ISSN 1743-3509 (on-line)
doi:10.2495/SU100071

the blast wave indicate the need for modified design parameters. The first step in obtaining these modified parameters is determining the pressure distribution acting on a round column. In order to accomplish this, the problem was modelled numerically using the commercial software ANSYS AUTODYN. Both the size of the blast and the size of the column were varied to determine the effect of these parameters on the distribution. The model was verified against the current design values and it showed good agreement.

2 Numerical model

The problem was modelled using the commercial software ANSYS AUTODYN, an explicit analysis tool for modelling the nonlinear dynamics and interactions of solids, fluids, and gases.

The model was necessarily constructed in 3D as the cylindrical curvature of the columns could not be captured using 2D axial symmetry, while none of the blast energy dissipated in the third dimension if 2D planar symmetry were used.

The AUTODYN material library properties of air and TNT were used. Air was given an internal energy of 206.8 J, which results in the standard atmospheric pressure of 101.325 kPa.

The blast was initially modelled using 2D axisymmetry from the explosion out to 1.95 m (just before the blast wave interacts with the column) in a multi-

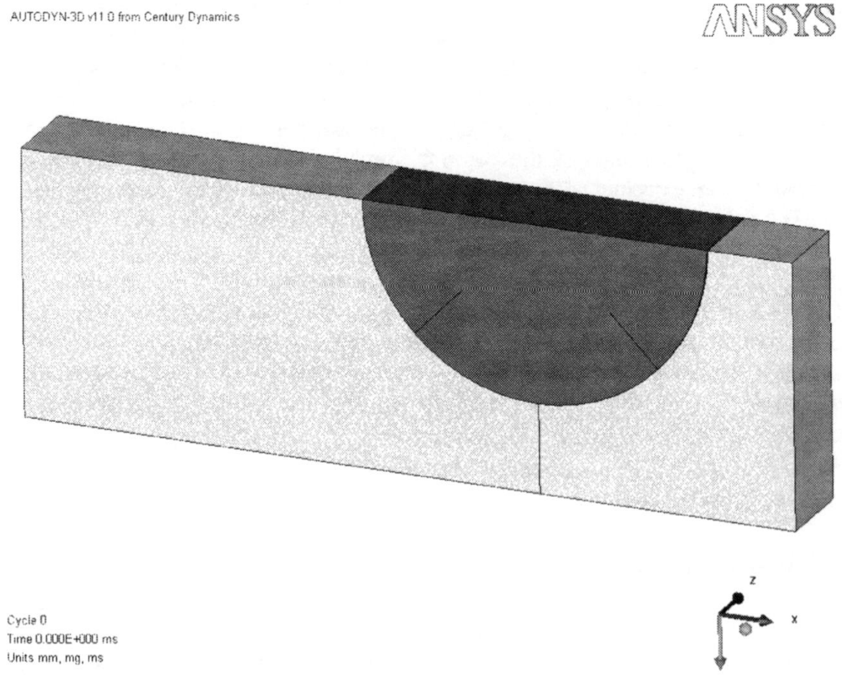

Figure 1: Euler parts and 1000 mm radius column.

WIT Transactions on The Built Environment, Vol 113, © 2010 WIT Press
www.witpress.com, ISSN 1743-3509 (on-line)

material Euler wedge. After which AUTODYN's remapping capabilities were used to set the results of the 2D model as initial conditions for the 3D model in an ideal gas Euler mesh. The remapping and the use of quarter symmetry reduced the computation time and memory requirements considerably.

The 2D blast was modelled as a multi-material Euler wedge divided into 1950 elements and filled with air. The radius of the central TNT sphere was calculated using the predetermined TNT mass and a density of 1657 kg/m^3.

The 3D model consisted of an Euler part surrounding a cylindrical Lagrange fill part. Fill parts are used to define the interaction boundaries in rigid coupling. The Euler gird was the same in all the models and only the column fill parts were varied. The dimensions of the Euler part were dictated by the range of the blasts (2000 mm), the largest column section (radius = 1000 mm), and the influence of the approximate outflow boundary condition (≈ 500 mm). Therefore the dimensions of the Euler part were x = 4500 mm, y = 1500 mm, and z = 500 mm. The Euler parts and a 1000 mm radius Lagrange fill part are shown in Figure 1.

The area of interest in the Euler part, where the blast interacted with the column, consisted of 10 mm cubic elements. Beyond the area of interest, AUTODYN's mesh grading capabilities were used to increase the element sizes geometrically by the maximum recommended rate of 1.2 in all three directions.

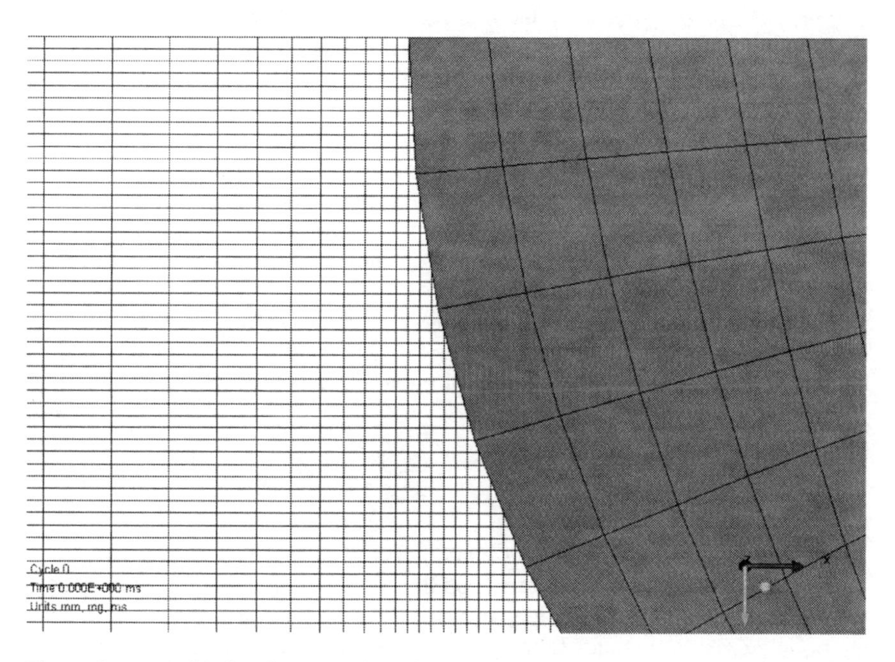

Figure 2: Grid detail at the interface between the 1000 mm radius column and the Euler part.

The column radii to be studied were chosen to be divisible by 10 mm to minimize the merging of Euler cells, which AUTODYN does automatically once a certain proportion of a cell is covered. AUTODYN also approximates circles using polygons, therefore, the number of elements in the cylindrical fill parts needed to be chosen carefully because they determined the precision of the circle. However, the restriction that the side of a Lagrange element must be larger than the smallest Euler element had to be adhered to too.

It was decided that all the cylindrical fill parts were to approximate Pi to two decimal places. Thus, the area of a polygon in terms of the number of its sides was equated to Pi to two decimal places multiplied by the radius squared. The number of sides was then solved. The number of sides used for all the columns was 56, which was obtained by having 15 cells across the radius in a type 2 cylindrical Lagrange part. This also gives a polygon side length of 11.2 mm which is larger than the minimum Euler element size of 10 mm. The interface between the 1000 mm radius column and the Euler grid is shown in Figure 2.

The default reflection boundary was applied to all planes of symmetry, and the approximate out flow boundary was applied to all surfaces where the blast is free to expand.

3 Investigation

The problem investigated was the variation of the reflected pressure (Pr) on the surface of a round column. The two variables that affect this distribution are the size of the column and the size of the blast. Therefore, in order to obtain a clear picture of the solution both the size of the column and the amount of explosive were varied to cover a practical range. The 100 mm radius lower bound for the column size was determined based on the fineness of the Euler grid in the model. While the 1000 mm radius upper bound was chosen as a practical limit. Thus, the column radii investigated were: 100 mm, 250 mm, 500 mm, 750 mm and 1000 mm with the two limiting cases of an unobstructed blast and a flat rigid wall.

For each of these seven cases, the blast was varied between a Z value of 0.8 $m/kg^{-1/3}$ to 2.4 $m/kg^{-1/3}$ in 0.4 $m/kg^{-1/3}$ increments resulting in 35 total runs. Z is the scaled distance and is given by the formula in eqn (1) below.

$$Z = \frac{R}{\sqrt[3]{W}}. \tag{1}$$

where R is the range and W is the weight of TNT.

To capture pressure measurements on the surface of the columns, nine gauges were placed radially at 11.25 degree increments around the circumference from the point closest to the blast to the point where the radius is perpendicular to the original point.

4 Verification

The model was verified against the design charts in TM 5-855 [2]. The values verified were the two limiting cases of the side-on pressure of an unobstructed blast and the pressure reflected off of an infinite flat rigid obstacle. The chart and model values are shown in table 1 below. The results showed very good agreement. The average per cent difference of the values was 12% for the side-on pressure and 21% for the reflected pressure.

5 Results and discussion

The surface plot shown in Figure 3 shows the variation of reflected pressure with respect to the radius of the column and the radial location around the circumference for a given Z value.

Table 1: Verification of model pressure values from model and charts.

	Chart		Model	
Z ($m/kg^{-1/3}$)	**P_{so} (MPa)**	**P_r (MPa)**	**P_{so} (MPa)**	**P_r (MPa)**
0.8	1.75	10.0	1.5	8.9
1.2	0.7	3.5	0.67	2.9
1.6	0.4	1.5	0.37	1.15
2.0	0.2	0.6	0.21	0.44
2.4	0.15	0.3	0.197	0.38

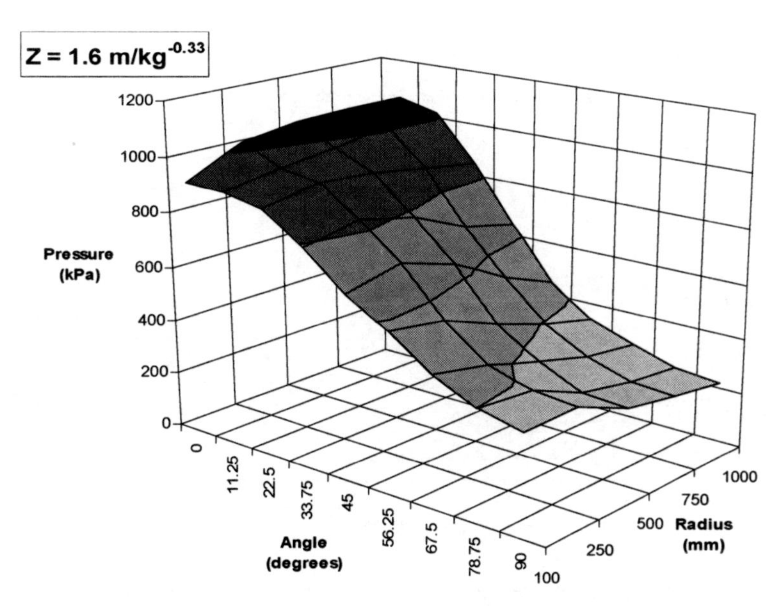

Figure 3: Surface plot of variation of pressure with respect to radius and angle for a given Z.

The effect of the radius and radial location on the reflected pressure can be isolated.

Examining the variation of pressure with respect to radius at the point of first contact shows that the pressure reaches its maximum value rapidly as the radius is increased. It should be noted that if the radius axis started at zero, then the pressure value would be the side on pressure. The maximum pressure reached was approximately 90% of the reflected pressure of a rigid wall.

The cross sections of the plot at an angle of 0 degrees for all the Z values studied are shown in Figure 4. The plots are a ratio of the reflected pressure and side on pressure for clarity, as the reflected pressure range would otherwise be too large to discern the plots for larger Z values. This plot shows that the trend of the initially rapidly increasing reflected pressure and the ceiling that is reached occurs for all the Z values studied. Also, that the asymptote is reached at radius of about 250 mm in all cases.

The pressure variation with respect to radius at the edge of the column was found to be approximately equal to the side-on pressure at that location. It tended to be lower for larger diameters because the pressure wave had to travel the longer distance around the circumference whereas the free wave travelled in a straight line.

Similarly to the plot above, the ratio of the reflected pressure to the side on pressure is shown in Figure 5. Again the plots show that the reflected pressure remains approximately equal to the side on pressure at that location for all the Z values studied.

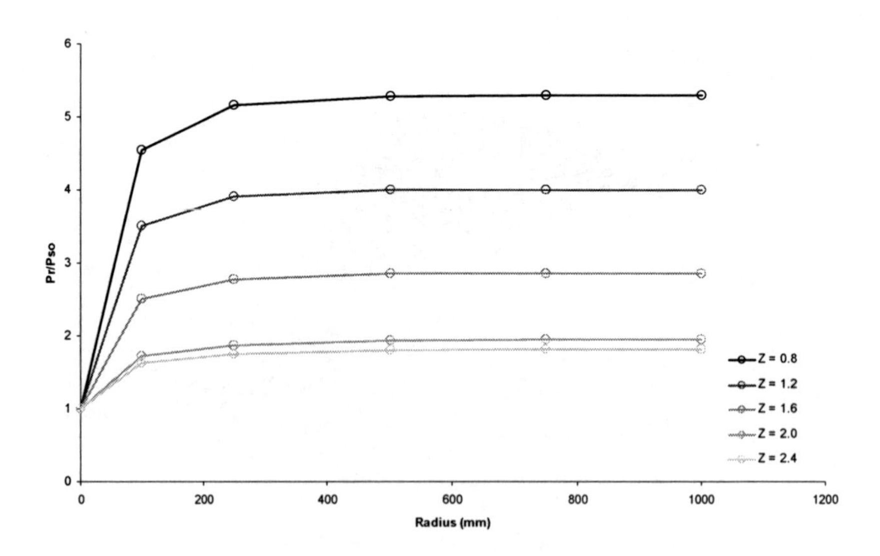

Figure 4: Variation of the ratio of reflected pressure to the side-on pressure with respect to the radius.

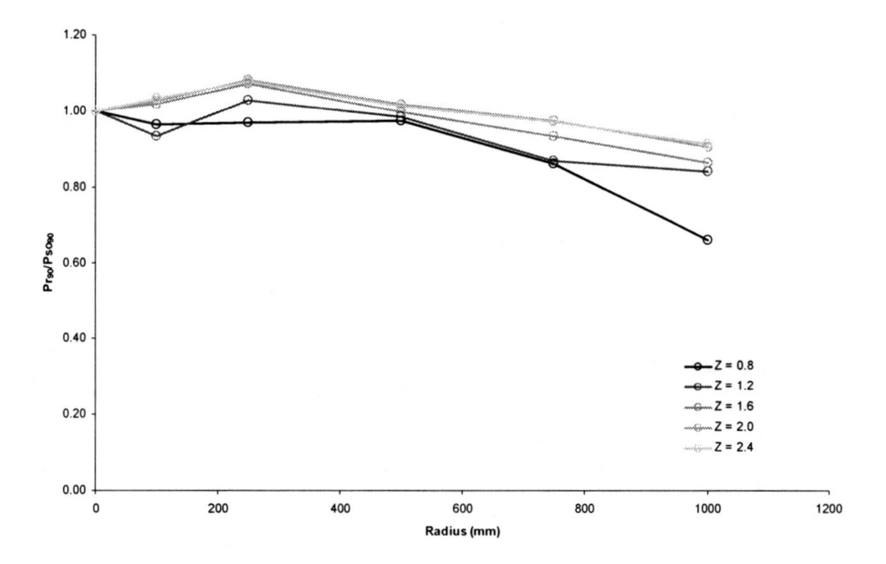

Figure 5: Variation of reflected pressure over the side-on pressure with respect to the radius at the side of the column.

Between these two radial extremes, the pressure increases initially as the radius increases, but then starts to decrease as the radius continues to increase. This is because as the radius increases, the pressure wave must travel further to reach the radial location on the circumference, thus dissipating more energy. This is also more pronounced the further around the circumference the location of interest is.

The variation of pressure with respect to angle can be approximated using a sinusoidal curve fit using eqn (2). The result of this fit is shown in Figure 4.

$$P_{rc} = P_{so} + \left(P_r - P_{so}\right)\left(\frac{\cos 2\phi}{2} + \frac{1}{2}\right). \tag{2}$$

This formula was used to find the sinusoidal curve fit for all the column size and Z combinations. The fit showed very good agreement with the numerical results. However, it did over estimate the pressure for the larger columns. Again, this is attributed to the increased distance the wave travels and that the equation does not take distance into account.

Thus, one could obtain the reflected and side on pressure for a given blast from the standard sources, and then use a sinusoidal fit to approximate the pressure distribution around a circular column.

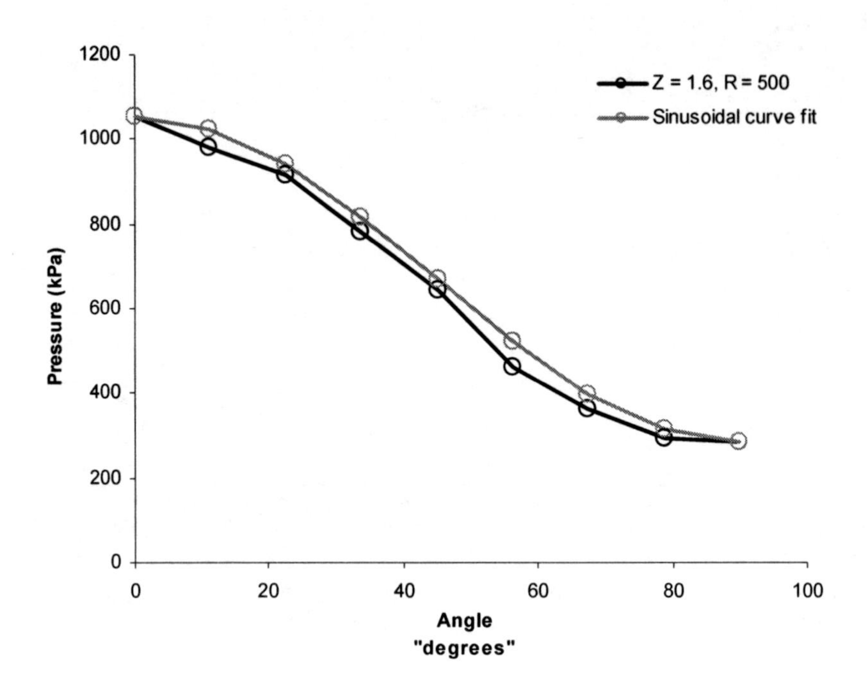

Figure 6: Variation of reflected pressure with radial location with sinusoidal curve fit.

6 Future work

The established sinusoidal fit can be used to find an average or equivalent pressure value to be used in design. A similar investigation needs to be conducted for determining the distribution of impulse around a circular column and finding an equivalent value that can be used in design. Perhaps most importantly, the determined equivalent pressure and impulse values need to be verified experimentally.

7 Conclusion

A numerical model was constructed in AUTODYN to investigate the pressure distribution around a circular column. The model was verified and showed good agreement with established values. It was found that as the column radius increased the maximum reflected pressure at the point closest to the blast approached a maximum of approximately 0.9 of the design value quickly. It was also found that the pressure varied sinusoidally from this maximum to a minimum at the side of the column approximately equal to the incident pressure. A sinusoidal function was used to fit the distribution around the column with good results and this curve fit can be used to find an equivalent design value.

References

[1] Fujikura, S., Bruneau, M., Lopez-Garcia, D., "Experimental Investigation of Multihazard Resistant Bridge Piers Having Concrete-Filled Steel Tube Under Blast Loading" ASCE Journal of Bridge Engineering, Vol. 13, No. 6 November 2008.
[2] US Department of the Army Technical Manual, TM 855. Design and Analysis of Hardened Structures to Conventional Weapons Effects, Washington, USA, 2002.

WIT Transactions on The Built Environment, Vol 113, © 2010 WIT Press
www.witpress.com, ISSN 1743-3509 (on-line)

Simplified evaluation of a building impacted by a terrorist explosion

D. Makovička[1] & D. Makovička Jr.[2]
[1]*Czech Technical University in Prague, Klokner Institute, Czech Republic*
[2]*Static and Dynamic Consulting, Czech Republic*

Abstract

This paper determines the parameters of the explosion wave excited by a charge planted by terrorists. A suitcase containing an industrial explosive (Danubit I, mass 6.45 kilogram) remotely controlled by mobile telephone was placed in the left-luggage office of a railway station. Certain simplified methods according to various publications, according to our own experimental results and according to 3D computations based on detailed calculation modelling of the interior of the room are compared to determine the explosive effects. Equivalent static analysis was applied to the dynamic response of the structural elements of the room (walls, floor, roof and windows). The damage caused to these structural elements is weighted on the basis of the angle of fracture of the central axis / surface, and on the basis of the limit stress state of these structures.

Keywords: explosion, room structure, simplified analysis, dynamic response, failure prognosis.

1 Introduction

When a small charge explodes in the internal space of a building, a pressure wave is formed by the explosion that applies a load on the surrounding internal elements of the structure (Figure 1). The pressure effects of even a small charge are usually high, and the primary consequence is that a window or a door structure may be broken and the pressure is released into the surrounding areas. Although the exhaust vents open, the load transmitted to the surrounding walls of the room, and to the ceiling and floors, is quite high, and the corresponding magnitude must first be estimated. This magnitude depends on numerous

 © 2010 WIT Press
www.witpress.com, ISSN 1743-3509 (on-line)
doi:10.2495/SU100081

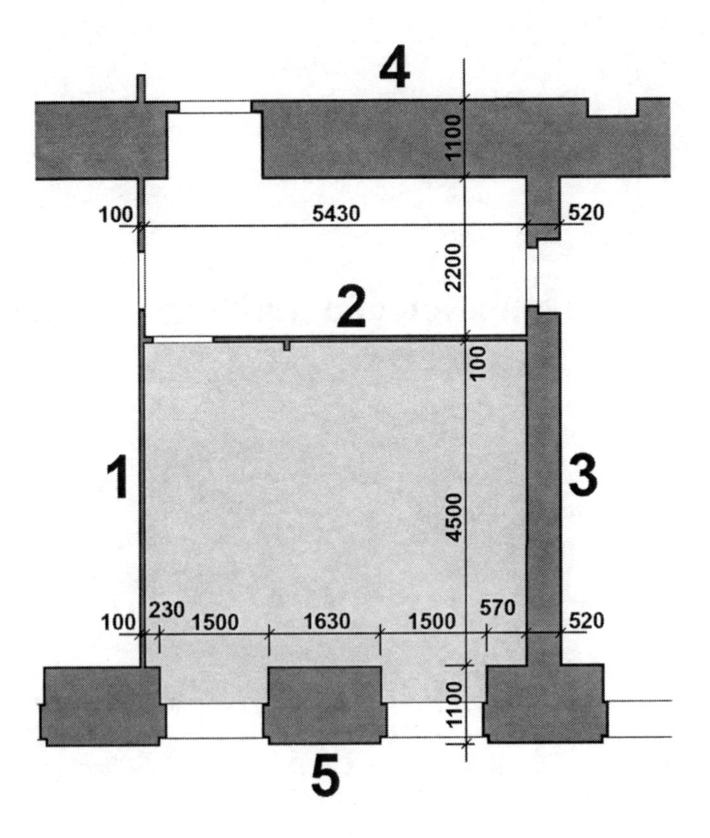

Figure 1: Ground plan of a room located on the ground floor of a building (the surrounding walls are numbered to provide easier identification of the response calculation).

parameters that have an impact on the load level, and therefore it is appropriate to adopt simplifying assumptions.

The load of the surrounding structure of the room can be determined either by means of relatively accurate calculations which take into account the internal space of the room, the composition of the explosive, and which deal with the interaction of the internal environment (air and combustion product mixture) with the structure of the room itself. Alternatively, simpler approximate procedures can be applied; these procedures are based on determining the parameters of the explosion load in a free space and then approximating them to the load in a semi-enclosed space (after the exhaust vents have opened).

2 Behaviour of masonry when loaded by an explosion

When dynamically loaded by an explosion, masonry displays almost linear elasticity up to the point of failure [6]. The real elastic modulus E is an important material quantity for calculating the response of masonry to the load due to an explosion. It can be estimated according to the deformability modulus E_{def} stated in the design standards, based on the experimentally verified experience of the authors of this paper:

a) To calculate the dynamic response of an undamaged structure near the failure limit:

$$E \approx 0.5 \cdot E_{def} \tag{1}$$

b) To calculate the dynamic response of a structure already damaged by visible cracks:

$$E \approx 0.1 \cdot E_{def} \tag{2}$$

The loading capacity of the brick masonry is decided in accordance with the standards used to design the bending compression / tensile strength of the masonry, with a certain margin of safety, given by the coefficients of the load, combination, etc.

If the loading capacity R_{tfd} is exceeded, a crack appears in the structure of the material. Thus the most unfavourable condition must apply to a safe structure, based on comparing the stress state or the deformation magnitude. The following relationship applies to the stress combination:

$$\min\,(\sigma_g \pm \sigma_{expl}) \geq -R_{tfd} \quad \text{or upon adjustment} \quad \sigma_{expl} - \sigma_g \leq R_{tfd} \tag{3}$$

where σ_{expl} is stress caused by the effects of the pressure wave when there is an explosion, σ_g is the normal stress at a given place (a joint) caused by the overburden weight itself.

In structural design based on limit state theory, it is usually more suitable to consider the carrying capacity strength moment, possibly in combination with the normal force, rather than the carrying capacity limit R_{tfd}. This stress criterion must be supplemented by an evaluation of the deformation of the structure. As a rule, the limit deformation value (shift or angular displacement) determines the actual destruction of the wall; the limit deformation value corresponds to the critical angle of the partial turning of the central line of the structure due to its bending. The limit angular displacement ψ at the failure limit is found in the range of approximately 2.3° to 5.7° for masonry [6, 9], a minimum of 6.5° for reinforced concrete [3], and a minimum of 10.5° for steel [3]:

$$\psi = 2\,\text{arctg}\,(2\,y\,/\,l\,) \tag{4}$$

where y is the maximum achieved deflection of the board (at the centre of the span), and l is the structure span along the shorter dimension.

3 Explosion load

When a charge explodes in an open space, the pressure effect of the impact wave on an obstacle (the load of the building structure) depends on the situation of the building with respect to the focus of the explosion, the impact wave parameters, etc. The entire phenomenon of the impact wave effect on the structure is then usually simplified for calculation purposes, using numerous assumptions, especially as regards the intensity and the time course of the impact wave effect and its distribution in contact with the given object [2, 7]. When an actual event takes place, the specific course of the load action depends on the swirl flow bypassing the structure surface, the atmospheric pressure, the temperature conditions and other factors that are usually neglected in a simplified analysis. The parameters of the explosive, too, are determined on the basis of average values; empirical formulas are used, and operate with mean (probable) coefficient levels. Thus the structure calculations concerning the impact wave effects are significantly burdened by these inaccuracies in the input quantities of the entire phenomenon.

Empirical formulas created by various authors [1, 3, 4, 7] are usually used for the time course of the pressure wave and subsequently the structure load. The structure of the formulas according to various authors is very similar, and they usually differ only in the magnitudes of the coefficients. Due to the variability of these coefficients, the uncertainty of the formulas is usually found to be in the range of ±20%, and possibly even more. The reliability of individual formulas improves with increasing distance of the pressure wave from the focus of the explosion.

The overpressure determined at the face of the air impact wave that spreads from the explosion site to the surroundings stems from the reduced distance [1, 3, 7, 8] is:

$$\overline{R} = \frac{R}{\sqrt[3]{C_w}} \tag{5}$$

where $\overline{R}$ is the reduced separation distance from the epicentre of the explosion [m/kg$^{1/3}$], R is the distance from the explosion epicentre [m], and C_W is the equivalent mass of the charge [kg TNT].

It is assumed that the energy released by the explosion is proportional to the mass of the explosive, and the solution consists in introducing a reference charge chosen to be represented by tritol (trinitrotoluene, TNT). Therefore the mass of various explosives is expressed in terms of the so-called tritol equivalent (k_{TNT}). If this equivalent cannot be found in the specialized literature (for example [7]), it can be calculated with sufficient accuracy using the relationship

$$k_{TNT\text{-}p} = 0.3\ Q_v - 0.2 \quad (\text{for } 2\ \text{MJ/kg} \leq Q_v \leq 5\ \text{MJ/kg}) \tag{6}$$

where $k_{TNT\text{-}p}$ is the pressure tritol equivalent of the explosive (equal to 1 for TNT), Q_v is the calculated explosion heat [MJ/kg] and $Q_v = 4.2$ MJ/kg for TNT.

Then the total equivalent mass C_W can be determined using the relationship [7]

$$C_w = C_N \cdot k_{TNT\text{-}p} \cdot k_E \cdot k_G \tag{7}$$

where C_w is the mass of the equivalent charge [kg TNT], C_N is the mass of the used charge of the (actual) explosive [kg], $k_{TNT\text{-}p}$ is the pressure tritol equivalent,

k_E is the charge seal coefficient, and k_G is the geometry coefficient of the impact wave spreading in the space.

The seal coefficient can be determined using the relationship

$$k_E = 0.2 + 0.8 / (1 + k_B) \tag{8}$$

where k_B is the cover mass [kg] divided by the explosive mass [kg], and expresses the ballistic ratio. The following applies to the geometry coefficient k_G, 1 for detonation in a free air space, 2 for surface detonation (on the ground).

The explosion wave spreads in spherical wavefronts from the focus point of the explosion. When the explosion is on the ground, the explosion energy is roughly double because, after complete reflection from the surface of the terrain, the pressure wave spreads in hemispherical wavefronts. The spreading geometry coefficient k_G is not stated by some authors in the formulas for determining the total equivalent mass; in such cases, and in the case of a ground explosion, the equivalent charge mass C_w is as a rule substituted by twice its value in empirical formulas.

In the simplified calculation [7], a ground explosion is represented by a situation when the explosive is located directly on the surface of the terrain ($h = 0$ m thus $k_G = 2$). An explosion in an open air space is a situation when the delay of the reflected wave from the surface of the terrain to the pressure wave front is higher than the duration of the overpressure phase of the pressure wave ($k_G = 1$). A linear interpolation is made between the values.

On the basis of comparing various resources in the literature (namely [1, 3, 4]) and on the basis of tests of bricked structures [2, 6] and window glass [5] during explosions of small charges, the authors of this paper proposed the application of realistic formulas. The empirical formulas below were verified in experiments using small charges (Semtex) in the vicinity of the loaded structure. Their resulting form then corresponds to the impact wave effects from a small solid charge in an external environment and during a ground explosion. Maximum overpressure p_+ and underpressure p_- at the face of the air impact wave and their durations τ_+ and τ_- are applicable both to ground explosions (C_W is replaced by the double value of the equivalent charge) and above-ground explosions in a free (air) environment:

$$p_+ = \frac{1{,}07}{\overline{R}^3} - 0{,}1 \quad \text{[MPa]} \quad \text{for } \overline{R} \le 1 \text{ m/kg}^{1/3} \tag{9}$$

$$p_+ = \frac{0{,}0932}{\overline{R}} + \frac{0{,}383}{\overline{R}^2} + \frac{1{,}275}{\overline{R}^3} \quad \text{[MPa]} \quad \text{for } 1 < \overline{R} \le 15 \text{ m/kg}^{1/3} \tag{10}$$

$$p_- = \frac{0{,}035}{\overline{R}} \quad \text{[MPa]} \tag{11}$$

$$\tau_+ = 1{,}6 \cdot 10^{-3} \cdot \sqrt[6]{C_w} \cdot \sqrt{R} \quad \text{[s]} \tag{12}$$

$$\tau_{-} = 1{,}6 \cdot 10^{-2} \cdot \sqrt[3]{C_w} \quad [\mathrm{s}] \tag{13}$$

After a normal (perpendicular) impact of the explosion wave on a solid obstacle, a reflected wave is formed with the reflection overpressure p_{ref} that loads the building structure from the front side (Figure 2). The overpressure value in the reflected wave corresponds to approximately twice the value of the overpressure for low overpressure values p_{+} of approximately up to 5 MPa (up to eight times the value for high overpressures of the order of several MPa) in the incident wave for the given distance R [7].

$$p_{ref+} \approx 2\, p_{+} \tag{14}$$

$$p_{ref-} \approx 2\, p_{-} \tag{15}$$

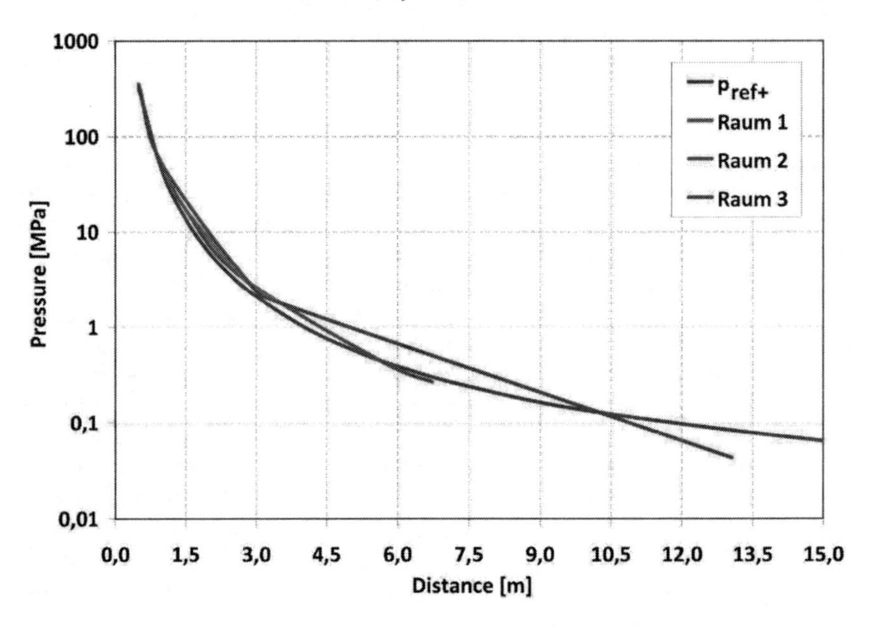

Figure 2: Magnitude of overpressure p_{ref+} in dependence on distance R of the charge location compared to the reflection overpressure p_{ref}^{f} for specific volumes of rooms 1 to 3.

After an explosion in the enclosed space of rooms in a building structure, with closed relieving vents, the load is increased by approximately 50% due to reflection from the surface of the walls, the ceiling and the floor of the room; the duration of the overpressure is thus roughly double. The resulting load of the surrounding structures and its duration can be expressed approximately as follows:

$$p_{load+} \approx 1.5 \cdot p_{ref+} \tag{16}$$

$$t_{load+} \approx 2\, \tau_{+} \tag{17}$$

Formulas similar to those for the overpressure phase of the load also apply approximately to the underpressure phase.

The reflective overpressure $p_{ref} = p_{ref+}$ in the rooms can also be calculated directly $p_{ref} = p_{ref}^{f}$ according to a method described in [9].

To determine the reflective overpressures and impulses, their values in the band $\overline{R} < 2$ m/kg$^{1/3}$ must either be read from the published curves [9], or their approximate values must be determined using the derived exponential relationships:

a) Reflective overpressure:

$$p_{ref}^{f} = 14{,}554 \times \overline{R}^{-1{,}4587} \quad \text{[MPa]} \quad \text{for } 0{,}05 < \overline{R} \leq 0{,}5 \text{ m/kg}^{1/3} \tag{18}$$

$$p_{ref}^{f} = 5{,}76 \times \overline{R}^{-2{,}762} \quad \text{[MPa]} \quad \text{for } 0{,}5 < \overline{R} \leq 5 \text{ m/kg}^{1/3} \tag{19}$$

b) Reflective impulse:

$$I_{ref}^{f} = 0{,}345 \times \sqrt[3]{C_w} \times \overline{R}^{-1{,}857} \quad \text{[kPa.s]} \quad \text{for } 0{,}05 < \overline{R} \leq 0{,}5 \text{ m/kg}^{1/3} \tag{20}$$

$$I_{ref}^{f} = 0{,}5823 \times \sqrt[3]{C_w} \times \overline{R}^{-1{,}0976} \quad \text{[kPa.s]} \quad \text{for } 0{,}5 < \overline{R} \leq 5 \text{ m/kg}^{1/3} \tag{21}$$

When substituting input values C_W into the relationships above, two differences from the calculations of the reflective overpressure and the reflective impulse in the open space outside the building must be taken into account:

a) Here, the indices f denote the detonation conditions in the free air space in the room, and the following is substituted for the charge size C_W (different from formula (3) above):

b)
$$C_w = C_N \cdot k_{TNT-p} \cdot k_E \cdot k_G = C_N \cdot k_{TNT-p} \cdot k_E \tag{22}$$

where the impact wave spreading geometry coefficient $k_G = 1.0$.

4 Overpressure calculation for specific rooms

Figure 2 shows the calculation values of the reflection overpressure p_{ref+} calculated using two different simplified procedures, namely according to formulas (9) to (15) derived for an explosion wave spreading in an external space, and furthermore, in order to compare the results of a direct calculation of the reflective overpressure in the rooms, according to formulas (18) and (19) for three selected rooms according to the methodology in [15]:

Room 1 (Figure 1): Volume 69 m^3, exhaust vents 1.7 m^2, area of the walls, floor and ceiling 104.9 m^2;

Room 2: Volume 69 m^3, exhaust vents 1.7 m^2, area of the walls, floor and ceiling 91.5 m^2;

Room 3: Volume 255.8 m^3, exhaust vents 24.7 m^2, area of the walls, floor and ceiling 283.1 m^2.

A comparison of the results obtained by the two simplified procedures shows that both methodologies provide sufficient accuracy for the usual volume of medium-size and large rooms, and can be applied to any position in which the charge is deposited in the internal areas of the rooms.

Now the calculated overpressures for selected wall 1 (using the marking based on Figure 1) will be compared, as calculated by the LS DYNA program [10], taking into account the interaction of the room environment with its walls (Fluid-Structure Interaction) in Figure 3. The calculated wall is divided into 16 fields, and the mean overpressure for the suitcase containing the explosive, deposited at the centre of the room, is calculated and shown in each of the fields. This figure shows that, due to reflections from the walls of the room, the explosion peaks are superimposed on each other and their coverage curve roughly corresponds to formulas (16) and (17). The peak pressure coverage is dominant for the global structure response. So that an equivalent "static" calculation is justified for the analysis of this problem.

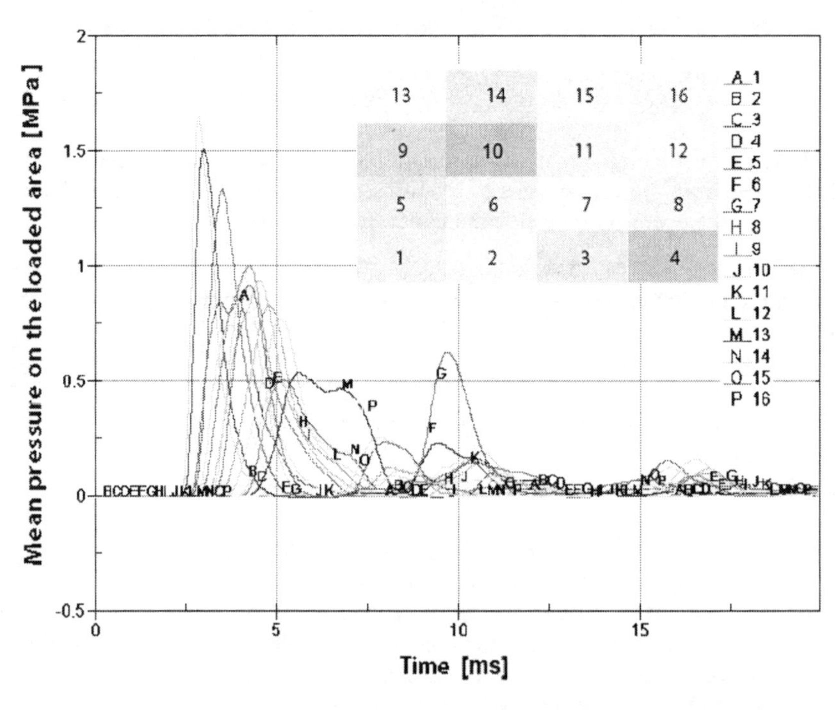

Figure 3: Time course of the mean pressure on the basis of a 3D calculation of the interaction that occurs between the pressure wave from the explosion of a charge in the centre of the room, and the wall at points corresponding to the wall as divided into rectangles.

5 Calculating the damage to the walls of the rooms

In order to evaluate the masonry of the walls and pillars of the room structure, we can use the load estimate p_{load+} and the duration of its effect t_{load+}, calculated for the possible distance R of the charge position from the evaluated wall, window, door or interwindow or inside pillars.

The equivalent static calculation of the wall uniformly and continuously loaded by load p_{load+} with duration of its effect t_{load+} was used to determine the failure hazard. In this procedure, the nature of the boundary conditions (support of the wall board) is chosen. As concerns the partition walls of a specific building, the individual wall board elements were considered to be independent, for the sake of simplification, hinge-mounted along their entire circumference.

From the viewpoint of uncertainty in the simplified effect of the pressure wave from the explosion, even such a simplification is acceptable and justified for an engineering estimate of the explosion effects.

As a rule, the dynamic coefficient δ is derived for the equivalent static calculation, and for a system with one degree of freedom it is the function of the natural period of dominant oscillation T of the structure and the pressure wave effect duration τ_+ or τ_-, based on whether the overpressure or underpressure wave is considered.

For the elastic-plastic system, dynamic coefficient δ is the function of the ratio of the impact wave effect duration τ_+ or τ_- on the natural structure oscillation period $T_{(i)} = T$ and on the ductility of the structure:

$$k_m = \frac{y_m}{y_{el}} \tag{23}$$

where y_m is the total elastic + plastic deflection (shift) of the structure, and y_{el} is the elastic part of the deflection (shift).

As for impact phenomena (very rapid) during bending stress of the structure, the ductility coefficient k_m can usually be considered to be equal to 3 to 5 for masonry, and from 5 to 10 for reinforced concrete, steel and wood. As for the load due to the impact wave, the dynamic coefficient including consideration of the ductile behaviour of the structure is found to be in the range $\delta = 1 \sim 2$. This magnitude was derived by N. M. Newmark (see [3]) for a simplified system with one degree of freedom in the following form:

$$\frac{1}{\delta} = \frac{T_{(i)} \cdot \sqrt{2 \cdot k_m - 1}}{\pi \cdot \tau_+} + \frac{1 - \dfrac{1}{2 \cdot k_m}}{1 + 0{,}7 \cdot \dfrac{T_{(i)}}{\tau_+}} \tag{24}$$

6 Evaluating the failure probability for a specific room

When calculating the load level p_{load+}, the load is found to be in the range of units of MPa or hundreds of kPa inside a room (Figure 1), based on the position of the charge inside the room. When comparing such high loads with the carrying capacity of the windows and doors, it reaches several units of kPa. It is apparent that such window and door openings will be smashed (destroyed) and will enable the pressure to be released into the surrounding (external or internal) areas.

Table 1 shows the calculated bending moments in the middle part of the wall board in vertical and horizontal directions, maximum deflection y at the centre of the wall board and the angle φ of angular displacement of the centre line of the wall board. The angle of 5° was chosen as the limit angle ψ at which the wall board masonry breakdown occurs (fracture, sweeping out of brick fragments, etc.). For the sake of transparency, the individual walls of the room are

Table 1: Failure risk estimation of structural parts (Figure 1).

Structural element	Distance of charge	Explosion load	Load duration	Vertical moment	Horizontal moment	Displacement	Rotation	Failure estimation
	R	p_{load+}	t_{load+}	M_{ver}	M_{hor}	y	φ	
	[m]	[MPa]	[s]	[kNm]	[kNm]	[mm]	[deg]	
1	Wall 6700×2800×100 mm							
	1.5	20.53	0.006	360	141	7088	157.7	Expected
	2.5	5.09	0.008	116	45	2282	117.0	Expected
2	Wall 5430×2800×100 mm							
	2	9.27	0.007	188	76	3697	138.5	Expected
	5	0.89	0.011	28	11	557	43.4	Expected
3	Wall 6700×2800×520 mm							
	1	64.73	0.005	4902	1924	688	52.3	Expected
	3	3.15	0.009	409	160	57	4.7	Partial failure
	6	0.58	0.012	105	41	15	1.2	Improbable
4	Wall 5430×2800×1100 mm							
	1	64.74	0.005	10202	4122	151	12.3	Probable
	2	9.27	0.007	2 030	820	30	2.5	Improbable
	4	1.52	0.010	460	186	7	0.6	Improbable
5	Pillar 1630×2800×1100 mm							
	1	64.74	0.005	9227	4050	44	6.3	Partial failure
	2	9.27	0.007	1751	769	8	1.2	Improbable
	4	1.52	0.010	369	162	2	0.3	Improbable

numbered and these numbers are shown in the ground plan of the room in Figure 1. It follows clearly from Table 1 that thin partition walls up to 150 mm in thickness of will be destroyed by the explosion.

As the explosion pressures markedly exceed the carrying capacity of such thin partition walls, the ruins of the partition walls will be swept into the surrounding areas. Thick bricked walls and interwindow pillars 900 mm and more in thickness will be destroyed only if the charge is placed in their vicinity, at a distance of about 1 m. For distances of the charge of more than 2 m, such a massive structure will transfer the explosion pressures without collapsing and without any other serious defects. Of course, the plaster will be damaged, cracks will appear in the walls, brick fragments may fall out, etc., but the structure will not collapse.

If a massive carrying wall or pillar (more than 900 mm in thickness) collapses under this ceiling, it is likely that the ceiling structure will fall through and damage will also occur to higher floors.

7 Conclusions

An example of a specific building was used to discuss the explosion and the building safety hazard when a terrorist charge is brought into the building in a suitcase and is equipped with a system for initiating the charge after it has been placed in the building and the terrorist has left.

Due to uncertainties in all parameters of the explosion load, a simplified methodology has been presented here. This methodology enables the parameters to be determined sufficiently concisely and the natural building structure to be evaluated on the basis of the parameters. The uncertainty in determining the explosion load parameters can be determined on the basis the results of a calculation using empirical formulas derived by the authors for small charges. The response of the structure is evaluated on the basis of the results of the equivalent static calculation, using the dynamic coefficient for the elastic-plastic system. The explosion hazard of the structure is evaluated on the basis of the maximum moments and deflections of the structure.

The example of a specific room is used to analyze its exposure and also the hazard to the entire building based on various possible placements of a charge a short distance or a longer distance away from the carrying structure and the partition walls.

Acknowledgement

This research was supported as a part of the research projects in GAČR 103/08/0859 "Structure response under static and dynamic loads caused by natural and man induced activity", for which the authors would like to thank the Agency.

References

[1] Henrych, J., *The Dynamics of Explosion and Its Use,* Academia: Prague 1979.

[2] Janovský, B., Šelešovský, P., Horkel, J. & Vejs, L., Vented Confined Explosions in Stramberk Experimental Mine and AutoReaGas Simulation. *J. Loss Prevention in the Process Industries*, Vol. 19, pp. 280–287, 2006.

[3] Koloušek, V. et al., *Building Structures Under Dynamic Effects* (in Slovak), SVTL: Bratislava 1967.

[4] Korenev, B.G. et al., *Dynamic Calculation of Structures Under Special Effects* (in Russian), Strojizdat: Moskva 1981.

[5] Makovička, D., Shock Wave Load of Window Glass Plate Structure and Hypothesis of Its Failure. In: *Structures Under Shock and Impact '98,* pp. 43–52, WIT Press: Southampton 1998.

[6] Makovička, D., Failure of Masonry Under Impact Load Generated by an Explosion. *Acta Polytechnica*, Vol. 39, No. 1/1999, pp. 63–91.

[7] Makovička, D. & Janovský, B., *Handbook of Explosion Protection for Buildings* (in Czech), CTU Publishing House in Prague, 2008.

[8] Makovička, D., Makovička, D., Janovský, B. & Adamík, V., Exposure of building structure to charge explosion in interior (in Czech), *Stavební obzor*, Vol. 18, No. 9/2009, pp. 257–265.

[9] Baker, W.E., Westine, P.S., Cox, P.A. et al., *Explosion Hazards and Evaluation*, Elsevier: Amsterdam 1983.

[10] LS-DYNA User's Manual, *Nonlinear Dynamic Analysis of Structures*, Version 950, Livermore Software Technology Corporation, May 1999.

A comparison of hydrodynamic and analytic predicted blast pressure profiles

G. M. Stunzenas & E. L. Baker
US Army Armament Research Development and Engineering Center, USA

Abstract

Modeling the structural response to blast relies on accurate descriptions of the blast loading pressure profiles. Traditionally, empirically based blast pressure histories are used for this modeling. However, the structural response and geometric configuration can strongly affect the blast loading profile, particularly for close-in blast loading configurations. As a result, high rate continuum modeling is being increasingly applied to directly resolve both the blast profiles and structural response. A variety of computer models exist for the purpose of analyzing blast pressures associated with different types of explosive charges and ranges. These computer models range from simple empirically based analytic models based off of the Hopkinson cube root scaling to multi-physics high rate finite element approaches, commonly known as "hydrocodes", which are capable of tracking shocks through the conservation equations of continuum mechanics. The purpose of this paper is to provide comparisons of blast profiles predicted by analytic models with a hydrodynamic model at various standoffs. Three computer models; BlastX, Conwep, and ALE3D, were used to model the detonation of five pounds of TNT. Pressure profiles for various standoffs were gathered from each computer model and compared. The ALE3D result is greatly dependant on the mesh size and appears to converge to the BlastX and Conwep solutions with increased mesh resolution.

Keywords: blast, explosives, modelling.

1 Introduction

Modeling the structural response to blast relies on accurate descriptions of the blast loading pressure profiles. Traditionally, empirically based blast pressure

WIT Transactions on The Built Environment, Vol 113, © 2010 WIT Press
www.witpress.com, ISSN 1743-3509 (on-line)
doi:10.2495/SU100091

histories are used for this modeling. However, the structural response and geometric configuration can strongly affect the blast loading profile, particularly for close-in blast loading configurations [1–3]. As a result, high rate continuum modeling is being increasingly applied to directly resolve both the blast profiles and structural response. An explosion produces shock waves in air, which extend outward from the point of detonation. This shock wave is composed of a highly nonlinear shock front, which decays as the distance from the source increases. This nonlinearity is characterized by a sharp, instantaneous increase in pressure, called the peak incident overpressure. The velocity of the shock is supersonic in the medium in which it travels. The gas molecules behind the shock travel at a lower particle velocity and generally make up what is referred to as the shock wind. As the volume in which the shock travels increases, the peak pressures associated with the shock decrease [2]. A variety of computer models exist for the purpose of analyzing blast pressures associated with different types of explosive charges and ranges. These computer models range from simple empirically based analytic models based off of the Hopkinson cube root scaling to multi-physics high rate finite element approaches, commonly known as "hydrocodes", which are capable of tracking shocks through the conservation equations of continuum mechanics. The purpose of this paper is to provide comparisons of blast wave profiles in open air predicted by analytic models with a high rate continuum model at various standoffs. Open air blast profiles were predicted using two analytic models: Conwep, BlastX, and a high rate continuum model: ALE3D. A comparison of the peak incident overpressure was made for a 5 pound spherical TNT charge at various distances from the charge.

2 Analytic blast modeling

Conwep blast calculations are conducted using a set of empirically based equations and curves. These empirically based scaled blast curves have been generated for spherical and hemispherical TNT charges [2]. These curves are used to predict blast variables including time of arrival, impulse, peak incident and peak reflected pressures. These curves were created based on Hopkinson cube root scaling [3], which relates the characteristic properties of the blast wave from an explosion of one energy level to that of another. According to this scaling law, the pressure at a certain distance from the charge is proportional to the cube root of the energy yield. Figure 1 presents scaled blast relationship curves [2] and is representative of how Conwep computes peak blast pressures and other blast characteristics for an open air detonation using a spherical TNT charge.

A scaled distance is calculated, which is the ratio of the distance from the charge to the cube root of the explosive weight. This scaled distance is then used to compute all of the variables associated with the particular blast wave of interest. Table 1 shows the peak incident overpressure computed by Conwep at various distances from the charge. This table also shows the results of using the chart alone with the scaled distance calculations. Based on the similar results,

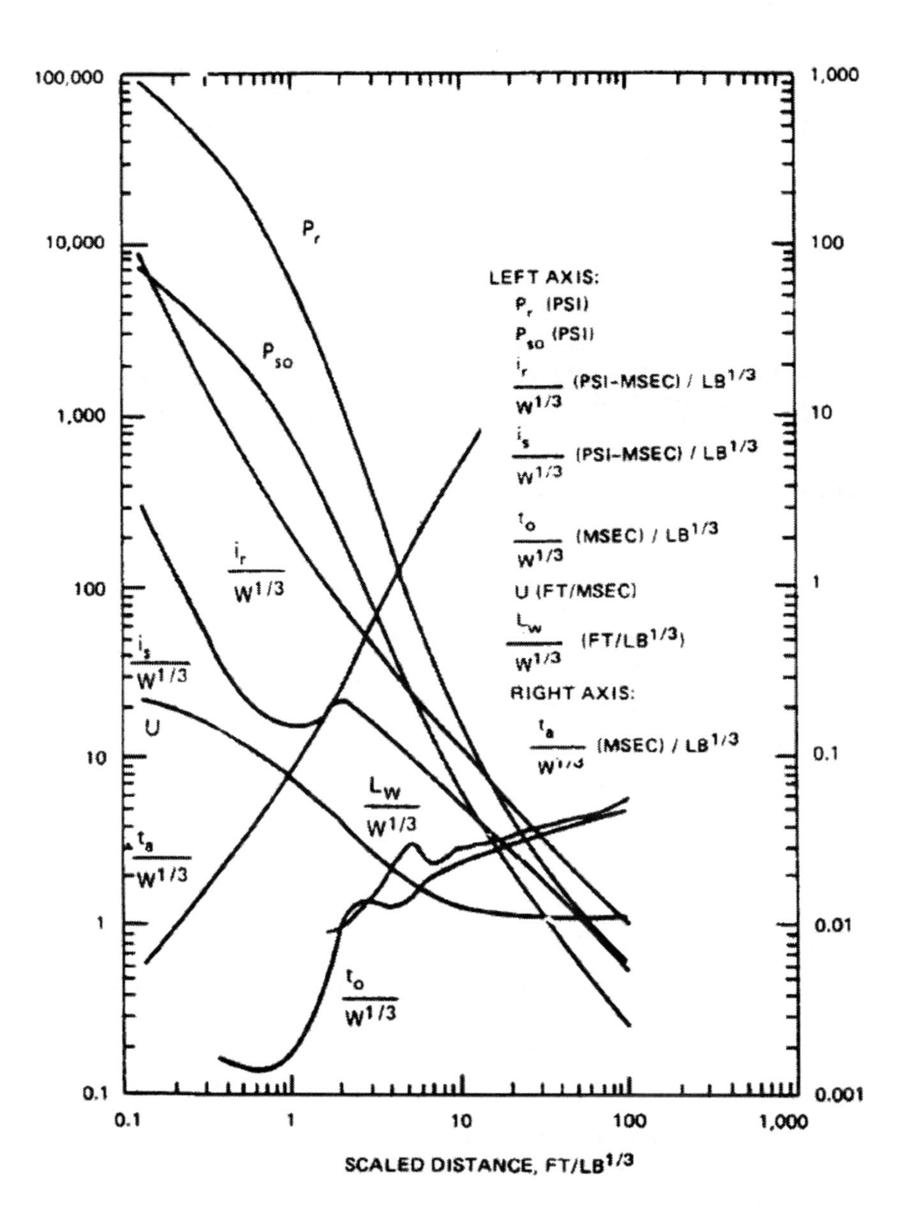

Figure 1: Blast wave characteristics vs. scaled distance for spherical TNT charge.

this demonstrates how Conwep uses this particular curve and scaling law to calculate air blast properties at various ranges.

One thing worth noting is that Figure 1 is on a log-log scale, which tends to makes accurate values difficult to assess from the graphical representations. This is likely the cause of any discrepancy between using the chart alone and the Conwep results. BlastX treats shock wave effects with a ray-based semi-empirical model. Similar to Conwep, it uses tabular blast data for spherical and cylindrical explosive charges. The blast data tables are based on hydrocode calculations for a 1 kg charge (of various explosives) at a standard set of atmospheric conditions. Results for other charge weights and atmospheric conditions are obtained similar to Conwep using Hopkinson scaling, as discussed above. BlastX uses the tabular values to calculate wave forms by interpolation of blast pressures, particle velocity, and density that were computed using the 1 kg spherical charge [5, 6].

Table 1: Comparison between peak pressures obtained from Conwep and Fig. 1.

Distance from Charge (ft)	Conwep Pressure Results (PSI)	Using Chart and Scaled Distance (PSI)
2.5	425.7	410
3	295.5	298
3.5	213.8	200
4	160	150
4.5	123.2	110
5	97.17	95
5.5	78.26	80
6	64.18	68
6.5	53.46	60

3 High rate continuum blast modeling

ALE3D is an arbitrary Lagrangian/Eulerian high rate finite difference hydrocode which is used to model fluid and solid elastic-plastic response of materials. A mesh is used to define a volume in space, and the conservation equations (mass, momentum, and energy) of continuum mechanics are applied and integrated through time, giving an updated nodal response to different forces, pressures, stresses, and strains [4]. The TNT charge was modeled using a standard Jones-Wilkins-Lee (JWL) detonation products equation of state. The air was modeled using a constant gamma equation of state. Table 2 presents the explosive products and air equations of state parameters.

Table 2: TNT JWL parameters and air constant gamma parameters used for the ALE-3D calculations.

	Density (g/cc)	Gamma	CJ Pressure (GPa)	Det Velocity (cm/microsecond)	EOS Coefficients
TNT JWL Parameters	1.63	2.66	17.7	.689	A = 3.712 Mbar B = .03231 Mbar R1 = 4.15 R2 = .95 Omega = .30
Air Constant Gamma Parameters	.000129	1.4	N/A	N/A	N/A

4 Results

Conwep, BlastX, and ALE3D were used to compute the peak pressure profiles of a 5 lb spherical charge of TNT, detonated in open air. The distance from the charge was varied, and the results are summarized in the Table 3. Figure 2 shows some images that were captured from the ALE3D simulation for the open airblast. Certain parameters were modified in ALE3D, while holding the mesh resolution constant to examine affect on the produced blast profiles and discrepancy between the different model outputs. The mesh resolution used for this study was approximately .12 cells/mm, resulting in 5.8 million total cells.

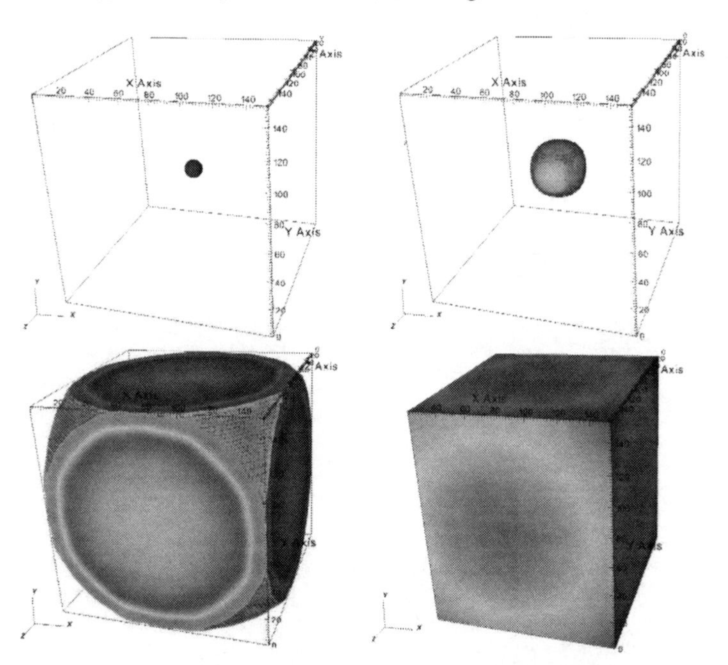

Figure 2: ALE3D open airblast simulation images – Blast front position.

The initial purpose of this comparison was to study how the kinetic energy advection method affected the results. Since this term is quadratic in nature, rather than linear, energy can sometimes be lost if the velocity varies strongly in a calculation. The kinetic energy advection method can be set so that the energy lost computationally during integration for a shock is added to the internal energy (fracke = 1). There has been some debate about what this does to the resulting calculation, as the objective is to keep a strong shock without putting the material on a wrong adiabat [4]. ALE3D was run in Eulerian mode for the purposes of this comparison. Table 3 summarizes the results for BlastX, Conwep, and ALE3D at various distances with the kinetic energy advection term (fracke) set to 0 and 1. The information in Figure 3 depicts the results in this table:

Table 3: Comparison of peak pressures for Conwep, BlastX, and ALE3D.

Distance from charge (ft)	Conwep Pressure (psi)	BlastX Pressure (psi)	ALE3D - Eulerian (fracke = 1) Pressure (psi)	ALE3D - Eulerian (fracke = 0) Pressure (psi)
2.5	425.7	400.3	226	220
3	295.5	278.4	175	171
3.5	213.8	212.8	142	144
4	160	161.5	119	121
4.5	123.2	123.6	99.4	103
5	97.17	96.57	90.2	86.1
5.5	78.26	76.78	82	82.6
6	64.18	62.7	73	77.5
6.5	53.46	52.25	69.7	69.9

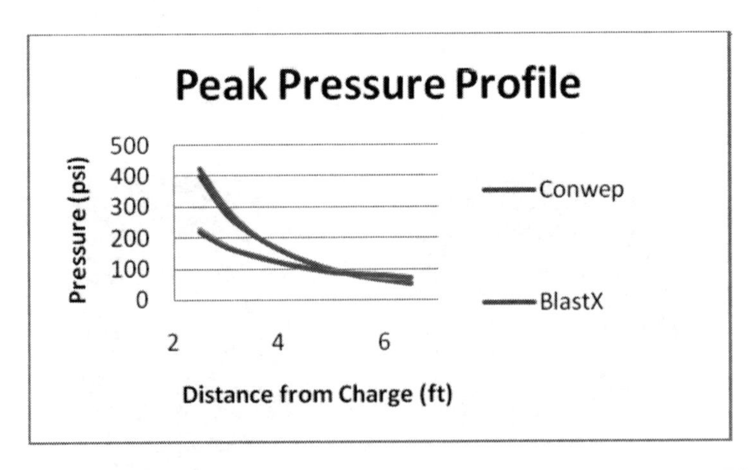

Figure 3: Graphical comparison of peak pressures between Conwep, BlastX, and ALE3D at various standoffs.

It can be seen that the kinetic energy advection term did not have a significant effect on the calculations, but the results using ALE3D were much different than the empirically-based Conwep and BlastX. However, the Conwep and BlastX calculations produce similar blast results. In order to further investigate the differences between the high rate continuum modeling and the analytic blast models, several further modifications were made to the ALE3D computations. In an attempt to better track the shock wave propagation, monotonic artificial viscosity was used, rather than the default linear-quadrate rate dependant artificial viscosity. Subsequently, different mesh ratios were used to allow the shock to expand as it travelled away from the charge. This approach was investigated, as it is well known that large changes in mesh size affect shock propagation calculations. Finally, the resolution of the calculation was investigated, by increasing the number of elements to determine dependency on mesh size. Table 4 summarizes the results of the analysis at a distance of 2.5 feet from the charge. The results indicate that convergence to the analytic results appears to be occurring with increased mesh resolution.

Table 4: Peak pressure results using ALE3D with modified input parameters.

2.5 Feet From Charge	**Analysis Description**	**Resulting Pressure (psi)**
Run 1	Eulerian, q weighting, slight increase in mesh size	260
Run 2	q weighting Modequipotential rather than Eulerian	262
Run 3	Eulerian, q weighting, different mesh ratios	281
Run 4	Eulerian, q weighting, different mesh ratios	284
Run 5	Eulerian, q weighting, finer mesh	300
Run 6	Eulerian, q weighting, finer mesh	326
Run 7	10 million elements	335
Run 8	18 million elements	339

5 Conclusion

More investigation is required if the true discrepancy between these computer models is to be determined, but based on the above analysis, it is quite evident that tracking shocks in ALE3D is greatly dependant on the mesh size. For the 5 LBS charge investigated, it appears that a mesh resolution of at least .15 cells/mm is required. The artificial viscosity term also had some impact on the calculations. Conwep and BlastX can not solely be depended upon when conducting blast analyses on structures simply due to the fact that they are empirical in nature. Much of the data calculated is based on curves and interpolated from existing databases. In the future, hydrodynamic codes, such as ALE3D will become increasingly necessary to solve these highly nonlinear and dynamic problems due to their capability of resolving both shock and structural response at the same time.

References

[1] Ngo, T., Mendis, P., Gupta, A. & Ramsay, J., Blast Loading and Blast Effects on Structures – An Overview, *EJSE Special Issue: Loading on Structures*, pp. 76 - 91, 2007.

[2] Headquarters, Department of the Army, Fundamentals of Protective Design for Conventional Weapons, *Technical Manual TM 5-855-1*, 3 November, 1986.

[3] Headquarters, Department of the Army, Engineering Design Handbook, Explosions in Air, *AMC Pamphlet AMCP 706-181*, 15 July 1974.

[4] Lawrence Livermore National Laboratories, *ALE3D High Performance Multi-Physics Simulations*, LLNL-MI-413853.

[5] Commission on Engineering and Technical Systems (CETS), *Protecting Buildings from Bomb Damage: Transfer of Blast-Effects Mitigation Technologies from Military to Civilian Applications*, National Academy Press: Washington, D.C., 1995.

[6] Ray J.C., Armstrong, B.J. & Slawson, T.R., Airblast Environment Beneath a Bridge Overpass, *Journal of the Transportation Research Board of the National Academies*, vol. 1827, pp. 63 - 68, 2003.

Transient response of a laminated sandwich plate with viscoelastic core subjected to air blast: theory and experiment

D. Balkan, O. Acar, H. S. Türkmen & Z. Mecitoğlu
Faculty of Aeronautics and Astronautics, Istanbul Technical University, Turkey

Abstract

This paper is focused on the theoretical and experimental study of the dynamic behavior of sandwich composite plates with a viscoelastic core subjected to blast load. The sandwich plate has carbon/epoxy face sheets and an aramid honeycomb core. The clamped boundary conditions are considered for all edges of the plate. The sandwich plate is modeled using first order shear deformation theory considering the geometrical nonlinearities. The equations of motion are derived by the use of the virtual work principle for the sandwich plate. The viscoelastic behavior is modeled by using the Kelvin-Voigt theory. The equations of motion are reduced into the time domain using Galerkin's method. The nonlinear-coupled equation system is solved by Mathematica Software. In the experimental study, the pressurized air in a shock tube is impinged on the sandwich composite plate by rupturing a membrane in front of the shock tube. The strain-time histories obtained from the theoretical analysis are compared with the experimental ones and a good agreement is found.
Keywords: blast load, Kelvin-Voigt model, Galerkin's method.

1 Introduction

Sandwich plates have been used extensively in the aircraft industry, both in civil and military aircraft, since 1940 [1]. They have now begun to be used in the structure of missiles and satellites. The use of sandwich materials in military structures is also growing rapidly.

WIT Transactions on The Built Environment, Vol 113, © 2010 WIT Press
www.witpress.com, ISSN 1743-3509 (on-line)
doi:10.2495/SU100101

Composite materials themselves, which are used as faced sheets of the sandwich plate, have damping properties due to fiber–matrix interactions. However, these damping levels are not high enough to safeguard the structure. Therefore, in the construction of sandwich composite plate, a core material with high viscoelastic damping capability can be used or some viscoelastic layers can be placed in addition to the conventional core layer.

Several studies for sandwich constructions are reviewed in the literature [3–5]. However, only a few studies in the nonlinear response of laminated sandwich composite plates subjected to blast load are investigated. Hause and Librescu [6] considered the response of anisotropic sandwich flat panels to explosive pressure pulses. Librescu et al. [7] investigated the linear and nonlinear dynamic response of sandwich panels under blast load. Dynamic responses of sandwich flat panels to time-dependent loads generated by an underwater and by a shock wave are considered by Librescu et al. [8]. Hause and Librescu [9] investigated the dynamic response of doubly curved anisotropic sandwich panels exposed to time-dependent loads. The nonlinear dynamic response of a sandwich plate subjected to blast load is investigated by Baş et al. [10].

Although there are a numerous studies considering the blast load effects on sandwich composite plates with various core configurations in the open literature, no study is found that examines the effects of a viscoelastic core on the dynamic behavior of sandwich plates under blast load. Furthermore, the no experimental works related to the subject are found in the literature. Esmailzadeh and Jalali [16] studied nonlinear oscillations of viscoelastic simply supported rectangular plates by assuming the Voigt-Kelvin constitutive model. Yan Wei et al. [13] investigated the three dimensional solution of a laminated orthotropic rectangular plate with viscoelastic interfaces, described by the Kelvin-Voigt model. Mesquita and Coda [14] considered an alternative Kelvin Viscoelastic formulation for the finite element method. Lekszycki et al. [15] characterized the constitutive parameters of the Voigt model and described viscoelastic materials applied in sandwich beams.

In the theoretical study, the equations of motion are derived by the use of the virtual work principle for a sandwich composite plate with a viscoelastic core. The sandwich plate is modeled using first order shear deformation theory. The geometrical nonlinearities are considered in the derivations. The viscoelastic behavior is modeled by using the Kelvin-Voigt linear viscoelastic theory. A system of five nonlinear equations is obtained. The displacement field in the space domain is approximated by the trial functions and then the equations of motion are reduced into the time domain using Galerkin's method. The nonlinear-coupled equation system in the time domain is solved by Mathematica Software and the results of theoretical analyses are obtained.

In the experimental study, a special experimental setup is used for the blast test.

The pressurized air in a shock tube is impinged on the sandwich composite plate by rupturing a membrane in front of the shock tube. The pressure distribution on the plate is obtained by using the pressure sensors placed at specific points on the plate. The straingages are used to obtain strain time

histories at certain locations on the sandwich plate. The strain-time histories obtained from the theoretical analysis are compared with the experimental ones and a good agreement is found.

2 Equation of motion

The rectangular sandwich plate with the length a, the width b and the thickness h is shown in figure 1. Cartesian coordinate system is used to derive the equations.

As it is mentioned in the Weierstrass theorem [11], "Any function which is continuous in an interval may be approximated uniformly by polynomials in this interval". So displacement field can be represented as follows:

$$\begin{aligned} u(x,y,z,t) &= u_0(x,y,t) + z\alpha_1(x,y,t) + z^2\alpha_2(x,y,t) + \ldots \\ v(x,y,z,t) &= v_0(x,y,t) + z\beta_1(x,y,t) + z^2\beta_2(x,y,t) + \ldots \\ w(x,y,z,t) &= w_0(x,y,t) + z\gamma_1(x,y,t) + z^2\gamma_2(x,y,t) + \ldots \end{aligned} \tag{1}$$

The Kirchhoff-Love Hypothesis for linear elastic thin plates results in the linearly distributed tangential displacements and a constant normal displacement through the thickness of the plate. So Eq.1 can be written as follows:

$$\begin{aligned} u(x,y,z,t) &= u_0(x,y,t) + z\alpha_1(x,y,t) \\ v(x,y,z,t) &= v_0(x,y,t) + z\beta_1(x,y,t) \\ w(x,y,z,t) &= w_0(x,y,t) \end{aligned} \tag{2}$$

Here x, y, z represent Cartesian coordinates. u, v, w indicate the displacements in the x, y, z directions, respectively; u_0, v_0, w_0 denote the displacement of a point on the neutral surface of plate. α_1, β_1 are rotations about

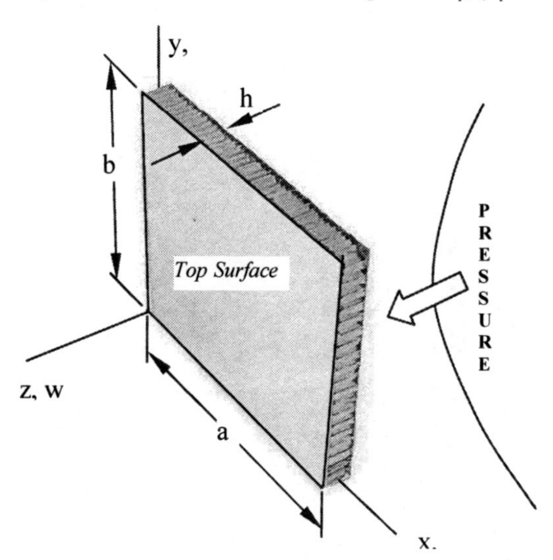

Figure 1: Sandwich composite plate and its coordinate system.

transverse normals. “The transverse normals do not remain perpendicular to the midsurface after deformation”. Kirchhoff’s assumption is not valid for the first order shear deformation theory [1].

The rotation functions for thin plates can be written as follows:

$$\alpha_1 = -\frac{\partial w_0}{\partial x}, \beta_1 = -\frac{\partial w_0}{\partial y}$$

The strain-displacement relations for the von Kármán Plate can be written as:

$$\begin{aligned}
\varepsilon_{xx} &= \frac{\partial u_0}{\partial x} + \frac{1}{2}\left(\frac{\partial w_0}{\partial x}\right)^2 + z\frac{\partial \alpha_1}{\partial x} \\
\varepsilon_{yy} &= \frac{\partial v_0}{\partial y} + \frac{1}{2}\left(\frac{\partial w_0}{\partial y}\right)^2 + z\frac{\partial \beta_1}{\partial y} \\
\varepsilon_{xy} &= \frac{\partial v_0}{\partial x} + \frac{\partial u_0}{\partial y} + \left(\frac{\partial w_0}{\partial x}\frac{\partial w_0}{\partial y}\right) + z\left(\frac{\partial \beta_1}{\partial x} + \frac{\partial \alpha_1}{\partial y}\right) \\
\gamma_{xz} &= \frac{\partial w_0}{\partial x} + \alpha_1 \\
\gamma_{yz} &= \frac{\partial w_0}{\partial y} + \beta_1 \\
\varepsilon_{zz} &= 0
\end{aligned} \tag{3}$$

ε_{xx}, ε_{yy}, ε_{zz} indicates strains in x, y and z directions, respectively.

Constitutive equations are derived considering linear viscoelastic behavior that is modeled by the Kelvin-Voigt linear viscoelastic theory [12].

$$\sigma(t) = E\varepsilon(t) + \eta\frac{\partial \varepsilon(t)}{\partial t}$$

For composite materials, stress-strain equations can be written as follows [16]:

$$\begin{gathered}
\begin{Bmatrix} \sigma_{xx} \\ \sigma_{yy} \\ \sigma_{xy} \end{Bmatrix}^{[k]} = \begin{bmatrix} \bar{Q}_{11} & \bar{Q}_{12} & \bar{Q}_{16} \\ \bar{Q}_{12} & \bar{Q}_{22} & \bar{Q}_{26} \\ \bar{Q}_{16} & \bar{Q}_{26} & \bar{Q}_{66} \end{bmatrix}^{[k]} \begin{Bmatrix} \varepsilon_{xx} \\ \varepsilon_{yy} \\ \varepsilon_{xy} \end{Bmatrix} + \eta\frac{\partial}{\partial t}\left\{ \begin{bmatrix} \bar{Q}_{11} & \bar{Q}_{12} & \bar{Q}_{16} \\ \bar{Q}_{12} & \bar{Q}_{22} & \bar{Q}_{26} \\ \bar{Q}_{16} & \bar{Q}_{26} & \bar{Q}_{66} \end{bmatrix}^{[k]} \begin{Bmatrix} \varepsilon_{xx} \\ \varepsilon_{yy} \\ \varepsilon_{xy} \end{Bmatrix} \right\} \\
\begin{Bmatrix} \sigma_{yz} \\ \sigma_{xz} \end{Bmatrix}^{[k]} = \begin{bmatrix} \bar{Q}_{44} & \bar{Q}_{45} \\ \bar{Q}_{45} & \bar{Q}_{55} \end{bmatrix}^{[k]} \begin{Bmatrix} \gamma_{yz} \\ \gamma_{xz} \end{Bmatrix} + \eta\frac{\partial}{\partial t}\left\{ \begin{bmatrix} \bar{Q}_{44} & \bar{Q}_{45} \\ \bar{Q}_{45} & \bar{Q}_{55} \end{bmatrix}^{[k]} \begin{Bmatrix} \gamma_{yz} \\ \gamma_{xz} \end{Bmatrix} \right\} \\
\begin{Bmatrix} Q_y \\ Q_x \end{Bmatrix} = \kappa\left\{ \begin{bmatrix} A_{44} & A_{45} \\ A_{45} & A_{55} \end{bmatrix} \begin{Bmatrix} \gamma_{yz} \\ \gamma_{xz} \end{Bmatrix} + \eta\frac{\partial}{\partial t}\left\{ \begin{bmatrix} A_{44} & A_{45} \\ A_{45} & A_{55} \end{bmatrix} \begin{Bmatrix} \gamma_{yz} \\ \gamma_{xz} \end{Bmatrix} \right\} \right\}
\end{gathered} \tag{4}$$

$\overline{Q}_{ij}$ and A_{ij} are stiffness matrix coefficients of sandwich materials.

For layered or sandwich composite plates, stress resultants can be integrated through the thickness of the each layer [17]. Then, stress and strain resultants can be obtained.

$\{\mathbf{N}\}$ and $\{\mathbf{M}\}$ are the vectors of force and moment components, respectively.

Coefficient in the matrices are given by [1],

$$D_{ij} = \frac{1}{3}\sum_{k=1}^{n}\left(\overline{Q}_{ij}\right)_k \left(h_k^{\ 3} - h_{k-1}^{\ 3}\right)$$

$$A_{ij} = \sum_{k=1}^{n}\left(\overline{Q}_{ij}\right)_k \left(h_k - h_{k-1}\right)$$

$$B_{ij} = \frac{1}{2}\sum_{k=1}^{n}\left(\overline{Q}_{ij}\right)_k \left(h_k^{\ 2} - h_{k-1}^{\ 2}\right)$$

3 Governing equation

The equations of motion are derived by the use of virtual work principle for a sandwich composite plate with a viscoelastic core.

$$\int_0^T \left(\delta U + \delta V + \delta K\right) dt = 0 \tag{5}$$

δU is virtual strain energy, δV is virtual work done by external applied forces, δK is virtual kinetic energy are given by

$$\begin{aligned} \delta U &= \int_{\Omega_0} \left\{ \int_{-h/2}^{h/2} \left[\sigma_{xx}\delta\varepsilon_{xx} + \sigma_{yy}\delta\varepsilon_{yy} + \sigma_{xy}\delta\gamma_{xy} + \sigma_{xz}\delta\gamma_{xz} + \sigma_{yz}\delta\gamma_{yz} \right] dz \right\} dxdy \\ \delta V &= -\int_{\Omega_0} \left[q_x \delta u + q_y \delta v + \left(q_b + q_t\right)\delta w \right] dxdy \\ \delta K &= \int_{\Omega_0} \overline{m}\left(\dot{u}\delta\dot{u} + \dot{v}\delta\dot{v} + \dot{w}\delta\dot{w}\right) dxdy \end{aligned} \tag{6}$$

q_b and q_t are the distributed forces at the bottom and at the top surfaces. q_x and q_y are distributed force components in the x and y directions. In virtual kinetic energy equation, exponentially used dot ($\dot{}$) represents time derivatives.

Substituting for the virtual strains from Eq. (5) into Eq. (6) the governing equations of the plate are obtained in integral form.

$$\frac{\partial N_{xx}}{\partial x} + \frac{\partial N_{xy}}{\partial y} = I_0 \frac{\partial^2 u_0}{\partial t^2} + I_1 \frac{\partial^2 \alpha_1}{\partial t^2} + q_x$$

$$\frac{\partial N_{xy}}{\partial x} + \frac{\partial N_{yy}}{\partial y} = I_0 \frac{\partial^2 v_0}{\partial t^2} + I_1 \frac{\partial^2 \beta_1}{\partial t^2} + q_y$$

$$\frac{\partial Q_x}{\partial x}+\frac{\partial Q_y}{\partial y}+\frac{\partial}{\partial x}\left(N_{xx}\frac{\partial w_0}{\partial x}+N_{xy}\frac{\partial w_0}{\partial y}\right)+\frac{\partial}{\partial y}\left(N_{xy}\frac{\partial w_0}{\partial x}+N_{yy}\frac{\partial w_0}{\partial y}\right)+q_z=I_0\frac{\partial^2 w_0}{\partial t^2}$$

$$\frac{\partial M_{xx}}{\partial x}+\frac{\partial M_{xy}}{\partial y}-Q_x=I_2\frac{\partial^2\alpha_1}{\partial t^2}+I_1\frac{\partial^2 u_0}{\partial t^2}$$

$$\frac{\partial M_{xy}}{\partial x}+\frac{\partial M_{yy}}{\partial y}-Q_y=I_2\frac{\partial^2\beta_1}{\partial t^2}+I_1\frac{\partial^2 v_0}{\partial t^2}$$

I_0 is normal moment of inertia coefficient, I_2 is rotary moment of inertia coefficient and I_1 is normal and rotary moment of inertia coefficient. Here, q_z is the distributed load in the z direction, which is summation of q_b and q_t. In this study q_z is the blast pressure and it is approximated by the following function,

$$q_z(x,y,t)=\left[1-\cos\left(\frac{2\pi t}{t_e}\right)\right]^f\left\{\frac{P_t+P_e}{2^f 2^g 2^s}\left[1-\cos\left(\frac{2\pi x}{a}\right)\right]^g\left[1-\cos\left(\frac{2\pi y}{b}\right)\right]^s-\frac{P_e}{2^f}\right\}$$

Here P_t is the peak pressure, P_e is the suction pressure measured at the side section of the plate, t_e is positive phase duration. f, g, s are the adjustable parameters in order to obtain and appropriate pressure function matching the experimental pressure data.

4 Solution method

The displacement functions are chosen by considering clamped boundary conditions and given as follows.

$$\begin{aligned}
u_0&=\sum_{m=1}^{M}\sum_{n=1}^{N}U_{mn}(t)\sin\frac{2m\pi x}{a}\left(1-\cos\frac{2n\pi y}{b}\right)\\
v_0&=\sum_{m=1}^{M}\sum_{n=1}^{N}V_{mn}(t)\left(1-\cos\frac{2m\pi x}{a}\right)\sin\frac{2n\pi y}{b}\\
w_0&=\sum_{m=1}^{M}\sum_{n=1}^{N}W_{mn}(t)\left(1-\cos\frac{2m\pi x}{a}\right)\left(1-\cos\frac{2n\pi y}{b}\right)\\
\alpha_0&=\sum_{m=1}^{M}\sum_{n=1}^{N}\Lambda_{mn}(t)\sin\frac{2m\pi x}{a}\left(1-\cos\frac{2n\pi y}{b}\right)\\
\beta_0&=\sum_{m=1}^{M}\sum_{n=1}^{N}\Psi_{mn}(t)\left(1-\cos\frac{2m\pi x}{a}\right)\sin\frac{2n\pi y}{b}
\end{aligned}\tag{7}$$

Accounting the first terms of the displacement functions, the Galerkin's Method is applied to Eq. (7) and time dependent nonlinear differential equations are obtained.

$$
\begin{aligned}
&a_0U + a_1V + a_2W^2 + a_3\dot{U} + a_4W\dot{W} + a_5\dot{V} + a_6\ddot{U} + a_7\ddot{\Lambda} = 0 \\
&b_0U + b_1V + b_2W^2 + b_3\dot{U} + b_4W\dot{W} + b_5\dot{V} + b_6\ddot{\Lambda} + b_7\ddot{\Psi} = 0 \\
&c_0UW + c_1W^3 + c_2VW + c_3W\dot{U} + c_4W\dot{V} + c_5W^2\dot{W} + c_6W + c_7 + c_8\ddot{W} + c_9\Lambda + \\
&c_{10}\Psi + c_{11}\dot{W} + c_{12}\dot{\Lambda} + c_{13}\dot{\Psi} = 0 \\
&d_0W + d_1\Lambda + d_2\Psi + d_3\dot{W} + d_4\dot{\Lambda} + d_5\dot{\Psi} + d_6\ddot{U} + a_7\ddot{\Lambda} = 0 \\
&e_0W + e_1\Lambda + e_2\Psi + e_3\dot{\Psi} + e_4\dot{\Lambda} + e_5\dot{\Psi} + e_6\ddot{V} + e_7\ddot{\Psi} = 0
\end{aligned} \tag{8}
$$

The nonlinear-coupled equation system in the time domain is solved by Mathematica Software, which uses the RKF45 solver, and the results of theoretical analyses are obtained. The results were compared with experimental results.

5 Experimental results

The sandwich plate used in the experiment has two layers carbon/epoxy face sheets on each side and an aramid honeycomb core. The clamped boundary conditions are considered for all edges of the plate. The dimensions of plate are a= b = 300 mm. The thickness of the plate core layer is 1.5 mm and the thickness of the face sheets are 0.332 mm.

In the experimental study, a special experimental setup is used for the blast test (Figure 2). The pressurized air in a shock tube is impinged on the sandwich composite plate by rupturing a membrane in front of the shock tube. The pressure distribution on the plate is obtained by using the pressure sensors placed at specific points on the plate. The straingages are used to obtain strain time histories at certain locations on the sandwich plate.

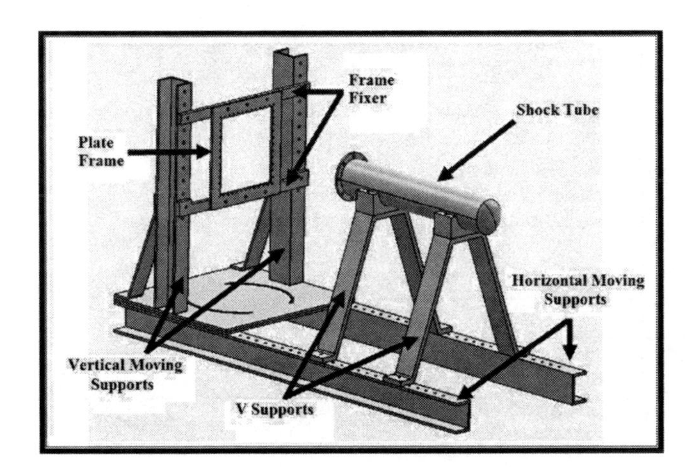

Figure 2: Experimental setup [24].

Three pressure sensors were placed in front side and two strain gauges were placed on the back side of sandwich plate. The location of the pressure sensors and strain gauges can be seen in Figure 3.

The sandwich plate has carbon/epoxy face sheets with two layers of ($0^o/90^o$) orientations and aramid core layer. Material properties of carbon/epoxy are given as E_1 = 45.5 GPa, E_2 = 45.5 GPa, G_{12} = 2.7 GPa, ν_{12} = 0.19 and ρ = 1340 kg/m^3. Material properties of aramid core layer are taken as damping coefficient η = 0.2, E = 25 MPa, G = 1.5 MPa, ν = 0.3 and ρ = 32 kg/m^3. The nonuniform blast pressure is used in the analysis. Parameters obtained from the experiment are as follows: P_t = 54 kPa P_e = 5 kPa, t_e = 0.01s.

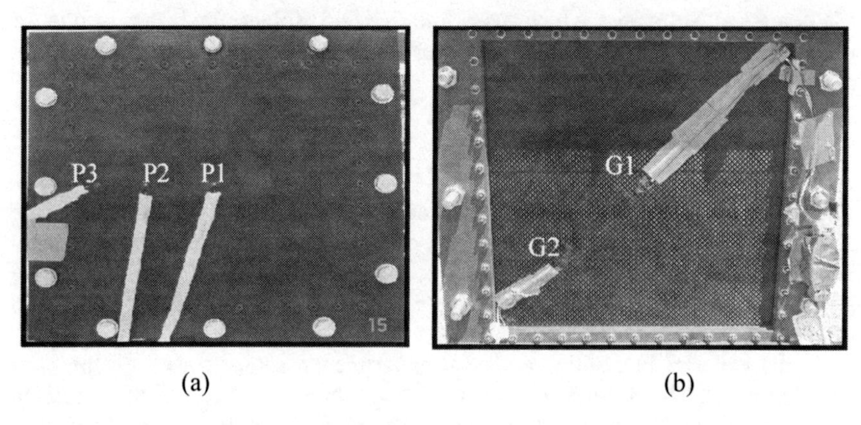

Figure 3: Location of sensors: (a) pressure sensors; (b) strain gauges.

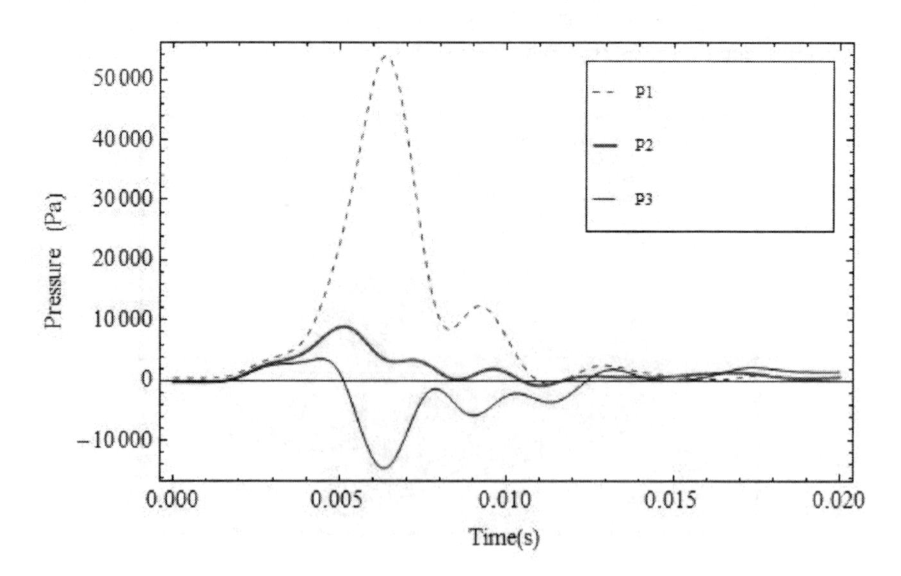

Figure 5: Pressure distribution at the location of pressure sensors.

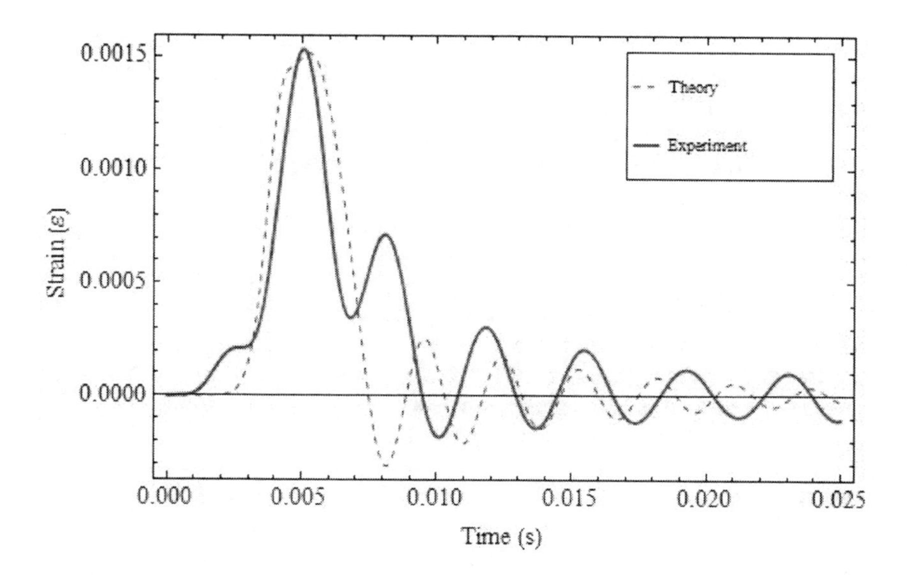

Figure 6: Time history of normal strain ε_x at the bottom center of the plate (P_e = 5 kPa).

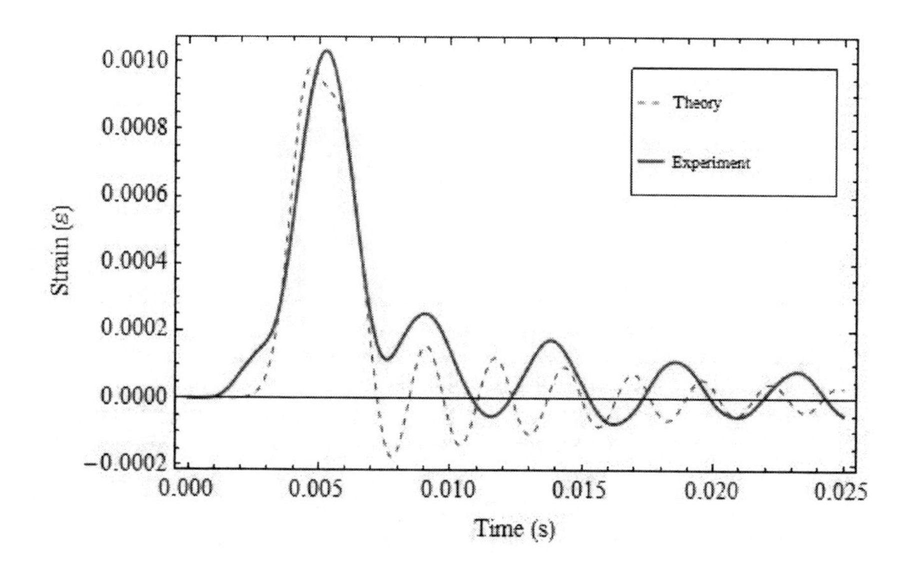

Figure 7: Time history of normal strain ε_x at the G2 location of the sandwich plate (P_e=5 kPa).

The pressure distribution obtained from blast test can be seen in Figure 5.

The results obtained from theoretical analysis are compared with experimental results in Figure 6.

The vibration frequencies are obtained from theoretical and numerical analysis are in good agreement but there is a phase shift. The peak amplitudes of normal strain are very approximate to each other.

6 Conclusion

In this study, transient dynamic behavior of sandwich composite plates with viscoelastic core under the blast load is studied. Sandwich plate has composite face layers and a viscoelastic core layer. The clamped boundary conditions are considered for all edges of the plate. The equations of motions are derived by the use of virtual work principle for a sandwich composite plate with a viscoelastic core.

In the experimental study, the pressurized air in a shock tube is impinged on the sandwich composite plate by rupturing a membrane in front of the shock tube. The strain-time histories obtained from the theoretical analysis are compared with the experimental ones and a good agreement is found for the peak values.

The blast pressure has peak value at the center of the plate and it decreases from center to the edges. The vibration amplitudes obtained from the experiment are higher than those of the approximate solution results. Moreover, the vibration frequencies obtained from the experiment is lower than the theoretical solution. It is observed that oscillations are damped instantly after the peak amplitude of the pressure.

Acknowledgement

This study was supported by The Support Programme for Scientific and Technological Research Project grant from TUBITAK.

References

[1] Allen, H. G., *Analysis and Design of Structural Sandwich Panels*, Pergamon Press: Oxford, 1969.

[2] Reddy, J. N., *Mechanics of Laminated Composite Plates and Shells Theory and Analysis*, CRC Press: Boca Raton, 2004b.

[3] Noor, A. K., Burton, W. S., and Bert, C. W., Computational Models for Sandwich Panels and Shells, *Applied Mechanics Reviews*, **4(3)**, pp. 155-199, 1996.

[4] Librescu, L., and Hause, T., Recent Development in the Modelling and Behavior of Advanced Sandwich Constructions: a survey, *Journal of Composite Structures*, **48(1-3)**, pp. 1-17, 2000.

[5] Vinson, J. R., Sandwich Structures, *Applied Mechanics Reviews*, **54(3)**, pp. 201-214, 2001.

[6] Hause, T., and Librescu, L., Dynamics Response of Anisotropic Sandwich Flat Panels to Explosive Pressure Pulses, *International Journal of Impact Engineering*, **31**, pp. 607-628, 2005.

[7] Librescu, L., Oh, S. Y., and Hohe, J., Linear and Non-linear dynamic response of sandwich panels to blast loading, *Composites: Part B*, **35**, pp. 673-683, 2004.

[8] Librescu, L., Oh, S. Y., and Hohe, J., Dynamic Response of Anisotropic Sandwich Flat Panels to Underwater and in-air Explosions, *International Journal of Solids and Structures*, **43**, pp. 3794-3816, 2006.

[9] Hause, T., and Librescu, L., Dynamic Response of Doubly-curved Anisotropic Sandwich Panels impacted by blast Loadings, *International Journal of Solids and Structures*, **43**, pp. 6678-6700, 2007.

[10] Baş, A., Kazancı, Z., and Mecitoğlu, Z., Nonlinear Response of A Sandwich Plate Subjected to Blast Load, *ASME International Mechanical Engineering Congress and Exposition*, November 11-15, (IMECE 2007), Seattle, Washington, USA, 2007.

[11] Weierstrass, K., Über die Analytische Darstellbarkeit Sogenannter Willkürlicher Functionen Einer Reellen Veränderlichen, *Sitzungsberichte de Akademie zu Berlin*, pp. 633-639, 789-805: also appeared in Weierstrass' Mathematische Werke, 1903, *Mayer and Muller*, Berlin, **3**, pp. 1-37, 1885.

[12] Shaw, M. T., MacNight, W. J, *Introduction to Polymer Viscoelasticity,* John Wiley & Sons, Inc.: Hoboken, New Jersey, 2005.

[13] Yan Wei, Ying Ji, Chen Weiqiu, A Three-Dimensional Solution for Laminated Orthotropic Rectangular Plates with Viscoelastic Interfaces, *Acta Mechanica Solida Sinica,* Vol. 19, No. 2, June 2006.

[14] Mesquita, A. D., Coda, H. B., Alternative Kelvin Viscoelastic Procedure for Finite Elements, *Applied Mathematical Modelling,* Vol. 26, 2002, pp. 501-516.

[15] Lekszycki, T., Olhoff, N., and Pedersen, J. J., Modelling and Identification of Viscoelastic Properties of Vibrating Sandwich Beams, *Composite Structures,* Vol. 22, pp. 15-31, 1992.

[16] Esmailzadeh, E., Jalali, M. A., Nonlinear Oscillation of Viscoelastic Rectangular Plates, *Nonlinear Dynamics*, **18**, 311-319, 1999.

[17] Vinson, J. R., *The Behavior of Sandwich Structures of Isotropic and Composite Materials*, Technomic Publishing Company: USA, 1999.

[18] Ward, I. M., Hadley, D. W., *An Introduction to the Mechanical Properties of Solid Polymers*, John Wiley & Sons, England, 1993.

[19] Zenkert, D., *An Introduction to Sandwich Construction*, Chameleon Press Ltd: London, UK, 1995.

[20] Türkmen, H. S., *Katmanlı Kompozit Panellerin Anlık Basınç Yüküne Dinamik Cevabı*, PhD. Thesis, p. Xvii+70, İstanbul Technical University, İstanbul, 1997.

[21] Bertholet, J., *Composite Materials: Mechanical Behavior and Structural Analysis*, Springer: New York, 1999.

[22] Jones, R., *Mechanics of Composite Materials*, Taylor & Francis Inc., Philadelphia, 1998.

[23] Gupta, A. D., Gregory, F. H., Bitting, R. L., and Bhattacharya, S., Dynamic Analysis of an Explosively Loaded Hinged Rectangular Plate, *Computers and Structures*, Vol. 26, No. 1-2 Aug. 987, pp. 339-344.
[24] Baş, A., Hibrit Katmanlı Kompozit Plakların Anlık Basınç YüküAltındaki Davranışının Deneysel ve Sayısal İncelenmesi, MS. Thesis, p. Iv+85, İstanbul Technical University, İstanbul, 2009.

Laboratory scale tests for internal blast loading

S. Kevorkian, N. Duriez & O. Loiseau
Institut de Radioprotection et Sûreté Nucléaire, France

Abstract

The definition of blast loads applying on a complex geometry structure is still nowadays a hard task when numerical simulation is used, essentially because of the different scales involved. As a matter of fact, modelling the detonation of a charge and its resulting load on a structure requires one to model the charge itself, the structure and the surrounding air, which rapidly leads to large size models on which parametrical studies may become unaffordable. Thus, on the basis of Crank-Hopkinson's law, an experimental set-up has been developed to support reduced scale structures as well as reduced scale detonating solid charges. As a final objective, the set-up must be used to produce the entry data for numerical assessments of the structural resistance.

The set-up is composed of two mock-ups equipped with sensors and has been designed to conduct non destructive studies. In the context of security, the general aim is to study the effects of detonation shock waves inside the test installation and to test the influence of various openings. This set-up offers the possibility of measuring the loading in terms of pressure-time curves.

The present paper summarizes the campaign of experiments performed in the year 2009 and gives the main features of the mock-up, the instrumentation and the pyrotechnics. During the campaign, internal blast tests have been conducted. Profiles of pressure versus time history are presented, taking into account relative positions of the explosive charge versus the gauges. The results obtained allow one to check that Crank-Hopkinson's law is verified and shows the gas pressure influence.

Keywords: blast waves, detonation, pressure measurements, reflections, gas pressure, safety.

WIT Transactions on The Built Environment, Vol 113, © 2010 WIT Press
www.witpress.com, ISSN 1743-3509 (on-line)
doi:10.2495/SU100111

1 Introduction

Although important developments have taken place during the last decade, the definition of blast loads applying on a complex geometry structure is still nowadays a hard task when numerical simulation is used, essentially because of the different scales involved (both in space and time). As a matter of fact, modelling the detonation of a charge and its resulting load on a structure requires modelling the charge itself, the structure and the air surrounding the charge and the structure, which rapidly leads to large size models on which parametrical studies may become unaffordable.

Because full-scale testing of realistic target geometries and realistic effects of charge position are often prohibitively expensive and time consuming, as far as detonation is involved, small-scale testing is a well proven means to assess blast loading. The most widely used method of blast scaling is Hopkinson's "cube-root" law for scaled distance, time and impulse.

This method has been used by IRSN to assess the pressure evolution in space in time though a free-field campaign of measurements [1]. In order to assess the pressure evolution due to small-scale detonations, IRSN realized a new campaign concerning the internal blast, which is described in this paper. Various authors used the method of blast scaling, in order to assess internal detonations. For instance, [2] shows a holographic interferometry system that permits the visualization and the measures of the propagation of an explosion of 10 mg of silver azide cylinder inside a small-scale closed room. Reference [3] presents results of explosions of a cylindrical charge made of composition B explosive inside several small-scale 3 and 4-wall cubicles of different sizes and shapes, these tests were made to establish method and criteria for blast effects prediction.

Always relating to internal explosion studies, [5] presents the experimental measurements of pressure due to detonation of an explosive gaseous mixture (1 g eq TNT) confined in a hemispherical soap bubble inside an unvented small-scale structure. For solid explosives, [4] presents a comparison of an experimental explosion of 1 lb of C4 in a rectangular bunker and numerical calculations (Method of Images). Reference [6] shows numerical simulations done in order to study the influence of the building geometry, positions of explosion vent and ignition point. All of these studies confirm the interest of studying explosion effects at small scale, mainly allowing one to capture a better understanding of the phenomena involved.

The experimental set-up described in the present paper is a laboratory scale set-up, constituted by a mock-up, able to bear the effects of detonations of solid explosives up to 16 g of TNT equivalent. In the context of security, the general aim is to study the effects of blast waves inside test installations and the influence of the openings.

2 Experimental set-up

The experimental set-up is composed of two mock-ups and sensors; it has been designed to conduct non destructive studies. Experimental campaigns are performed at the SNPE's Research Centre located at Le Bouchet (Vert-le-Petit, France). SNPE ensures all the pyrotechnics handling aspects of the experiments and also provide a spare data recording system.

2.1 Mock-ups

For this campaign, IRSN used two mock-ups made of steel, representing a parallelepipedic room.

In order to study the scale factor in a small-scale configuration for internal blast, two mock-ups with homothetic dimensions were made. The largest mock-up is 40 cm wide, 80 cm large and 40 cm high. The scale factor between the two different mock-ups was determined in such a way that a factor of four between the masses of explosives used in the two mock-up sizes was employed (i.e.: the scale factor λ is equal to 0.63 for the length).
For each size, three configurations are available; the first one is the mock-up without the front face, the second one with the front face with one opening and the third one with a full front face.

Threaded holes were made through the walls, soil and roof in order to allow insertion of air blast pressure gauges. Charges are supposed to be placed at the geometrical centre of the mock-up thanks to a guiding cap inserted from the top of the box.

2.2 Gauges

Eleven piezoelectric pressure transducers (Kistler, reference: 603B, range: 0-200 bars) are mounted on the mock-up. In the absence of pressure gauges, the holes in the walls are filled with a specific screw, so that they do not constitute unexpected venting or opening surfaces. Each pressure transducer is statically calibrated prior to the test campaign. The transducers are connected to an amplifier which is connected to the data acquisition system with electric microdot cables.

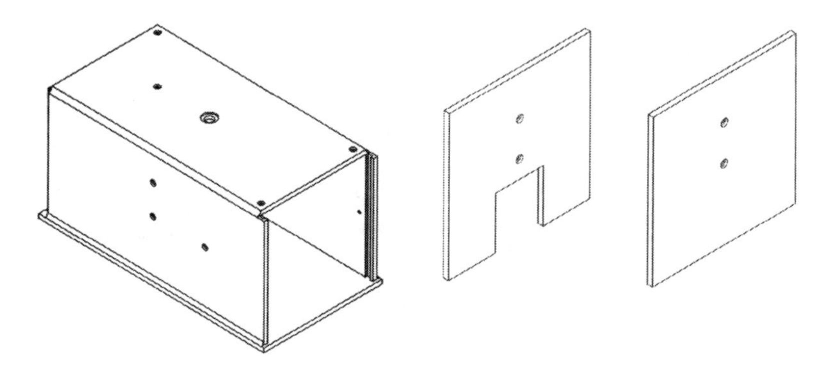

Figure 1: The mock-up, the front faces with and without openings.

2.3 Detonating charges

A cylindrical charge of hexocire is initiated from one of its extremity using an electrical detonator. The masses used for the campaign described in the present paper are 1, 2, 3 and 4 g of TNT equivalent for the smallest mock-up and 4, 8, 12 and 16 g of TNT equivalent for the largest one.

2.4 Acquisition system

A LTT-186 data acquisition and transient recorder system connected to a PC has been used for the data acquisition in parallel with a NICOLET Odyssey.

3 Physical phenomena

The pressures observed after the explosion of a charge in a confined space is composed of two distinct phases.

The first phase is the reflected blast load impulse and the second one is due to the pressure of the gases created during the explosion [7, 8]. These phases are illustrated by Figure 2.

3.1 Reflections

During an explosion, the shock wave is expanding in air up to the walls of the room, and then it is reflected (Figure 3).

For these experiments, an a priori estimate of the peak pressure was obtained using the software SHOCK [9].

In the literature, the maxima of pressure on a structure are most commonly estimated from scaled blast data or theoretical analyses of normal blast wave reflection from a rigid wall [8]. The subsequent shocks due to reflections are supposed to be attenuated.

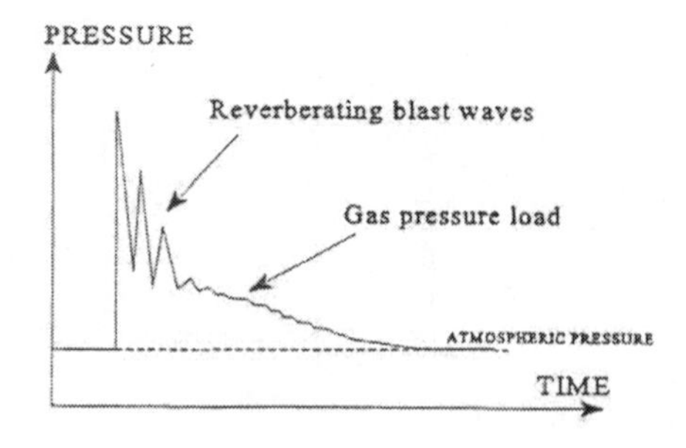

Figure 2: Profile of pressure vs. time history.

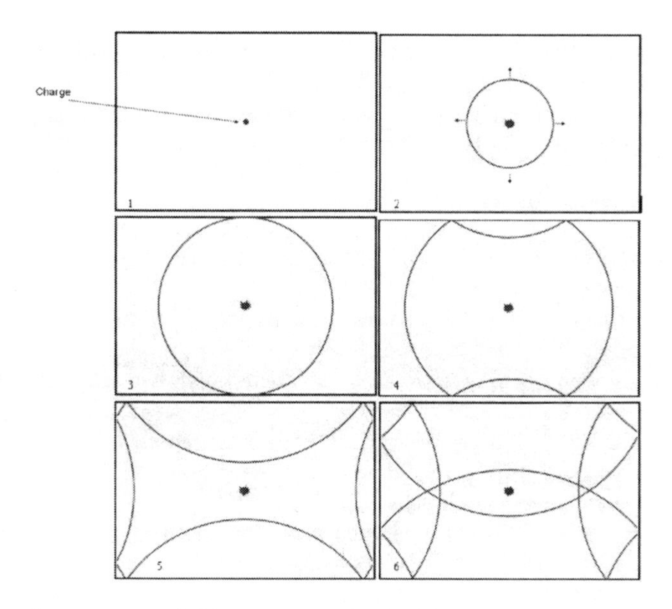

Figure 3: Air blast propagation inside a rectangular room.

3.2 Gas pressure

When an explosion is produced, the detonation gas generated induces an increase of pressure inside the room. This pressure decreases more or less quickly depending of the number and the size of openings in the room. This phenomenon occurs after the propagation of the air blast.

The pressure and impulse due to the detonation gases have been estimated by using the software FRANG [10].

4 Internal blast campaign

4.1 Objectives

The aim of this campaign is to validate the reduced scale experimental concept for an internal blast.

The records of pressure vs. time history are used in order to:

- Check the scaling law or Crank-Hopkinson's law in an internal blast configuration;
- Determine the characteristic parameters of the blast waves – pressure, pulse, time of arrival – identified by the time evolution of pressure and compare the values obtained with values from abacuses available in the literature or out of calculation codes [9, 10];
- Observe experimentally the different physical phenomena composing the air blast : shock wave with reflections and gas pressure;

- Verify the trials reproducibility and the symmetry of measurements from symmetrically placed gauges;
- Show the difference of air blast pressure considering the position of the gauges in the room.

4.2 Description of the trials

For each size of the two mock-ups, four series of trials have been performed with four different masses of explosive.

For each masse of explosive, trials were performed with the three different types of opening.

For each type of opening, trials were conducted with two sets of positions for the gauges.

As a whole, 48 trials were performed, each one with eleven gauges in place. All the recordings of pressure vs. time history have been studied. Only the most representative are included in this article.

At first, the charge was placed at the exact geometrical centre of the mock-up, at the same height as the half-height gauges. By doing so, fragments of the detonator envelope came and hit the gauges after the charge initiation, causing several damages. To avoid this inconvenient, the charge was positioned at a height equal to one quarter of the overall mock-up height.

4.3 Position of charge and gauges

The positions of the gauges were chosen in order to verify:

- The reproducibility of the trials — gauges placed at the centre of the faces are the same for the two sets of gauges positions (K4 & K6, K9 & K11) and two gauges of the floor are the same (K1 & K3);

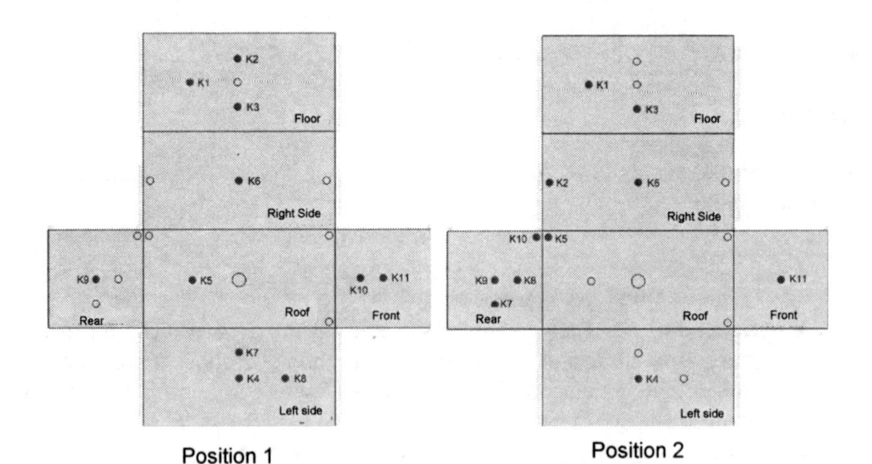

Figure 4: Positions of the gauges for the mock-up with front face.

- The symmetry of the propagation of the air blast — considering the different axes, pairs of symmetrical gauges were studied (K4 and K6, K9 and K11, K2 and K3, K1 and K5, for the position 1);
- The profile of pressure for the largest number of geometrical positions inside the box:
    - o centre of faces, nearest from the charge (K4 and K6), farthest from the charge (K9 and K11)
    - o corners (K5 and K10 for the position 2)
    - o edges (K2 for the position 2)
    - o intermediate positions (K1,K2,K3, K7,K8, K10)

5 Experimental results

Experimental results are compared with estimates obtained from SHOCK [9]. A relatively large difference has been observed between the estimates and the experimental results notably regarding the values of peak overpressure. For instance, discrepancies as large as 70% have been observed. These major discrepancies have been imputed to the shape of the explosive charge. Indeed, the SHOCK software [9] considers a spherical charge whereas the trials were performed with cylindrical charges. In the configuration used for the trials, the reduced distance between the explosive charge and the gauges is rather small (less than 1 m/(TNT kg)$^{1/3}$) and the influence of the shape of an explosive charge is more important for the smaller the distances, see Figure 5 taken from references [8] and [11]. The comparison of the impulse values between trials and a priori estimates is about 27%, showing less influence of the charge shape itself.

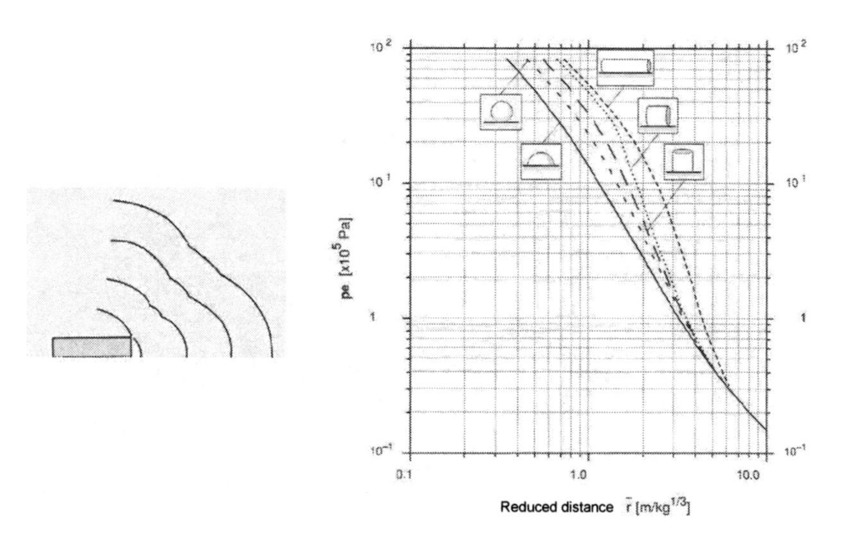

Figure 5: Influence of the shape of the charge.

5.1 Position of the gauges

In this section, the different pressure profiles are presented considering the position of the gauge.

Figure 6 is the record of pressure vs. time history of the gauge K4, placed in the middle of the largest face of the experimental box. The maxima of pressure are obtained for this gauge, which is the closest gauge to the charge. It can be seen that the first overpressure peak is rather high, and followed by several peaks due to the different reflections.

Figure 7 is the record of pressure vs. time history of the gauge K9, placed in the middle of the smallest face of the experimental box. On the results, it has been observed that the largest overpressure peak appears a few microseconds after a first one, smaller, and is then followed by several peaks. This second

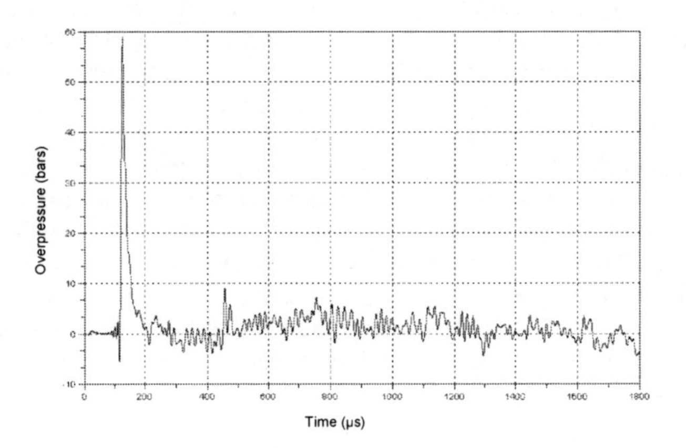

Figure 6: Centre of the largest face (K4, 16 g eq TNT).

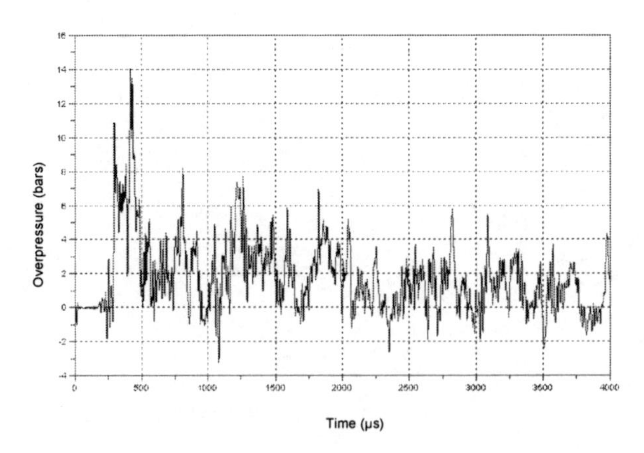

Figure 7: Centre of the smallest face (K9, 16 g eq TNT).

overpressure peak is induced by the four simultaneous reflections on the walls surrounding the smallest face, which are at equal distance from the gauge.

Figure 8 is the record of pressure vs. time history of the gauge K5, placed in one corner of the room. It can be noticed that in this case, the first peak to appear is much less intense than the immediate second one. This particular gauge being near and equidistant from three surfaces of reflection, the first peak corresponds to the incident wave and the second to the combination of all the reflected waves on the adjacent faces. The following peaks are induced by the following reflections.

5.2 Scale factor

Considering the Crank-Hopkinson's scaling law, the pressure ratio for the two scale mock-ups should be equal to 1, and the impulse ratio equal to the scale factor, i.e. 0.63.

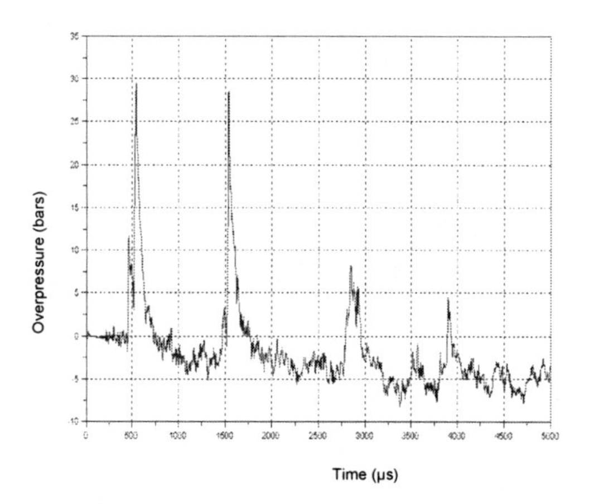

Figure 8: Corner (K5, 16 g eq TNT).

Table 1: Ratio between peak overpressure at two different scales. The average was obtained considering all the gauges and all the trials.

gauges / masses	K1	K2	K3	K4	K5	K6	K7	K8	K9	K10	K11	Average
2g/8g	0.934	1.555	0.882	0.721	1.370	0.900	0.445	0.911	1.018	1.035	0.763	0.958
4g/16g	0.636	1.142	0.677	1.438	0.957	0.903	0.591	0.713	1.510	0.926	1.044	0.958
4g/16g	1.201	1.386	1.173	0.684	0.581	1.354	1.330	0.857	1.176	0.865	0.739	1.186
4g/16g	1.215	1.104	0.828	0.629	0.908	1.145	0.389	1.179	1.244	0.738	0.767	0.922
4g/16g	0.870	0.810	0.867	1.192	0.894	1.115	0.715	0.938	0.980	0.615	0.608	0.873
4g/16g	1.306	0.884	1.445	0.740	0.730	1.004	x	0.763	0.621	x	0.870	0.929
3g/12g	0.878	1.904	1.049	0.377	1.122	0.747	0.475	0.798	1.010	0.826	0.796	0.908
3g/12g	0.934	1.530	0.782	0.599	1.403	0.645	0.577	1.077	0.550	0.693	1.012	0.891
1g/4g	1.041	1.291	0.573	0.729	1.245	1.311	0.891	0.864	0.977	0.837	1.719	1.043
2g/8g	1.180	0.647	0.875	0.488	x	0.477	x	0.748	0.845	0.792	0.754	0.756
3g/12g	1.268	1.097	1.000	1.271	1.133	x	0.641	1.429	1.086	0.838	x	1.085
4g/16g	0.803	0.942	0.660	0.816	1.485	1.602	0.669	1.128	0.979	0.576	x	0.966
									Average			0.956

These results gives good insurance that the Crank-Hopkinson's scaling law has been verified through the present internal blast loading experimental campaign.

The compared recordings of the profile pressures vs. reduced time, obtained for the explosion of 4 g (eq TNT) in the small model and 16 g (eq TNT) in the large model are drawn on Figures 9 and 10. The good correspondence between the curves obtained at one scale and at another, confirms the good reproducibility of the tests even at different scales.

5.3 Gas pressure

For each size of mock-up and each mass of explosive, trials were performed with different type of openings in order to emphasize the influence of the gas pressure on the pressure profile measured on the mock-up faces. The recordings obtained for these three different configurations are plotted in Figure 11. During the first

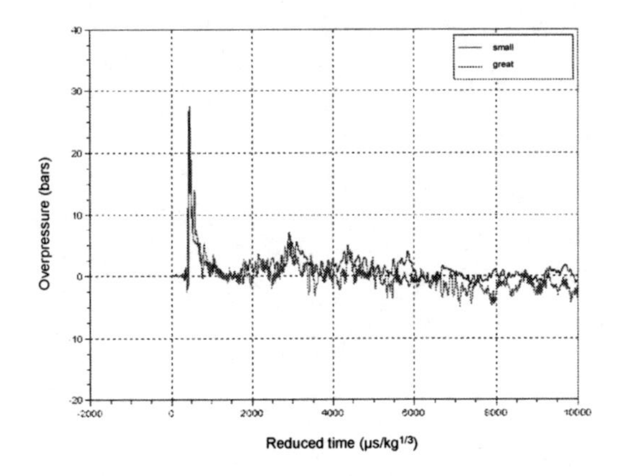

Figure 9: Centre of greatest face (K4, 4 and 16 g (eq TNT)).

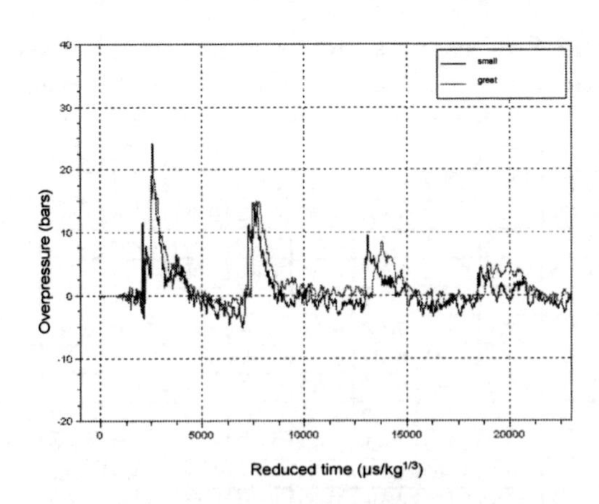

Figure 10: Corner (K5, 4 and 16 g (eq TNT)).

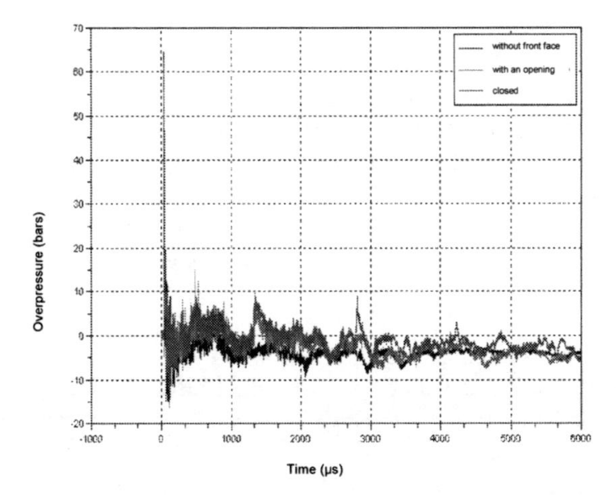

Figure 11: Pressure for the three type of opening (4 g eq TNT).

milliseconds of the test, the pressure profile is the same in the three configurations; this part corresponds to the air blast propagation and reflections; after that, differences between the profiles begin to appear. The pressure profiles corresponding to the closed mock-up and the mock-up with one opening present more intense peaks than in the pressure profile observed for the open mock-up. Furthermore, the amplitude of these peaks is more important in the case of closed mock-up. This observation confirms qualitatively that the detonation gas pressure has an influence on the pressures measured by the gauges. Nevertheless, considering the effect of the drift of the gauge in the (0;200 bar) range, the evaluation of the gas pressure amplitude remains difficult to obtain out of the present experiments.

6 Conclusions

In order to conduct security studies for which the effects of blast waves in the vicinity of industrial sensitive installations need to be investigated, IRSN has developed an experimental set-up composed of a modular table, mock-up and pressure transducers. This experimental set-up is a support for non-destructive studies and dedicated to testing various shock wave propagations. The first campaign performed in the end of 2006 allowed to qualify the measurement chain and validate the concept of small-scale experiments. This validation was conducted through a free-field campaign. This campaign performed in 2009 allowed to validate the concept of small-scale experiments for an internal blast. In this framework, the results obtained allowed to check that the Crank-Hopkinson's "cube-root" law is verified. This campaign has also shown that improvements should be made in order to quantify the effects of gas pressure, notably requiring the use of some other sensor technology or range.

References

[1] K. Cheval, O. Loiseau & V. Vala, *Laboratory scale test for the assessment of solid explosive blast effects*, Structures Under Shock and Impact X,2008
[2] A; Miura, T. Mizukaki, T. Shiraisji, A. Matsuo, K. Takayama, I. Nojiri, *Spread behaviour of explosion in closed space*, 2004, Journal of Loss Prevention in the Process Industries 17 (2004) 81-86
[3] W.A. Keenan, J.E. Tancreto, *Blast environment from fully and partially vented explosions in cubicles*, 1975, Civil Engineering Laboratory
[4] P. Chan, H.H. Klein, *A study of blast effects inside an enclosure*, 1994, Journal of fluids engineering,116, n°3, 450-455
[5] A. Zyskowski, I. Sochet, G, Mavrot, P. Bailly, J. Renard, *Study of the explosion process in a small scale experiment – Structural loading*, 2004, Journal of Loss Prevention in the Process Industries 17 (2004) 291-299.
[6] N. Sonoda, A. Hashimoto, A. Matsuo, Influence of vessel geometry on the effect of explosion vent, 5th International Seminar on Fire and Explosion Hazards, Edinburgh, UK, April 2007
[7] Kinney, Graham, Explosive Shocks in air, Springer Verlag (1962)
[8] Baker & al., *Explosion hazards and Evaluations*, Elsevier (1983)
[9] SHOCK, Naval Civil Engineering Laboratory, USA
[10] FRANG, Naval Civil Engineering Laboratory, USA
[11] B. Anet, E. Binggli, *LS2000 Luftsossphänomene infolge nuklearer und konventioneller explosionen*

An arbitrary Lagrangian Eulerian (ALE) based numerical method for the computation of gas-particle two phase flow

S. Zhang, M. Zhang, H. Zhou & J. Xiong
Beijing Institute of Applied Physics and Computational Mathematics, China

Abstract

In this paper, an arbitrary Lagrangian Eulerian (ALE) based numerical method has been presented for the numerical simulation of gas-particle two phase flow with moving boundaries. The main stages for the implementation of the algorithm have been discussed. The numerical results of cylindrical implosion and 2D dusty gas explosion have shown the effectiveness of the method.
Keywords: ALE method, compatible Lagrangian method, gas-particle two phase flow.

1 Introduction

Gas-particle two phase flow is a complex flow phenomenon in the natural environment and industrial engineering. With the development of physical modeling, numerical investigation has been playing a more and more important role in the research field. For most numerical investigation, the physical domain is fixed, thus the Eulerian numerical methods are generally used with fixed grid and fixed boundaries. However, for the investigation of the strong interaction of gas-particle two phase flow with structure where the solid structure deformation is severe and the physical boundary is moving, the numerical method should have the ability to track the moving boundary (also the fluid-structure interface) and a moving grid is need. In this paper, an arbitrary Lagrangian Eulerian (ALE) based numerical method is presented for gas-particle two phase flow with a trajectory model, which satisfies the above requirements [1, 2].

WIT Transactions on The Built Environment, Vol 113, © 2010 WIT Press
www.witpress.com, ISSN 1743-3509 (on-line)
doi:10.2495/SU100121

For the investigation of transient dynamic response of complex engineering systems, such as laser and explosion driven problems and fluid-structure interaction problems, the ALE method is the most valuable numerical method. The grid can move with arbitrary speed, which takes advantage of tracking the material interface with the grid. When the grid moves with material speed, the ALE method is just a pure Lagrangian method that is suitable for computational solid mechanics. In this mode, the grid may be distorted under the circumstances of severe shear deformation of the flow field, especially for fluid mechanics; thus, the grid should move at an appropriate speed to keep the grid in good quality to support the computation. A special case is that the grid is fixed without moving, which is just the Eulerian mode of the ALE method.

The object of this paper is the application of the ALE method to gas-particle two phase flow with a two-way coupling trajectory model. After a brief introduction of the ALE method, some key problems are presented for the application of the method in multiphase flow. Two numerical examples have been given to show the effectiveness and robustness of the method.

2 Numerical method

In this section, the general ALE method is discussed at first, followed by its application to gas-particle simulation.

The explicit ALE algorithm of this paper can be divided into two stages in general. The first stage is the Lagrangian stage, followed by the rezone and remap stage.

In the first stage the Lagrangian method is the compatible Lagrangian method [3]. The new method is a kind of physically motivated discretization method that can achieve the goal of representing as faithfully and accurately as possible the mathematic and physical property underlying dynamic differential equations while keeping the accuracy and efficiency of the algorithm in a simple and generic manner. The theoretical basis of the method is the support operator method, utilizing the vector identities of differential calculus to derive compatible sets of the fundamental vector differential operators, such as gradient divergence and curl in discrete form. This is first done by first specifying one of these operators (called the prime operator) and then using the vector identities written in discrete summation form to consistently determine the others. Although a finite truncation error is inherent to the method, certain global physical properties, such as the conservation of total energy, can be satisfied to numerical roundoff error, although only the internal equation is to be solved.

The differential equations to be solved can be written in the following form in general for continuous mechanics.

$$\frac{1}{V}\frac{dV}{dt} = \nabla \cdot \vec{v} \tag{1}$$

$$\rho\frac{d\vec{v}}{dt} = \nabla \cdot \vec{\vec{\sigma}} \tag{2}$$

$$\rho \frac{de}{dt} = \vec{\vec{\sigma}} : \vec{\vec{\varepsilon}} \tag{3}$$

$$\vec{\vec{\sigma}} = \vec{\vec{\vec{\vec{C}}}} : \nabla \vec{v} \tag{4}$$

where $V(=1/\rho)$, ρ, e, $\vec{v}$, $\vec{\vec{\sigma}}$, $\vec{\vec{\varepsilon}}$ are the specific volume, density, internal energy, velocity, stress and strain tensor. Equation (4) is the general constitutive equation.

The compatible Lagrangian method utilizes a staggered grid for space discretization and predictor-corrector two steps for time discretization. For shock problems, artificial viscosity should be used [4, 5]. The contact and impact algorithm is a better choice if there are complex interactions at the interfaces between different materials.

The second stage is the rezone and remap stage, which is necessary when the grid is in severe distortion [6, 7]. The object of the stage is to construct a new grid with much better geometric quality than the old one to conduct the computation smoothly and to project the known flow field in the old grid to the new grid with high accuracy [1].

When the ALE method above is applied to the numerical simulation of gas-particle two phase flow with a two-way coupling trajectory model [8], things are similar to the Eulerian methods for the calculation of the particle phase; the main difference is in the computation of the continuous carrier phase described above. However, two key problems should be addressed with much care [8]. The first one is the searching problem, which requires a fast search technique to determine which cell of the moving unstructured grid a particle with known coordinates locates. The second one is the interaction of the particles with the moving wall of a condensed matter, which may occur in the problem of the interaction of a gas-particle two phase flow with a structure. The particles may bounce from the wall, or slide on the wall, so appropriate models and algorithms are needed.

3 Numerical method

Two examples are presented to show the effectiveness of the ALE method of this paper. The first one in fact is a simple example of the interaction of a gas-particle two phase flow with a structure with moving boundaries. The second one is a fixed boundary gas-particle flow, which may be simulated traditionally with the pure Eulerian method, but here is simulated with the ALE method.

3.1 1D cylindrical dusty gas implosion

Initially, the particle distribution in space is uniform, and the particles with a uniform diameter of 10μm are at rest in the static air at a pressure of 1atm. The dusty gas is driven by a thin metal shell toward the center with an initial high velocity. The dynamic behavior of the shell is described with an elastic–plastic model. The numerical results are shown in figures 1 to 3 in non-dimensional style.

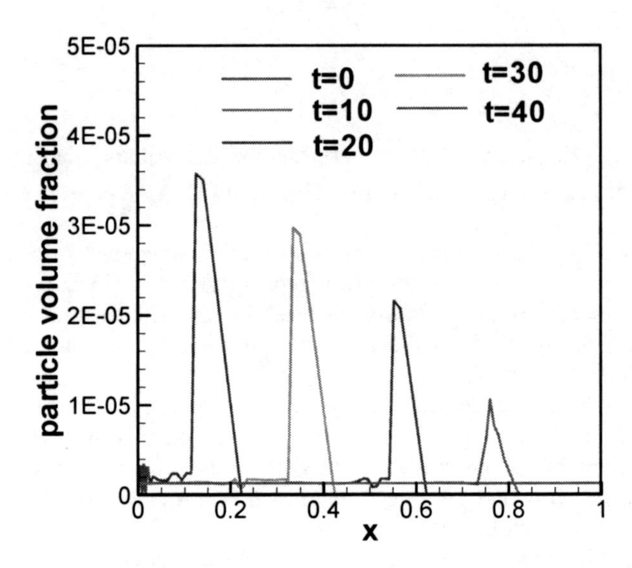

Figure 1: The distribution of the particle volume fractions.

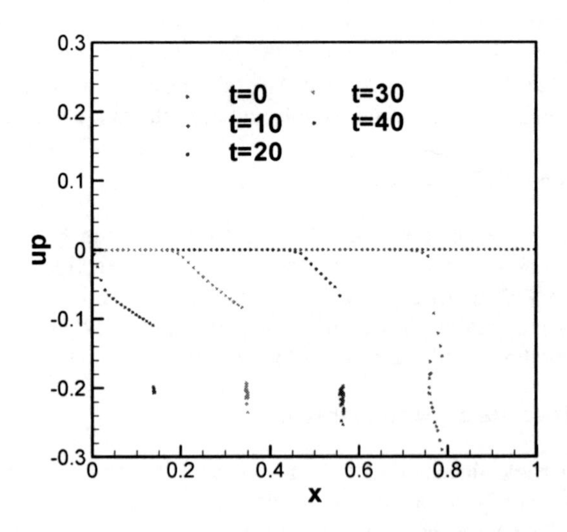

Figure 2: The distribution of the particle velocities.

Figure 1 shows the distribution of the particle volume fraction in a 1D dusty gas implosion at t = 0, 10, 20, 30, and 40. With the moving of the interface, the gas is compressed and the particles move with the gas. The particles near the interface collide with the interface and are accelerated. The particle is accumulated at the region near the surface. The maximal value of the particle volume fraction increases with time during the period of compression. Figure 2 shows the distribution of particle velocity in a 1D dusty gas implosion at t = 0, 10, 20, 30, and 40. The particles near the interface have the maximal velocity. The maximal particle velocity decreases with time. The compressed gas decreases the particle and it results in the increasing of the maximal value of the particle volume fraction with time, as shown in figure 1. Figure 3 shows the distribution of particle temperature in a 1D dusty gas implosion at t = 0, 10, 20, 30, and 40. Due to the compression, the gas near the interface has a higher temperature. Therefore, the temperature of the particles near the interface increases by adsorbing the heat from the surrounding gas.

3.2 2D explosion of dusty gas

At t = 0, the ratio of the gas density and pressure is 10, between the inner and the outer region of the column. The particle concentration is 1 inside the column and 0 outside the column. Two phases are at rest in the beginning. Figure 4 gives the particle concentrations at t = 0, 2, and 5. The particle concentration in the center and periphery is high. In addition, the particle concentration is high in the direction perpendicular to the coordinate axis.

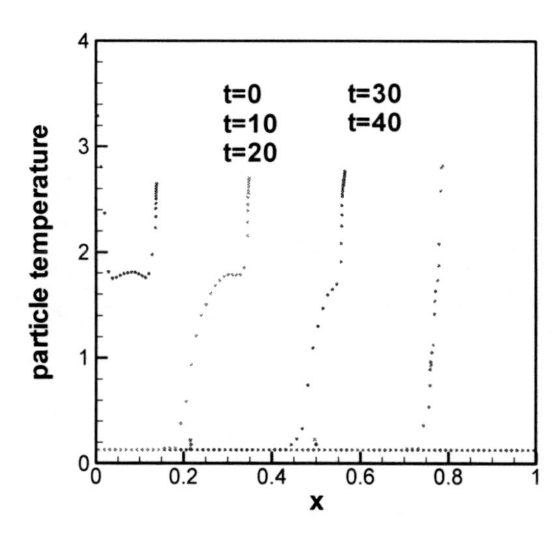

Figure 3: The distribution of the particle temperatures.

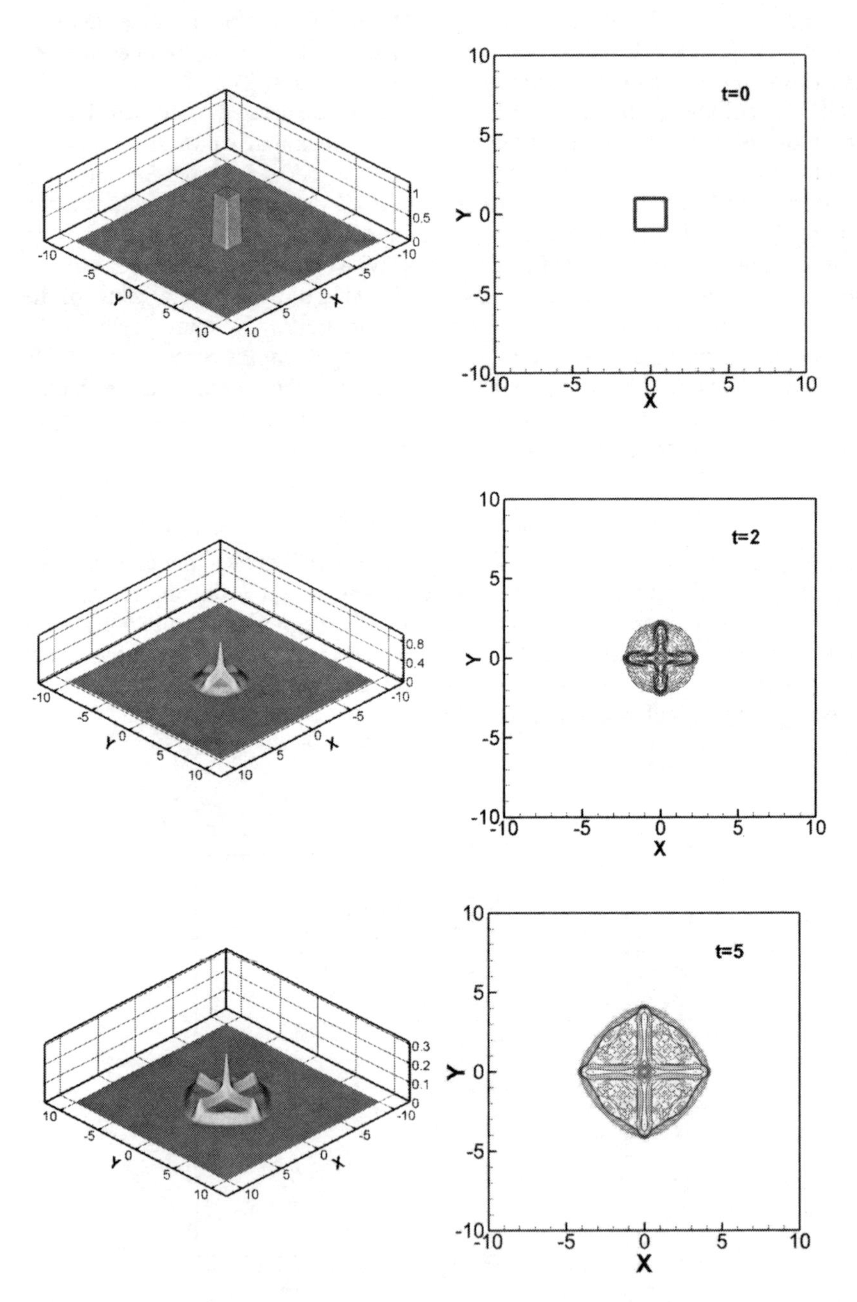

Figure 4: Particle concentration of the 2D dusty gas explosion: t=0.0 (top); t=2.0 (middle); t=5.0 (bottom).

4 Conclusion

An arbitrary Lagrangian Eulerian (ALE) based numerical method has been presented for the numerical simulation of a gas-particle two phase flow with moving boundaries. This ALE method takes advantage of the recently developed compatible Lagrangian method. A brief introduction was given for the implementation of this algorithm. The numerical results have proved that the new method is effective and robust for the transient analysis of a strong driven gas-particle two phase flow.

References

[1] Hirt, C.W., Amsden, A.A. and, Cook, J., An arbitrary Lagrangian-Eulerian computing method for all flow speeds. *J. Comp. Phys.*, 14:227–253, 1974. reprinted in 1997, 135: 203–216.

[2] Benson, D.J., Computational methods in Lagrangian and Eulerian hydrycodes. *Comput. Methods Appl. Mech. Engrg.*, 99:235–394, 1992.

[3] Caramana, E.J., Burton, D.E. Shashkov, MJ. and Whalen, P.P., The construction of compatible hydrodynamics algorithms utilizing conservation of total energy. *J. Comp. Phys.* 146(1):227-262, 1998.

[4] Caramana, E.J., Shashkov, M. J. and Whalen, P.P., Formulations of artificial viscosity for multi-dimensional shock wave computations. *J. Comp. Phys.*, 144(1):70-97, 1998.

[5] Haibing Zhou, Jun Xiong and Shudao Zhang., Formulation of Artificial Viscosity for Lagrangian Analysis of Shocks. *The 8th World Congress on Computational Mechanics (WCCM8) and the 5th European Congress on Computational Methods in Applied Sciences and Engineering* (ECCOMAS 2008), June 30 -July 5, 2008, Venice, Italy.

[6] Canann, S.A., Tristano, J.R., Staten, M.L., An approach to combined Laplacian and optimization-based smoothing for triangular, quadrilateral, and quad-dominant meshes, in: *Proceedings of the Seventh International Meshing Roundtable*, Dearborn, MI, pp. 479-494. 1998.

[7] Margolin, L. G. and Shashkov, M. J., Remapping, recovery and repair on a staggered grid. *Comput. Methods Appl. Mech. Engrg.*, 193:4139–4155, 2004.

[8] Mingyu Zhang, Shudao Zhang, Haibin Zhou, Jun Xiong, Numerical Investigation of the Interaction between Shock and the Particles. The paper in *this symposium*.

Section 4
Interaction between computational and experimental results

Theory and calibration of JWL and JWLB thermodynamic equations of state

E. L. Baker[1], D. Murphy[1], L. I. Stiel[2] & E. Wrobel[1]
[1]*US Army Armament Research Development and Engineering Center, USA*
[2]*New York Polytechnic University, USA*

Abstract

Structure geometric configuration and response can be strongly coupled to blast loading particularly for close-in blast loading configurations. As a result, high rate continuum modeling is being increasingly applied to directly resolve both the blast profiles and structural response. In this modeling, the equation of state for the detonation products is the primary modeling description of the work output from the explosive that causes the subsequent air blast. The Jones-Wilkins-Lee (JWL) equation of state for detonation products is probably the currently most used equation of state for detonation and blast modeling. The Jones-Wilkins-Lee-Baker (JWLB) equation of state is an extension of the JWL equation of state that we commonly use. This paper provides a thermodynamic and mathematical background of the JWL and JWLB equations of state, as well as parameterization methodology. Two methods of parameter calibration have been used to date: empirical calibration to cylinder test data and formal optimization using JAGUAR thermo-chemical predictions. An analytic cylinder test model that uses JWL or JWLB equations of state has been developed, which provides excellent agreement with high rate continuum modeling. This analytic cylinder model is used either as part of the formal optimization or for post parameterization comparison to cylinder test data.
Keywords: blast, explosives, equation of state, modelling.

1 Introduction

Structure geometric configuration and response can be strongly coupled to blast loading particularly for close-in blast loading configurations. As a result, high

WIT Transactions on The Built Environment, Vol 113, © 2010 WIT Press
www.witpress.com, ISSN 1743-3509 (on-line)
doi:10.2495/SU100131

rate continuum modeling is being increasingly applied to directly resolve both the blast profiles and structural response. Modeling the structural response to blast relies on accurate descriptions of the blast loading pressure profiles. When high rate continuum modeling is directly applied for the blast calculation, the explosive produced blast profile is calculated using detonation modeling of the high explosive event. In this modeling, the equation of state for the detonation products is the primary modeling description of the work output from the explosive that causes the subsequent air blast. The Jones-Wilkins-Lee (JWL) equation of state for detonation products is probably the currently most used equation of state for detonation and blast modeling. The Jones-Wilkins-Lee-Baker (JWLB) equation of state is an extension of the JWL equation of state that we commonly use. The purpose of this paper is to provide a thermodynamic and mathematical background of the JWL and JWLB equations of state, as well as parameterization methodology.

2 JWL equation of state

The JWL thermodynamic equation of state [1] was developed to provide an accurate description of high explosive products expansion work output and detonation Chapman-Jouguet state. For blast applications, it is vital that the total work output from the detonation state to high expansion of the detonation products be accurate for the production of appropriate blast energy. The JWL mathematical form is:

$$P = \mathrm{A}\left(1-\frac{\omega}{\mathrm{R_1}V*}\right)e^{-\mathrm{R_1}V*}+\mathrm{B}\left(1-\frac{\omega}{\mathrm{R_2}V*}\right)e^{-\mathrm{R_2}V*}+\frac{\omega E}{V*} \qquad (1)$$

where $V*$ is the relative volume, E is the product of the initial density and specific internal energy and ω is the Gruneisen parameter. The equation of state is based upon a first order expansion in energy of the principle isentrope. The JWL principle isentrope form is:

$$Ps \equiv \mathrm{A}e^{-\mathrm{R_1}V*}+\mathrm{B}e^{-\mathrm{R_2}V*}+\mathrm{C}V*^{-(\omega+1)} \qquad (2)$$

For JWL, the Gruneisen parameter is defined to be a constant:

$$\omega \equiv \left.\frac{V*dP}{dE}\right|_{V*} \qquad (3)$$

Energy along the principle isentrope is calculated through the isentropic identity:

$$dE_s = -P_s dV* \Rightarrow E_s = \frac{A}{R_1}e^{-R_1V^*}+\frac{B}{R_2}e^{-R_2V^*}+\frac{C}{\omega V^{*\omega}} \qquad (4)$$

This relationship defines the internal energy referencing for consistency, so that the initial internal energy release is:

$$\Rightarrow E_0 = E_{CJ} - \frac{1}{2} P_{CJ} (V_0^* - V_{CJ}^*) \tag{5}$$

The general equation of state is derived from the first order expansion in energy of the principle isentrope:

$$P = P_S + \left. \frac{dP}{dE} \right|_{V*} (E - E_S) = P_S + \frac{\omega}{V^*} (E - E_S) \tag{6}$$

(2), (4), (6)

$$\Rightarrow P = \mathrm{A}\left(1 - \frac{\omega}{\mathrm{R}_1 V*}\right) e^{-\mathrm{R}_1 V*} + \mathrm{B}\left(1 - \frac{\omega}{\mathrm{R}_2 V*}\right) e^{-\mathrm{R}_2 V*} + \frac{\omega E}{V*} \tag{7}$$

From eqns (4) and (5) it can be seen the E0 represents the total work output along the principle isentrope. For blast, this would represent the total available blast energy from the explosive.

3 JWLB equation of state

The JWLB thermodynamic equation of state [2] is an extension of the JWL equation of state. JWLB was developed to more accurately describe overdriven detonation, while maintaining an accurate description of high explosive products expansion work output and detonation Chapman-Jouguet state. The equation of state is more mathematically complex than the Jones-Wilkins-Lee equation of state, as it includes an increased number of parameters to describe the principle isentrope, as well as a Gruneisen parameter formulation that is a function of specific volume. The increased mathematical complexity of the JWLB high explosive equations of state provides increased accuracy for practical problems of interest. The JWLB mathematical form is:

$$P = \sum_n A_i \left(1 - \frac{\omega}{R_i V^*}\right) e^{-R_i V^*} + \frac{\lambda E}{V^*} \tag{8}$$

$$\lambda = \sum_i (A_{\lambda i} V^* + B_{\lambda i}) e^{-R_{\lambda i} V^*} + \omega \tag{9}$$

where V^* is the relative volume, E is the product of the initial density and specific internal energy and λ is the Gruneisen parameter. The JWL equation of state may be viewed as a subset of the JWLB equation of state where two inverse exponentials are used to describe the principle isentrope (n=2) and the Gruneisen parameter is taken to be a constant ($\lambda = \omega$).

4 Analytic cylinder model

An analytic cylinder test model that uses JWL or JWLB equations of state has been developed, which provides excellent agreement with high rate continuum modeling. Gurney formulation has often been used for high explosive material acceleration modeling [3], particularly for liner acceleration applications. The work of Taylor [4] provides a more fundamental methodology for modeling exploding cylinders, including axial flow effects by Reynolds hydraulic formulation. A modification of this method includes radial detonation product flow effects and cylinder thinning. The modifications were found to give better agreement with cylinder expansion finite element modeling [5]. One method of including radial flow effects is to assume spherical surfaces of constant thermodynamic properties and mass flow in the detonation products. The detonation products mass flow is assumed to be in a perpendicular direction to the spherical surfaces. A diagram of a products constant spherical surfaces cylinder expansion due to high explosive detonation is presented in Figure 1.

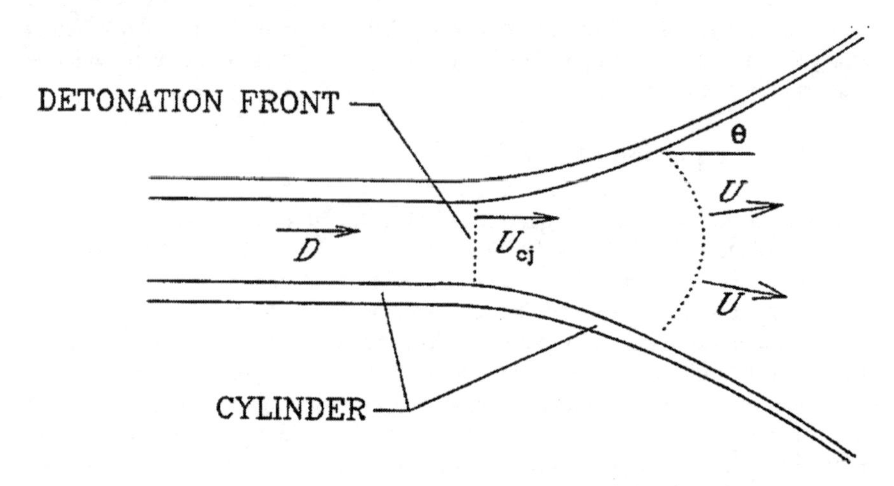

Figure 1: Analytic cylinder test model.

It should be noted that flow velocities are relative to the detonation velocity, D. If constant detonation product properties are assumed across spherical surfaces, the following model results using the JWLB thermodynamic equation of state

Mass:

$$\rho_{cj} U_{cj} A_0 = \rho U A \tag{10}$$

Axial Momentum:

$$P_{cj} r_0^2 - P r^2 = \frac{m}{\pi} D^2 \cos\Theta - \frac{m}{\pi} D^2 + \rho U^2 r^2 - \rho_{cj} U_{cj}^2 r_0^2 \quad .... (11)$$

Energy:

$$\rho_{cj} U_{cj} A_0 \left(\frac{U_{cj}^2}{2} + e_{cj} \right) + P_{cj} U_{cj} A_0 = \rho U A \left(\frac{U^2}{2} + e \right) + PUA \tag{12}$$

Principle Isentrope:

$$P = \sum_i A_i e^{\frac{-R_i \rho_0}{\rho}} + C \left(\frac{\rho_0}{\rho} \right)^{-(\omega+1)}, \quad de = -Pd\left(\frac{1}{\rho} \right) \tag{13}$$

Taylor Angle:

$$v = 2D sin \frac{\Theta}{2} \tag{14}$$

Spherical Area:

$$A = \pi r^2 \frac{2(1 - cos\Theta)}{\sin^2 \Theta} \tag{15}$$

The final equation set used for solution is:

$$(4) \Rightarrow P = \sum_i A_i e^{\frac{-R_i \rho_0}{\rho}} + C \left(\frac{\rho_0}{\rho} \right)^{-(\omega+1)} \tag{16}$$

$$(4) \Rightarrow e_{cj} - e = \sum_i \frac{A_i}{\rho_0 R_i} \left(e^{\frac{-R_i \rho_0}{\rho_{cj}}} - e^{\frac{-R_i \rho_0}{\rho}} \right)$$

$$+ \frac{C}{\omega \rho_0} \left[\left(\frac{\rho_0}{\rho_{cj}} \right)^{-\omega} - \left(\frac{\rho_0}{\rho} \right)^{-\omega} \right] \tag{17}$$

$$(3) \Rightarrow \frac{U^2}{2} = \frac{U_{cj}^2}{2} + \frac{P_{cj}}{\rho_{cj}} - \frac{P}{\rho} + e_{cj} - e \tag{9}$$

$$(2) \Rightarrow \frac{v^2}{2} = \left[P \left(\frac{r}{r_0} \right)^2 - P_{cj} + \rho \left(\frac{r}{r_0} \right)^2 U^2 - \rho_{cj} U_{cj}^2 \right] \frac{C}{m \rho_0} \tag{18}$$

$$(1), (5), (6) \Rightarrow \rho = \frac{\rho_{cj} U_{cj}}{U \left(\frac{r}{r_0} \right)^2} \left[1 - \left(\frac{v}{2D} \right)^2 \right] \tag{19}$$

This set of equations is solved for a given area expansion, $(r/r_0)^2$ using Brent's method [6]. The spherical surface approach has been shown to be more accurate for smaller charge to mass ratios without any loss of agreement at larger charge to mass ratios. It should be recognized that this analytic modeling approach neglects initial acceleration due to shock processes [7] and is therefore anticipated to be more accurate as the initial shock process damps out. The model as expressed does not consider the fact that the cylinders thin during radial expansion. One simple way to account for this wall thinning is to assume that the wall cross sectional area remains constant and r and v represents the inside radius and inside surface wall velocity.

$$v_{out} = v\frac{r_{in}}{r_{out}}\ ;\ r_{out}^2 = r_{in}^2 + r_{out_0}^2 - r_{in_0}^2 \tag{20}$$

5 Eigenvalue analytic cylinder model

High explosives are often aluminized for blast enhancement. Eigenvalue detonations are observed for some aluminized explosives [9]. For this reason, it was of interest to develop a modified analytic cylinder test model that provides a description of the detonation products isentropic expansion from the eigenvalue detonation weak point, rather than from the Chapman-Jouguet state. It was found that the most straight forward method of implementation of an eigenvalue detonation analytic cylinder model was to refit the isentrope associated with the eigenvalue weak point using eqn (13). In this way, equations 1-11 remain correct, except that eigenvalue weak point is used, rather than the Chapman-Jouguet state. With this approach, it is important to realize that the weak-point isentrope fit is not the same as the principle isentrope fit. The final form is:

$$P = \sum_i A_{wi} e^{\frac{-R_{wi}\rho_0}{\rho}} + C_w\left(\frac{\rho_0}{\rho}\right)^{-(\omega+1)} \tag{21}$$

$$e_w - e = \sum_i \frac{A_{wi}}{\rho_0 R_{wi}}\left(e^{\frac{-R_{wi}\rho_0}{\rho_w}} - e^{\frac{-R_{wi}\rho_0}{\rho}}\right) + \frac{C_w}{\omega\rho_0}\left[\left(\frac{\rho_0}{\rho_w}\right)^{-\omega} - \left(\frac{\rho_0}{\rho}\right)^{-\omega}\right] \tag{22}$$

$$\frac{U^2}{2} = \frac{U_w^2}{2} + \frac{P_w}{\rho_w} - \frac{P}{\rho} + e_w - e \tag{23}$$

$$\frac{v^2}{2} = \left[\begin{array}{c} P\left(\frac{r}{r_0}\right)^2 - P_w \\ +\rho\left(\frac{r}{r_0}\right)^2 U^2 - \rho_w U_w^2 \end{array}\right]\frac{C}{m\rho_0} \tag{24}$$

$$\rho = \frac{\rho_W U_W}{U\left(\frac{r}{r_0}\right)^2}\left[1 - \left(\frac{v}{2D_W}\right)^2\right] \tag{25}$$

6 High rate continuum modeling comparison

ALE3D high rate continuum modeling, Figure 2, was compared to analytic cylinder test modeling using identical JWLB equations of state for TNT, LX-14 and PAX-30 for 1 inch diameter charges and 0.1 inch and 0.2 inch thick copper cylinders.

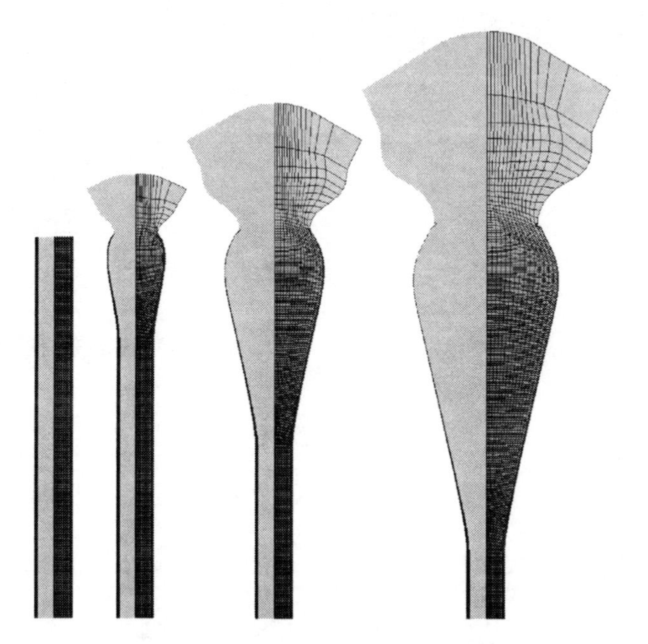

Figure 2: Modeling at 10μs intervals for 0.1" thick copper cylinder.

Figures 3, 4 and 5 present the comparison of the analytic cylinder test model to the ALE3D modeling for TNT, LX-14 and PAX-30 respectively. The analytic cylinder model slightly under predicts the velocities at 2 and 3 inside area expansions, but is in very close agreement by 6 and 7 inside area expansions. This is consistent with the fact that this analytic modeling approach neglects initial acceleration due to shock processes. Strong shock effects are typically observed in the 2 to 3 volume expansion region and are practically damped out by 6 volume expansions, where very close agreement between the analytic model and ALE3D results are observed.

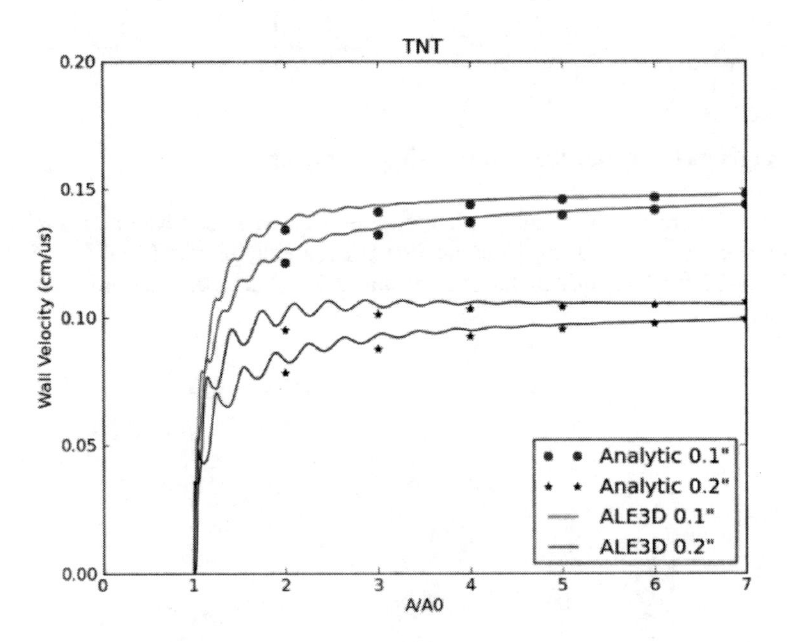

Figure 3: TNT cylinder analytic model versus ALE3D.

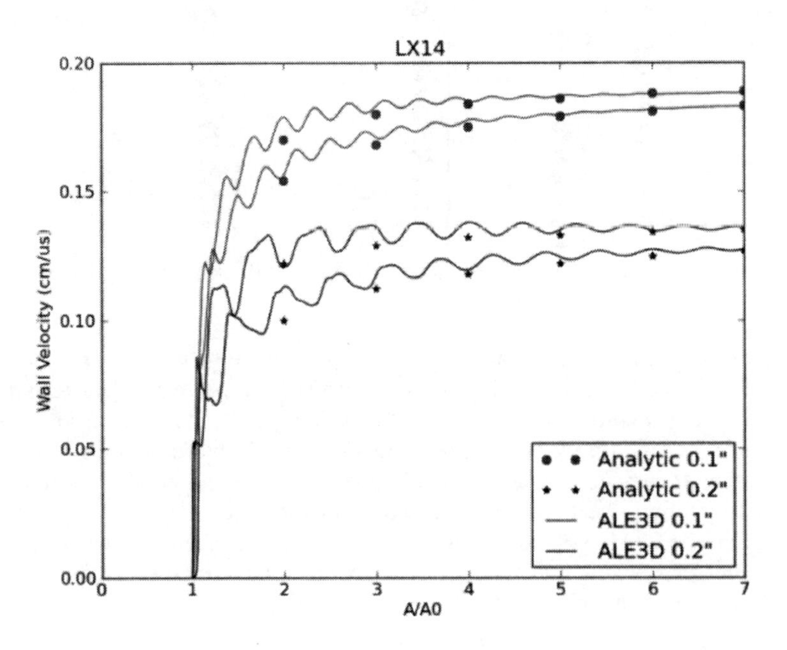

Figure 4: LX-14 cylinder analytic model versus ALE3D.

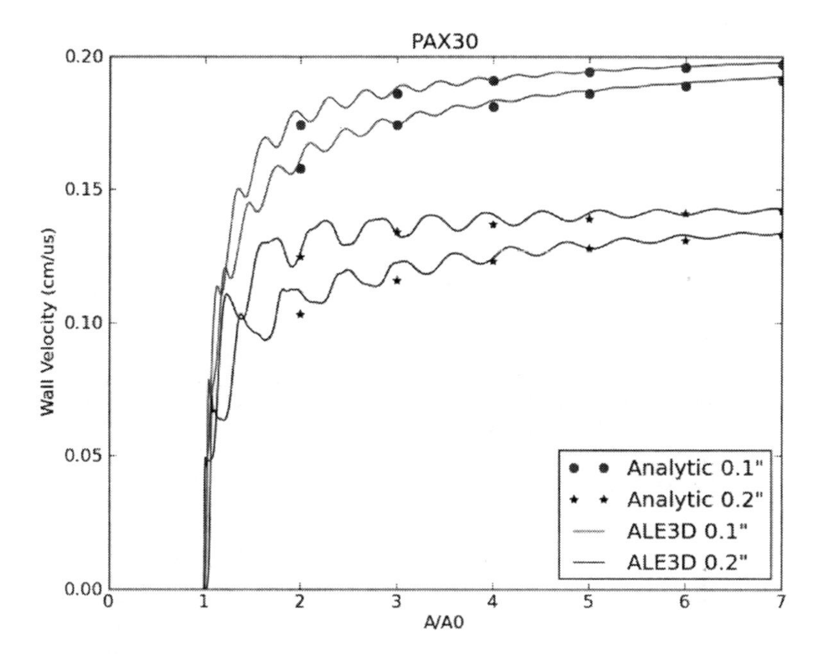

Figure 5: PAX-30 cylinder analytic model versus ALE3D.

7 Parameterization

We use two methods of parameterization are used to calibrate the JWL and JWLB constants. Both employ non-linear variable metric optimization techniques [2] for the parameterization process. In the first method [2], the equation of state parameters are optimized to reproduce the experimental cylinder velocities using the analytic cylinder test model, as well as to reproduce a desired Chapman-Jouguet detonation velocity and pressure. Typically, the total principle isentrope work output E_0 is also fixed to provide a desired total blast output. The cylinder velocities are used in a cost function to be minimized, whereas the Chapman-Jouguet state and E_0 are treated as equality constraints. The second method of parameterization [8] is to directly fit the predicted pressure and Gruneisen parameter versus specific volume behavior predicted by the thermo-chemical equation of state computer program JAGUAR. Formal non-linear optimization is used for the parameterization procedure. The LX-14 high energy explosive example presented in Figure 4 used the technique of parameterization for the JWLB equation of state. JWL and JWLB equation of states were parameterized for LX-14 using the JAGUAR predictions and non-linear optimization routines. The resulting JWL and JWLB equations of state were then used to model a standard 1.2 inch outside diameter and 1 inch inside diameter copper cylinder test (0.1" thick wall) and compared to experimental data using the analytic cylinder test model. Table 1 presents the resulting outside cylinder velocity results at different inside cylinder cross sectional areas. The results clearly show the improved agreement to experimental data obtained when using the more mathematically complex JWLB mathematical form. The improved agreement is attributed to the improved agreement to the JAGUAR predicted detonation products behavior that is achieved using the JWLB form.

Table 1: LX-14 JWL and JWLB cylinder test velocity predictions (Km/s) compared to experimental data.

		ANALYTIC CYLINDER	
A/A0	EXPERIMENTAL	JWL	JWLB
2	1.505	1.562	1.519
3	1.664	1.705	1.667
4	1.745	1.759	1.738
5	1.791	1.79	1.78
6	1.817	1.812	1.807
7	1.833	1.828	1.826
		% ERROR	
2		3.787	0.930
3		2.464	0.180
4		0.802	0.401
5		0.056	0.614
6		0.275	0.550
7		0.273	0.382
AVERAGED ERROR (%)		**1.276**	**0.510**

Table 2: PAX-30 JWL cylinder test predictions compared to experiments.

A/A0	EXPERIMENTAL	JWL	JWLB	JWLB w-point
2	1.499	1.599	1.55	1.541
3	1.682	1.759	1.702	1.703
4	1.774	1.823	1.780	1.779
5	1.827	1.862	1.831	1.825
6	1.859	1.89	1.868	1.856
7	1.883	1.911	1.897	1.879
			% ERROR	
2		6.6711	3.4023	2.8019
3		4.5779	1.1891	1.2485
4		2.7621	0.3157	0.2818
5		1.9157	0.2189	0.1095
6		1.6676	0.4841	0.1614
7		1.4870	0.7435	0.2124
AVERAGED ERROR (%)		**3.1802**	**1.0589**	**0.8026**

Similar to the LX-14, JWL and JWLB equation of states were also parameterized for PAX-30 using the JAGUAR predictions and non-linear optimization routines. The resulting JWL and JWLB equations of state were again used to model a standard 1.2 inch outside diameter and 1 inch inside diameter copper cylinder test (0.1" thick wall) and compared to experimental data using the analytic cylinder test model. However, PAX-30 is an aluminized explosive that is known to produce eigenvalue detonations [9]. Table 2 presents the resulting outside cylinder velocity results at different inside cylinder cross sectional areas. Again, the results clearly show the improved agreement to experimental data obtained when using the more mathematically complex JWLB mathematical form. The results also show a slight improvement by using the eigenvalue analytic cylinder model that represents expansion from the weak point (w-point). Table 3 presents JWLB equation of state parameters for TNT, LX-14 and PAX-30, which were used in this study.

Table 3: JWLB equation of state parameters for TNT, LX-14 and PAX-30.

	TNT	PAX-30		LX-14	
ρ (g/cc)	1.6300	1.885	1.909	1.820	1.8350
E0 (Mbar)	0.0657	0.13568	0.1376	0.102195	0.1032
D (cm/μs)	0.6817	0.8342*	0.8429*	0.86337	0.8691
P (Mbar)	0.1930	0.2419*	0.2464*	0.33529	0.3418
A1 (Mbar)	399.2140	406.224	405.3810	399.995	399.1910
A2 (Mbar)	56.2911	135.309	14.8887	20.1909	52.1951
A3 (Mbar)	0.8986	1.5312	1.49138	1.42441	1.59892
A4 (Mbar)	0.0092	0.006772	0.0076	0.02273	0.0249
R1	28.0876	26.9788	13.2982	13.93720	27.4041
R2	9.7325	10.6592	8.0204	7.230140	8.4331
R3	2.5309	2.52342	2.4942	2.558910	2.6293
R4	6.9817	0.335585	0.3566	0.736406	0.7498
C (Mbar)	0.0076544	0.013561	.0135749	0.011016	0.385366
ω	0.345920	0.234742	0.234664	0.384733	.0110204
A^{λ} 1	58.2649	72.6781	66.6542	41.71970	68.6476
A^{λ} 2	6.1981	5.64752	5.7776	6.83632	6.7497
B^{λ} 1	2.9036	2.8728	3.1440	6.42909	4.1338
B^{λ} 2	-3.2455	-3.10754	-3.2552	-4.47655	-4.4607
R^{λ} 1	25.5601	27.8109	25.5996	25.72540	26.2448
R^{λ} 2	1.7034	1.71375	1.7099	1.71081	1.6977

* Eigenvalue weak point detonation state (not the Chapman-Jouguet state).

8 Conclusions

An analytic cylinder test model has been developed by ARDEC for explosive equation of state calibration and verification. The analytic model was based on adiabatic expansion along the principle isentrope from the Chapman-Jouguet state. Additionally, an eigenvalue extended analytic cylinder expansion model has been developed based on isentropic expansion from the detonation eigenvalue weak point, rather than from the Chapman-Jouguet state. High explosives often include additive aluminium for blast effects. This eigenvalue model is applicable to Al based explosives, such as PAX-30, that exhibit eigenvalue detonations. The results for these explosives show only a very small reduction of explosive work output for eigenvalue detonations compared to Chapman-Jouguet detonations. This is due to the fact that the Chapman-Jouguet principle isentrope and eigenvalue weak point isentrope lie very close to each other. Excellent agreement between the analytic cylinder test and high rate continuum modeling predicted cylinder velocities is achieved when using the same JWL or JWLB parameters.

References

[1] Lee, E.L., Hornig, C. & Kury, J.W., "Adiabatic Expansion of High Explosive Detonation Products", Lawrence Livermore Laboratory, Rept. UCRL-50422 (1968).

[2] Baker, E.L. "An Application of Variable Metric Nonlinear Optimization to the Parameterization of an Extended Thermodynamic Equation of State", *Proceedings of the Tenth International Detonation Symposium*, Edited by J. M. Short and D. G. Tasker, Boston, MA, pp. 394-400, July 1993.

[3] Gurney, R. W., "The Initial Velocities of Fragments from Bombs, Shells, and Grenades," BRL Report 405, U.S. Army Ballistic Research Lab, 1943.

[4] Taylor, G. I., Analysis of the Explosion of a Long Cylindrical Bomb Detonated at One End, Scientific Papers of Sir G. I. Taylor, Vol 111:2770286, Cambridge University Press (1963), 1941.

[5] Baker, E.L., "Modeling and Optimization of Shaped Charge Liner Collapse and Jet Formation", Picatinny Arsenal Technical Report ARAED-TR-92017, January 1993.

[6] Brent, R., *Algorithms for Minimization without Derivatives.* Prentice-Hall, Englewood Cliffs, NJ, 1973. Reprinted by Dover Publications, Mineola, New York, January 2002.

[7] Backofen, J.E., "Modeling a Material's Instantaneous Velocity during Acceleration Driven by a Detonation's Gas-Push", 2005: Proceedings of the Conference of the American Physical Society Topical Group on Shock Compression of Condensed Matter, AIP Conf. Proc., July 28, 2006, Volume 845, pp. 936-939.

[8] Baker, E.L. & L.I. Stiel, "Improved Cylinder Test Agreement with JAGUAR Optimized Extended JCZ3 Procedures", *Proceedings of the International Workshop on New Models and Numerical Codes for Shock Wave Processes in Condensed Media*, St. Catherines College, Oxford, UK, September 1997.

[9] Baker, E.L., Stiel, L.I., Capellos, C., Balas, W. & Pincay, J., "Combined Effects Aluminized Explosives", Proceedings of the International Ballistics Symposium, New Orleans, LA, USA, 22-26 September 2008.

A study on the comparison of dynamic behaviour of high speed railway bridges

S. I. Kim, H. M. Kim & I. H. Yeo
Department of Track & Civil Engineering,
Korea Railroad Research Institute, Korea

Abstract

The dynamic behaviour of two steel bridges crossed by the Korean High Speed Train (KTX) has been investigated experimentally and the results are compared with the specification requirement and other typical prestressed concrete box girder bridge's responses. The investigated bridges are 2-girder steel bridges of 1@40m span length, 2@50m span length, and a PSC Box girder bridge of 2@40m span length. A set of experimental tests were performed during the operation of KTX, and a number of accelerometers, LVDTs and ring-type displacement transducers were utilized for measurement of three kinds of dynamic responses (acceleration, deflection, and end-rotation angle). The measured responses show that the vertical deflections and end-rotation angles of the three bridges all satisfy the specification requirement with a large margin, but acceleration responses that are very close or exceed the limit value were also found. Most of the excessive acceleration responses were found when the passing speed of the train is close to the critical velocity, which causes resonance. No noticeable differences of dynamic responses due to the different materials (steel or concrete) could be found within these experimental results.
Keywords: high speed railway bridge, bridge-train interaction, moving load.

1 Introduction

A high speed railway bridge must support trains travelling at speeds of up to 350 km/h. The possibility of resonance occurring due to the cyclic dynamic load is extremely high. For this reason, safety evaluation and dynamic movement evaluation are both critical elements in the development of a high speed railway bridge.

WIT Transactions on The Built Environment, Vol 113, © 2010 WIT Press
www.witpress.com, ISSN 1743-3509 (on-line)
doi:10.2495/SU100141

Most of the bridges along Korea's Gyeong-bu High Speed Railway consist of 2-span continuous PSC box girder bridges, with each span at a length of 40 meters. In parts where the railway passes through the city or roads, the bridges are mostly 2-girder steel plate girder bridges with 40 m to 50 m spans. In some parts, arch bridges have been set up. PSC box girder bridges are dominant because there is a general consensus that concrete bridges are better than steel bridges in terms of vibration and noise. Both theoretical and experimental studies have been continuously done on the 2-span continuous PSC box girder bridges, which account for the majority of the bridges along the Gyeong-bu High Speed Railway. Recently, research has also been underway into approaches to reducing excessive vibration [1].

At the same time, research on the dynamic behaviour of the bridge caused by the moving train load has been pursued since the 1960s. In the study of railway bridges, the emphasis on this type of research has been increasing because of the possibility of resonance caused by the cyclic load of the train (Yang et al. [2]). Resonances occur in bridges when the exciting frequency of the train is equal to the natural frequency of the bridge. This resonance is directly related to instability in dynamic behaviour, and can have a very bad influence on the ride comfort of the train passengers (Sogabe et al. [3]). However, thus far there has been little research on the behaviour of actual high speed rail bridges while a train is moving along the bridge. Furthermore, there have as of yet been no research results presented comparing the dynamic behaviour of concrete and steel high speed railway bridges.

In this paper, we performed field experiments on the dynamic characteristics and behaviour of Ewon Bridge and Jitan Bridge. Both bridges are 2-girder steel composite bridges on Gyeong-bu High Speed Railway. We then compared the results of our experiments with the behaviour of Yeonjae Bridge, a PSC box girder bridge. We also compared the results with the standard dynamic behaviour requirements set forth by Eurocode and Korean High Speed Railway. Through interpretative analysis, we quantitatively compared the steel bridge to the PSC box girder bridge.

2 Bridges

We used three bridges as the subjects of this research. The first bridge was the 40 m simple span bridge between the 59th pier and the abutment of Ewon Bridge. This bridge is located at the 174 km mark of Gyeong-bu High Speed Railway's Daejon - Gimcheon line. The second bridge is the Jitan Bridge (2@50m), which is located on the 177 km mark of the same railway line. The bridges are both 2-girder steel composite bridges, and are shown in fig. 1. Girder depth is 3 m for Ewon Bridge, and 4m for Jitan Bridge; distance between girders is 6.5 m, and total bridge width is 14.0 m. We chose Yeonjae Bridge as our comparison bridge. This bridge is a PSC box girder bridge with 2@40 m span, and is like the majority of the bridges on the Gyeong-bu High Speed Railway.

Figure 1: Ewon Bridge, Jitan Bridge and Yeonjae Bridge.

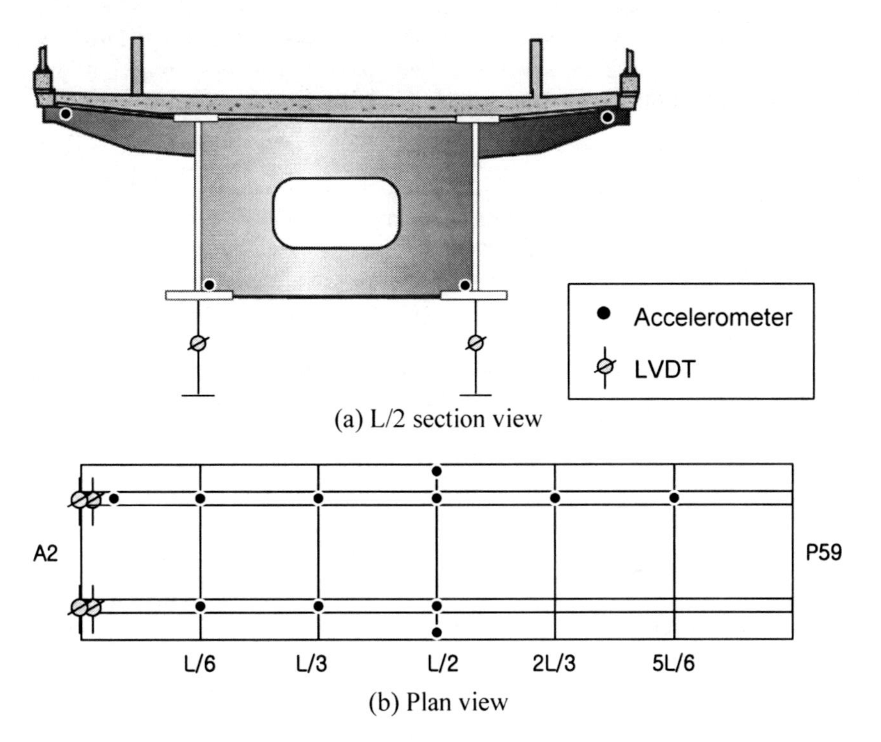

(a) L/2 section view

(b) Plan view

Figure 2: Sensor disposition map (Ewon Bridge).

3 On-site experiments

3.1 Selection of measurement categories and placement of sensors

Eurocode [4] and the specification of Korean High Speed Railway [5] used to evaluate the traffic safety and ride comfort criteria. Vertical deflection and vertical acceleration of the bridge deck are two major criteria.

To determine whether the dynamic performances of our test bridges satisfied the requirements of the standard, we measured the acceleration and displacement at the centre of the span, and also measured the rotation angle of the end deck. To obtain the rotation angle, we used two displacement sensors (LVDT or ring type sensors) placed at a constant distance from each other, and first obtained two vertical displacements. Then, we calculated the rotation angle from those measurements. In addition, we placed vertical direction accelerometers at constant intervals to measure specific dynamic characteristics (natural frequency, mode shape, damping ratio). The sensors used and their placements are shown in table 1 and fig. 2.

3.2 Measurement of dynamic response

Measurements of dynamic response on our subject bridges are performed on three separate dates. We used the previously-installed sensors and DAQ, and measured dynamic responses while a train passed over our subject bridges. Measurement data for each bridge is shown in table 2.

3.2.1 Time domain data

Figs. 3–5 below show examples of the data obtained from each bridge. In terms of the rotation angles of the end decks, we have shown the value calculated from the two measurements that we obtained from displacement sensors. The S/N characteristic is relatively low in comparison to other measurements. However, we have concluded that it will not pose as much of a problem in investigating the overall dynamic behaviour characteristic of the bridge.

The sampling rate and the low-pass filter were set to 1000 Hz and 200 Hz, respectively, to maintain consistency with our previous research.

Table 1: List of the installed sensors (Ewon Bridge).

<table>
<tr><th>Sensor type</th><th>Q’ty</th><th>Location</th><th>Response</th></tr>
<tr><td>accelero- meter</td><td>11</td><td>A2, L/6, L/3, L/2, 2L/3, 5L/6</td><td>natural frq.
mode-shape
acceleration</td></tr>
<tr><td rowspan="2">displacement transducer (LVDT)</td><td>2</td><td>L/2</td><td>vertical deflection</td></tr>
<tr><td>4</td><td>A2</td><td>end-rotation</td></tr>
</table>

Table 2: Quantity of the measured data set.

<table>
<tr><th rowspan="2">Bridge Name</th><th rowspan="2">Date</th><th colspan="2">Quantity of Data Set</th></tr>
<tr><th>North Bound</th><th>South Bound</th></tr>
<tr><td>Ewon</td><td>2007.2.27</td><td>14</td><td>13</td></tr>
<tr><td>Jitan</td><td>2007.4.13</td><td>5</td><td>5</td></tr>
<tr><td>Yeonjae</td><td>2007.4.11</td><td>22</td><td>20</td></tr>
</table>

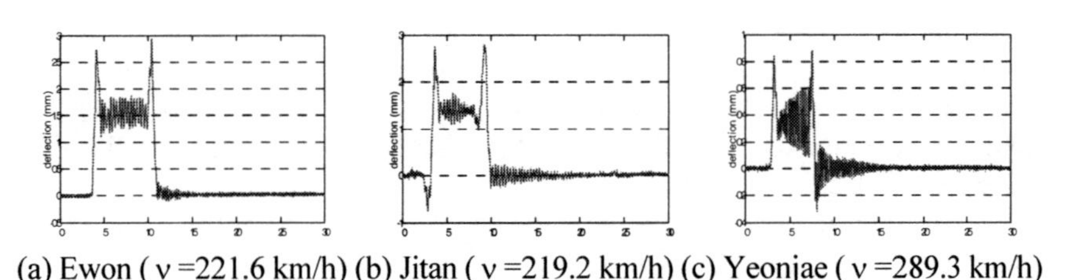

(a) Ewon (ν =221.6 km/h) (b) Jitan (ν =219.2 km/h) (c) Yeonjae (ν =289.3 km/h)

Figure 3: Vertical displacement of mid-span section.

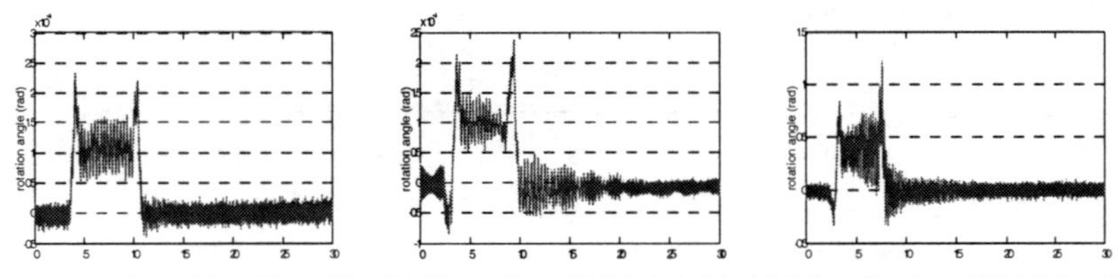

(a) Ewon (ν =221.6 km/h) (b) Jitan (ν =219.2 km/h) (c) Yeonjae(ν =289.3 km/h)

Figure 4: End-rotation angle.

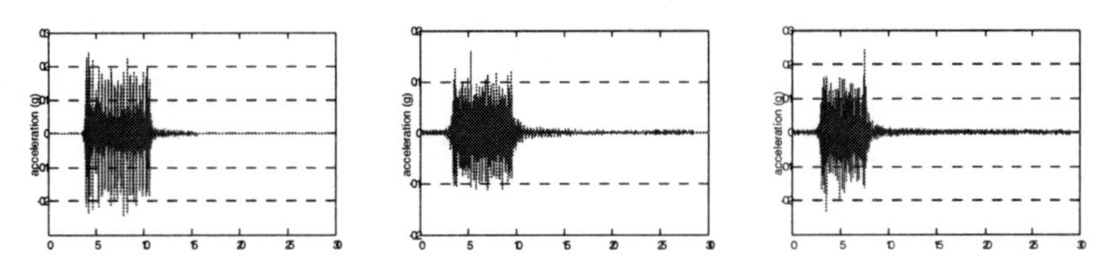

(a) Ewon (ν =221.6 km/h) (b) Jitan (ν =219.2 km/h) (c) Yeonjae (ν =289.3 km/h)

Figure 5: Vertical acceleration of mid-span section.

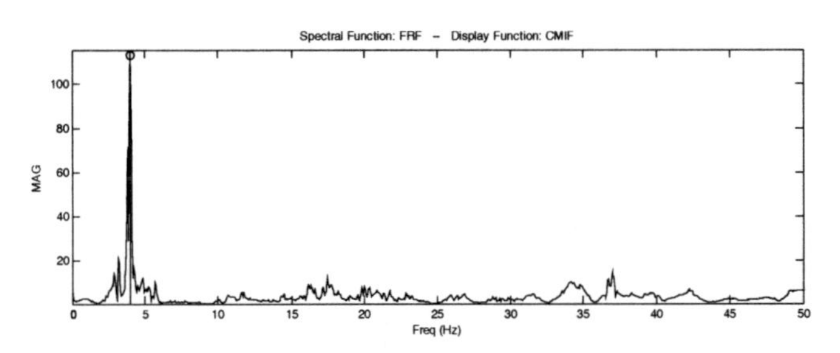

Figure 6: Acceleration spectrum of Ewon Bridge.

3.2.2 Natural frequency, mode shape and damping ratio

When the train is passing on the bridge, it is difficult to measure the natural frequency of the bridge independently, because of the interaction between the train and the bridge and the extra load of the train. Consequently, of the total measured acceleration signal, we only took the part after the train had left the bridge, and then extracted the natural frequency and mode shape.

Fig. 6 shows the signal spectrum obtained from the nine accelerometers installed on the lower girder parts of the Ewon Bridge. Fig. 7 shows the mode shape extracted from the spectrum. As can be seen from the image, there are almost no excitations in modes other than the 1st mode because of the train movement. The 1st natural frequency was 4.0 Hz.

To calculate the damping ratio of the subject bridges, we first obtained the frequency domain from the FFT spectrum of the corresponding mode. Then, we performed a curve fitting by using SDOF theory, and applied the least squares method. The following eqn. (1) is the formula that we used to obtain the frequency response function of SDOF.

$$\frac{A(\omega/\omega_n)^2}{\sqrt{\left\{1-\left(\frac{\omega}{\omega_n}\right)^2\right\}^2+\left\{2\xi\left(\frac{\omega}{\omega_n}\right)\right\}^2}} \tag{1}$$

Here, ω_n is the natural frequency, ξ is the damping ratio, and A is the proportional constant.

Fig. 8 shows the curve fitting results of the 1st mode from Jitan Bridge. We determined that the natural frequency was 3.08 Hz, and the damping ratio was 1.2%.

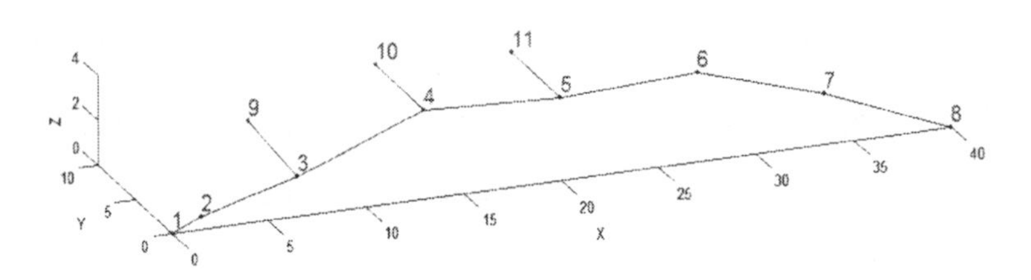

Figure 7: The 1st vibration mode shape of Ewon Bridge.

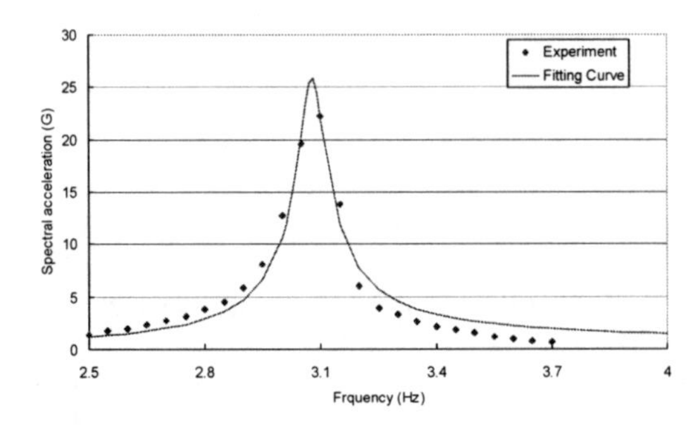

Figure 8: The 1st mode curve fitting of Jitan Bridge.

Table 3: Dynamic characteristics of the tested bridges.

Bridge	The 1st Natural frequency(Hz)	Damping Ratio(%)	Critical Speed(km/h)
Ewon	4.0	2.4	269.3
Jitan	3.08	1.2	207.3
Yeonjae	4.35	2.4	292.8

Table 4: Requirements of dynamic performances.

Response	Reference Value
Acceleration	0.35g
Vertical Deflection (for Passenger Comfort)	L/1900=21.05mm (for L=40m), L/2200=22.73mm (for L=50m),
End-rotation	5×10^{-4} rad

Table 3 shows the natural frequencies and the damping ratios of each bridge.

Bridge resonance caused by the train load occurs when the 1st bending natural frequency of the bridge is equal to the frequency of the passing train, as shown in eqn. (2) below.

$$\Omega(Hz) = V / S_{eff} = \omega_1 (Hz) \tag{2}$$

Here, Ω is the train's frequency, V is the train's velocity, S_{eff} is the effective beating interval of the train, and ω_1 is the 1st natural frequency of the bridge.

Therefore, the critical speed of the train that causes the resonance can be expressed as in eqn. (3). If the natural frequency of the bridge and the effective beating interval of the train are known, then the critical speed can be estimated in advance. Unlike ordinary highway bridges, over which all types of vehicles pass at random intervals, only specific trains are allowed to pass over railway bridges. Consequently, we can analyze the critical speed that generates resonance by using eqn. (3). The natural frequency of the bridge, and 18.7 m, the effective beating interval of the KTX, can be substituted into the formula to calculate the critical speeds. The calculated speeds are shown in table 3.

$$V_{cr} = \omega_1 \times S_{eff} \tag{3}$$

4 Dynamic responses

4.1 Criteria for dynamic performance

The evaluation standard for vertical deflection, vertical acceleration and rotation angle of the end deck are listed in Eurocode and in the design guideline for Korean High Speed Railway. All of these standards are shown in table 4.

4.2 Investigation of dynamic performance measurements

4.2.1 Maximum deflection

In fig. 9, the maximum displacements obtained from the three bridges are shown in proportion to the standard value. The standard value in accordance with the standard set in table 4 is 21.05 mm for Ewon and Yeonjae Bridges, and 22.73 for Jitan Bridge.

As can be seen, the maximum deflection was greatest for Ewon, followed by Jitan and Yeonjae, respectively. All three bridges showed safe responsive values in comparison to standard values. When compared with the PSC box girder bridge, vertical displacements of steel bridges are great. However, the deflection is still within 15% of the standard, and is therefore sufficient.

4.2.2 Maximum vertical acceleration

Fig. 10 shows the measurements of the maximum vertical acceleration of the bridge deck.

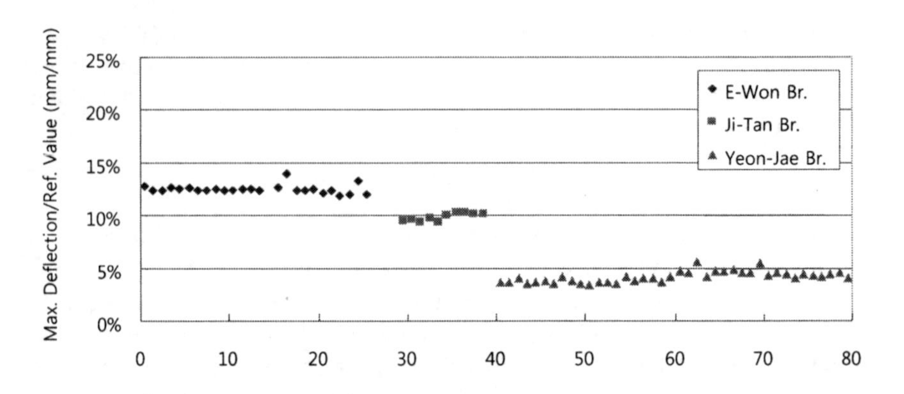

Figure 9: Measured maximum vertical deflections.

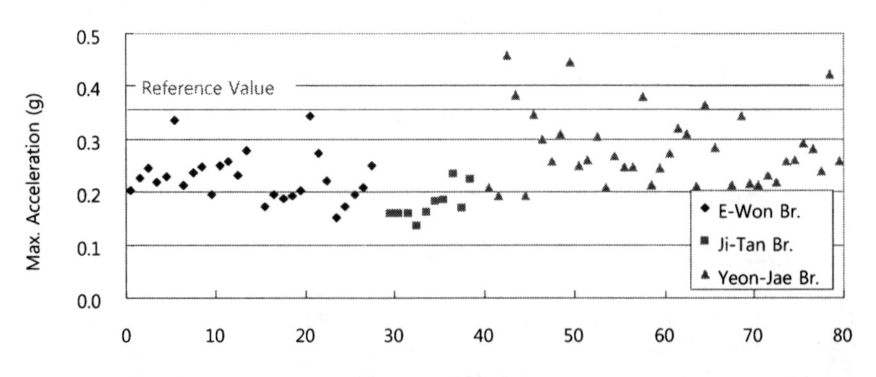

Figure 10: Measured maximum vertical accelerations.

The results obtained in the measurement of vertical acceleration were very different from the vertical displacement results. In other words, responses exceeding the criteria were generated from the Yeonjae Bridge - the PSC box girder. This is because when the train is moving on Yeonjae Bridge, the actual train speed approaches the critical speed, and thus generates resonance. On Ewon Bridge, the acceleration response is relatively higher than that of Jitan, but no responses exceeding the standard were observed.

4.2.3 Rotation angle of the end deck

Fig. 11 shows the maximum values for the rotation angles. The angles closely resemble the results of the vertical deflections. However, in the case of Ewon Bridge, a greater angle was measured in the northbound lane than in the southbound lane. Yeonjae Bridge showed similar results. We concluded that the results are related to differences in passenger numbers and running characteristics between the two lanes. As in the case of the displacements, the greatest rotation angle was observed on Ewon Bridge, followed by Jitan and Yeonjae Bridges, respectively. All of the angles were within the criteria.

4.3 The importance of actual running speed

When a high speed railway bridge has a high likelihood of resonance when the train is operating at its normal speeds, then the train may cause the dynamic instability of the bridge when operating at critical speeds. The behaviour at this time is controlled by the damping ratio. Therefore, the actual running speed of the train is very important in relation to the dynamic characteristics of the bridge. Fig. 12 shows train speeds on the Yeonjae Bridge, a PSC box girder (2@40) and on the steel composite bridges Ewon (1@40) and Jitan Bridge (2@50).

As it can be seen in the figure, the actual running speed on the Yeonjae Bridge is very close to the critical speed, while for Ewon and Jitan Bridges, the running speed is quite far from the critical speed.

Fig. 3 shows the time history curve of vertical displacements on Yeonjae and Ewon Bridges. As can be seen in the figure, the curve of Yeonjae Bridge is

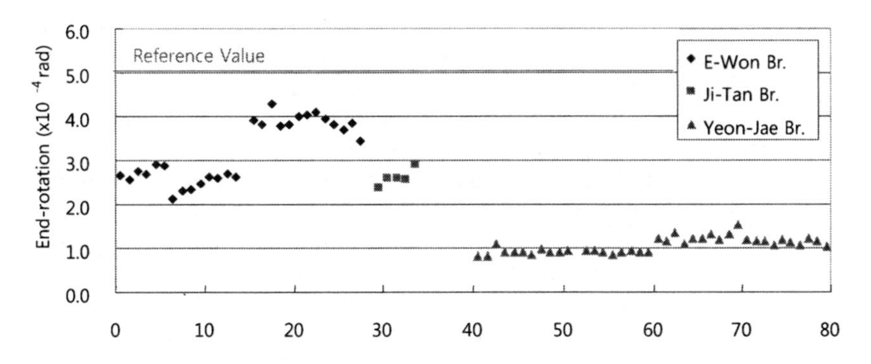

Figure 11: Measured maximum end-rotation angles.

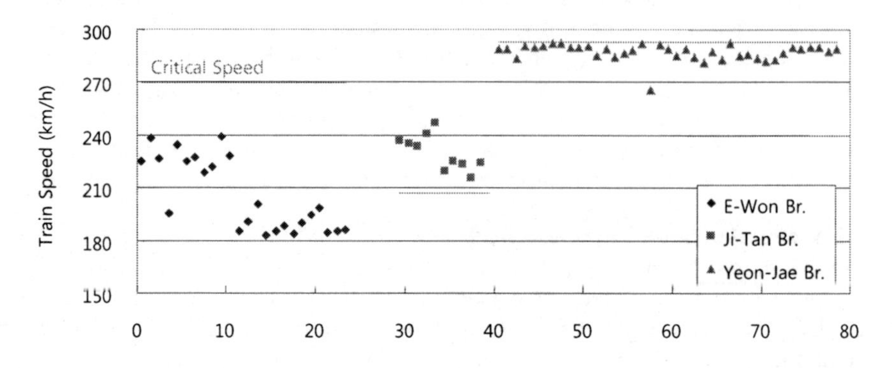

Figure 12: Critical speeds and train-passing speeds.

similar to that of a resonance wave, while for Ewon, the curve is a typical vertical displacement time history curve caused by the KTX.

All three bridges have sufficient surplus displacements. In the case of the Yeonjae Bridge, because of the high rigidity of the PSC box girder bridge, it has sufficient surplus displacement, even if resonance does occur. However, continuous resonance waves like those shown in fig. 3(c) can shorten bridge life. In addition, in the case of vertical acceleration, the Yeonjae Bridge is shown to have higher acceleration results than the steel composite bridges because of the generation of resonance.

Generally, steel composite girders are known to be more vulnerable to noise and vibration than concrete bridges. Our results, however, show that when the bridges have the same level of static/dynamic performance, what matters more than the form of the bridge are its dynamic characteristics. We also discovered that the actual running speed of the train can have a great influence on the bridge.

5 Conclusion

In this paper, we performed dynamic measurements on part of the Ewon Bridge, and on Jitan Bridge. Both are bridges along the Gyeong-bu High Speed Railway. We then compared the results with the dynamic performance standard values set forth by Eurocode and Korean High Speed Railway, as well as with response values from Yeonjae Bridge, a PSC box girder bridge.

To identify the dynamic performance of the bridges, we measured and analyzed the following elements: maximum vertical acceleration, maximum vertical displacement, and rotation angle of the end decks. Most of the measurements were within the regulation values. On Yeonjae Bridge, however, some acceleration responses exceeded regulation values.

The steel composite bridge showed higher values for displacement and rotation angles in comparison to the PSC box girder bridge. However, for acceleration response it showed lower values. This result shows that although the steel composite bridge is lower in rigidity than the PSC box girder bridge, there

are lower acceleration responses, because no resonance occurs when the train passes through the bridge.

Therefore, the component that controls the dynamic performance of the bridge is whether or not resonance is generated. Differences in bridge material did not result in differences in the dynamic performance of the bridges.

The dynamic responses obtained and analyzed in this paper were all obtained when the train was running at an operating speed. Therefore, we are conducting further numerical analysis of the influence of resonance caused by changes in speed.

Acknowledgement

This research was supported by a grant from Railroad Technology Development Program (RTDP) funded by Ministry of Land, Transport and Maritime Affairs of Korean government.

References

[1] E.S. Choi, W.J. Chin, J.W. Lee, J.W. Kwark, J.Y. Kang, B.S. Kim, Diminution of bridge vibration for high-speed trains, *Journal of the Korean Society for Railway*, **9(1)**, pp. 125~130. 2006.

[2] Yang Y.B., Yau J. D. and Hsu L.C., Vibration of simple beams due to trains moving at high speeds, *Engineering Structures*, **19(11)**, 936-944, 1997.

[3] Sogabe, M., Matsumoto, N., Kanamori, M., Sato, M. & Wakui, H., Impact factors of concrete girders coping with train speed-up", *Quarterly Report of RTRI*, **46(1)**, 46-52, 2005.

[4] EUROCODE 1 Pt. 2, *Actions on Structures: General Actions - Traffic Loads on Bridges*, European Committee for Standardization, 2003.

[5] Korea Rail Network Authority, *Design Specification for Honam High Speed Railway*, Korea Rail Network Authority (in Korean), 2007.

Section 5
Protection of structures from blast loads

Modelling the response of UHPFRC panels to explosive loading

G. K. Schleyer[1], S. J. Barnett[1], S. G. Millard[1] & G. Wight[2]
[1]*School of Engineering, University of Liverpool, UK*
[2]*VSL Australia Pty Ltd, Melbourne, Australia*

Abstract

Explosive testing of full-size fibre-reinforced concrete panels was conducted at GL Industrial Services at Spadeadam test site, Cumbria, England in 2008. The panels were manufactured by VSL Australia and shipped to Spadeadam for testing. This paper reports these tests and a simplified analysis of the response of the panels. Each panel measured 3.5m by 1.3m by 100mm thick. The panels were contained within a large concrete enclosure to minimise clearing around the sides from the blast wave and placed between 7m and 12m from a 100 kg TNT equivalent explosive charge. Two of the panels were fabricated with different levels of steel fibre dosage. The remaining two panels were fabricated with steel fibres together with supplementary steel bar reinforcement. Numerical computer modelling was carried out using the Autodyn package to predict the behaviour of the four panels before testing. Based on the predictive modelling, each panel was placed a suitable distance from the explosive charge so as to cause permanent damage but not total structural collapse. The maximum flexural tensile strain rate evaluated on the back face of the panel was in the region of $1.0s^{-1}$. Simplified modelling of the panels was also carried out using a single-degree-of-freedom representation together with a resistance-deflection relationship that took account of characteristic brittle cracking and ductile softening behaviour following ultimate capacity. An outline of the method with results is given in the paper.

Keywords: fibre-reinforced concrete, explosive testing, SDOF modelling, ductile softening behaviour.

WIT Transactions on The Built Environment, Vol 113, © 2010 WIT Press
www.witpress.com, ISSN 1743-3509 (on-line)
doi:10.2495/SU100151

1 Introduction

Ultra high performance fibre-reinforced concrete (UHPFRC) has been recently developed (Richard and Cheyrezy [1]) and has material properties which are much improved from those of conventional concrete. UHPFRC contains a very high cement content and very low water-cement ratio (typically 0.15-0.18), which is achieved using a super-plasticiser. In addition, the only aggregate used is fine silica sand, together with a high dosage of fine high tensile steel fibres of the order of 0.2 mm in diameter. Elevated temperature curing at 90 °C enables early-age compressive cube strength of between 150-200 MPa to be achieved. In addition, a flexural tensile strength in the range of 25-50 MPa is provided by the inclusion of the steel fibres. UHPFRC also has a high fracture energy of around 20,000-40,000 J/m^2.

These properties suggest that UHPFRC could be a suitable material to resist blast and impact loading. Studies were carried out in the UK and Australia, (Rebentrost and Wight [2]) to investigate the properties of UHPFRC under impact and explosive loading. Preliminary results from the UK study were presented at PROTECT 2007 (Barnett et al. [3]) and showed that under quasi-static loading a high flexural tensile strength could be achieved which was sensitive to the dosage of steel fibres in a non-proportionate manner. Further results of the sensitivity of UHPFRC material properties to the strain rate applied during high-speed loading were reported at PROTECT 2009 (Millard et al. [4]) and focus primarily on the flexural tensile and shear strengths. Back face tensile spalling and shear punching are thought to be two critical modes of failure for conventional concrete under blast or high-speed impact loading (Magnusson and Hallgren [5]).

A collaborative research programme between the University of Liverpool and the University of Sheffield has made use of drop hammer and Hopkinson bar test facilities to investigate the dynamic increase factor (DIF) of UHPFRC in both flexural and shear behaviour. The results are described by Millard et al. [6] along with an independent study of the same. The results from both studies correlate well and reveal that a DIF of the flexural tensile strength rising from 1.0 at 1.0s^{-1} on a slope of $^1/_3$ on a log (strain rate) versus log (DIF) graph can be used for design purposes. The results also show that no significant increase in shear strength is to be gained at high loading rates. Results have been reported for conventional concrete at high strain rates in several publications [7–9].

The collaborative programme culminated in a series of small-scale and large-scale explosive tests using the University Sheffield's field test site at Buxton and the facilities at RAF Spadeadam test site. The results suggest that UHPFRC is suitable for blast and impact mitigation applications particularly in counteracting the threat of an urban terrorist attack. This paper focuses on the full-scale tests at RAF Spadeadam and simplified modelling using single-degree-of-freedom (SDOF) procedures.

2 Full-scale tests

Explosive testing of full-size panels was conducted at RAF Spadeadam, Cumbria. Four panels were manufactured by VSL Australia Pty Ltd, Melbourne and transported to Spadeadam for testing. Each panel measured approx. 3.5m by 1.3m by 100mm thick. The panel was positioned vertically and supported along the upper and lower edges so that it spanned one-way. Each panel was contained within a large concrete enclosure with steel side plates to minimise clearing. The enclosure prevented the blast pressure wave from reaching the back face of the panel and reducing the loading on the front face. A 100kg TNT equivalent explosive charge was used to load the panels with stand-offs between 7m and 12m.

Numerical computer modelling was also carried out using the Autodyn package to predict the behaviour of the four panels prior to testing. Some of the panels were instrumented with a contactless laser displacement transducer to measure the mid-span displacement during blast loading. In addition a simple mechanical friction "broomstick gauge" was used to measure the maximum displacement. The maximum flexural tensile strain rate evaluated on the back face of the panel during full-scale blast loading was estimated to be in the region of $1.0s^{-1}$.

2.1 VSL panels

Two of the full-size panels were fabricated with different levels of steel fibre dosage, 2% and 4% by volume, respectively. The fibre lengths were 13 mm and 25 mm by 0.2 mm in diameter. The remaining two panels were fabricated with steel fibres together with identical additional steel bar reinforcement. Based on predictive modelling, each panel was placed at a stand-off from the explosive charge designed to cause permanent damage but not structural collapse as given in table 1.

Table 1: Stand-off distances and fibre content by volume of test panels.

Panel #	Fibres	Additional steel bar reinforcement?	Stand-off (m)
1	2% 13mm long	Yes	9
2	2% 13mm long	No	12
3	2% 13mm + 2% 25mm	No	12
4	2% 13mm long	Yes	7

VSL carried out a static analysis of the test panels and produced moment-curvature relationships as shown in fig. 1. This data is also summarised in table 2 and was used to develop a resistance-deflection function for the SDOF modelling described in section 3. The stiffness given in table 2 is based on a

Table 2: Summary of static analysis of test panels.

Panel #	M_{max} (kNm)	Curvature at M_{max} ($mm^{-1}\times10^{-5}$)	Deflection at M_{max} (mm)	Stiffness k (kPa/mm)	R_{max} (kPa)
1	194.5	8.582	39.4	2.48	97.7
2	123.8	7.161	27.4	2.27	62.2
3	154.3	7.637	31.2	2.48	77.5
4	194.5	8.582	39.4	2.48	97.7

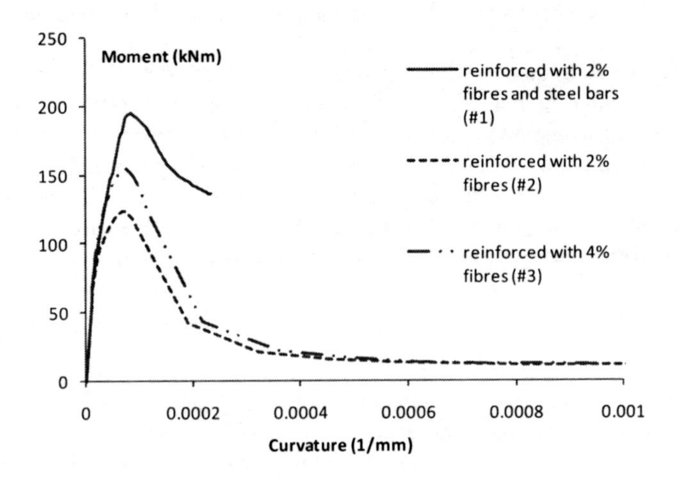

Figure 1: Moment-curvature relationship of VSL panels.

linear approximation to maximum capacity (deflection at M_{max}). The initial stiffness is higher.

2.2 Test results

The test panels were arranged in 'arena' style layout in which a number of targets are positioned at various stand-offs around a central explosive charge. The blast wave produced by detonating the explosive charge radiates in all directions and so can be used to load a certain number of test pieces at the same time according to their stand-off distance from the charge. The panels were all set up to receive the air blast loading with a normal angle of incidence to the propagating blast wave. The panels would therefore experience a reflected pressure and impulse. The loading can be readily estimated using air-blast data from TM 5-1300 [10].

The results of the tests are summarised in table 3. Fig. 2 shows the two unreinforced panels #2 and #3 after testing permanently deformed by 180mm and 90mm, respectively with no apparent elastic rebound.

Table 3: Test results.

Panel #	Stand-off (m)	Maximum deflection (mm)	Permanent deflection (mm)
1	9	110	20
2	12	180	180
3	12	90	90
4	7	210	50

Panel #2 (2% fibres)

Panel #3 (4% fibres)

Figure 2: Unreinforced panels #2 and #3 after testing at 12m stand-off.

3 SDOF modelling

SDOF methods are based on the representation of the actual structure by an equivalent spring-mass system that is constrained to have only one degree of freedom. The equivalent system parameters, mass, spring stiffness, spring yield load and applied force (M_e, k_e, R_{me} and F_e respectively) are selected such that the deflection of the equivalent concentrated mass, M_e, is the same as that for some significant point on the structure, usually the point of maximum deflection as shown in fig. 3. Forces and stresses are not directly equivalent. The constants of the spring-mass system are evaluated on the basis of an assumed deformed shape of the structure, normally the approximate shape as that resulting from the static application of the loading.

The load transformation factor, K_L, which relates the equivalent force on the spring-mass system to the total force on the actual structure, is evaluated from a consideration of the work done by the equivalent spring-mass system and the structural model. Similarly, the mass transformation factor, K_M, which relates the lumped mass on the spring-mass system to the total mass of the actual structure, is evaluated from a consideration of the kinetic energy of the equivalent spring-mass system and the structural model.

WIT Transactions on The Built Environment, Vol 113, © 2010 WIT Press
www.witpress.com, ISSN 1743-3509 (on-line)

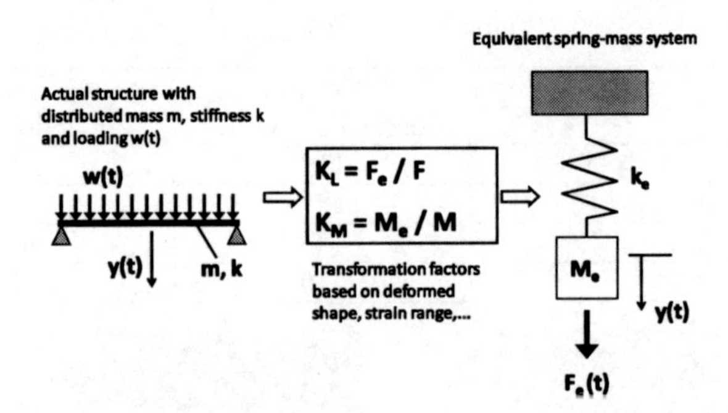

Figure 3: SDOF representation.

The load transformation factor, $K_L = F_e/F = k_e/k$ where k and F are the linear spring stiffness and total force for the actual structure respectively and k_e and F_e are the spring stiffness and equivalent force for the spring-mass system. The mass transformation factor $K_M = M_e/M$ where M is the total mass for the actual structure and M_e is the lumped mass for the spring-mass system.

SDOF analysis can be used for any duration of load and can be used for both elastic and elastic-plastic behaviour. The SDOF system is therefore a good model of the dynamics of the real system and allows non-linear behaviour to be modelled. Further details of the method and calculation of the equivalent spring-mass properties can be found in Biggs [11] and Baker et al. [12].

A structural member's capacity to resist a load (static or dynamic) can be defined by a resistance function which combines elastic stiffness with plastic energy absorption. Standard SDOF methods are based on an effective bi-linear resistance function as shown in fig. 4 for a beam with fixed ends and simple

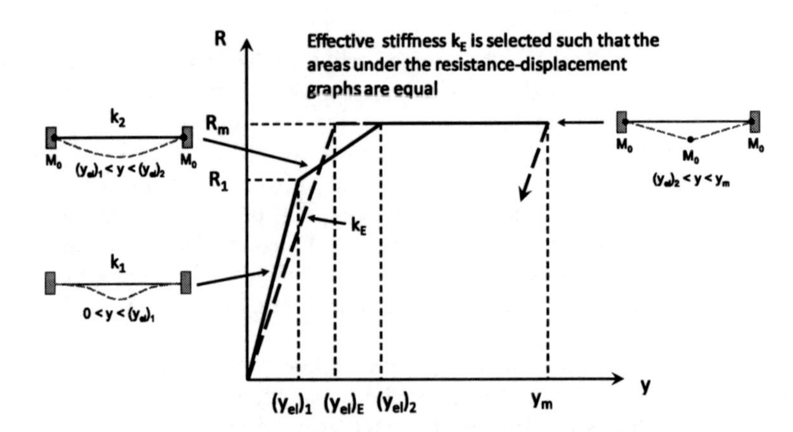

Figure 4: Effective bi-linear resistance-deflection relationship for a beam with fixed supports as used in the SDOF method.

elastic-perfectly plastic material behaviour. This representation is usually adequate in the context of rapid design screening for modelling the flexural behaviour of ductile steel beams and RC beams and slabs.

However, the FRC panels exhibit characteristic softening behaviour due to a combination of brittle cracking and ductile yielding after reaching their ultimate capacity. Standard graphical or numerical implementation of the SDOF method do not provide for definition of softening or hardening behaviour after yielding. Enhanced numerical implementation of SDOF methods have been developed to allow typical hardening or softening behaviour or a combination of both to be modelled in the analysis. More accurate definition of the resistance-deflection relationship leads to better predictions of structural response. SBEDS [13] is an engineering spreadsheet tool intended for use by structural engineers with knowledge of structural dynamics and blast effects. It is a product of the US Army Corps of Engineers (USACE) developed by several consulting engineering companies in the US contracted by the USACE protective Design Centre (PDC). It enables the user to define a resistance-deflection function with up to 5 points, shown typically in fig. 5. This tool was used for the SDOF analysis of the test panels.

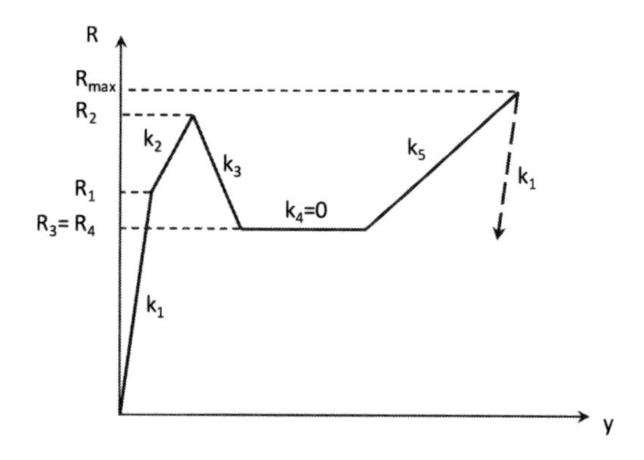

Figure 5: Typical resistance-deflection function as used in SBEDS [13].

3.1 SDOF parameters

The parameters for the SDOF analysis were determined from the static mechanical properties and moment-curvature relationships supplied by VSL, shown in fig. 1 and summarised in tables 4 and 5. A dynamic increase factor (DIF) of 1.2 and 5% critical damping was applied to panels #1 and #4 with additional steel bar reinforcement. This effectively raised the maximum resistance e.g. in panel #1 from 97.7 to 117.2 kPa. Subsequent points were simply raised by the same amount. A DIF=1.1 and no damping was applied to panels #2 and #3, reinforced only with steel fibres.

Table 4: SDOF model parameters.

Parameter	Value	Units
Length, L	3.5	m
Load-mass factor (elastic)	0.78	-
Load-mass factor (plastic)	0.66	-
Area, A	4.55	m^2
Density, ρ	2450	kg/m^3
Mass, M	1115	kg

Table 5: Resistance-deflection parameters.

Panel #	1	2	3	4	Units
k_1	46.9	2.27	2.48	46.9	kPa/mm
R_1	29.1	68.42	85.25	29.1	kPa
k_2	1.79	-1.0	-1.0	1.79	kPa/mm
R_2	117.2	20.0	20.0	117.2	kPa
k_3	-0.18	-0.1	-0.1	-0.18	kPa/mm
R_3	95.64	10.0	10.0	95.64	kPa
k_4	-0.05	-0.01	-0.01	-0.05	kPa/mm
R_4	87.74	5.02	5.02	87.74	kPa
k_5	0.0	0.0	0.0	0.0	kPa/mm
R_5	87.74	5.02	5.02	87.74	kPa

The load parameters given in table 6 were determined using the data from TM 5-1300 [10] based on cube root scaling of TNT equivalent surface bursts. The free field parameters of reflected pressure and impulse are well defined at a distant target. The equivalent load duration assumes a right-angled triangular load pulse shape for the blast load. Only the positive phase of the loading was considered in the analysis; the negative phase was ignored.

Table 6: Load parameters.

Stand-off (m)	Peak reflected pressure (kPa)	Peak reflected impulse (kPa-msec)	Equivalent load duration (msec)
7	2488	2400	1.9
9	1160	1754	3.0
12	498	1240	5.0

3.2 SDOF results

The analysis results along with the test results for comparison are summarised in table 7. The deflection vs. time graphs for each of the 4 panels are shown in figs 6–9 together with their respective resistance (kPa) vs. deflection (mm) graph. These results are discussed in section 4.

Table 7: Comparison of results.

Panel #	Maximum deflection (mm)		Permanent deflection (mm)	
	Test	SDOF	Test	SDOF
1	110	106	20	46
2	180	179	180	175
3	90	90	90	78
4	210	200	50	147

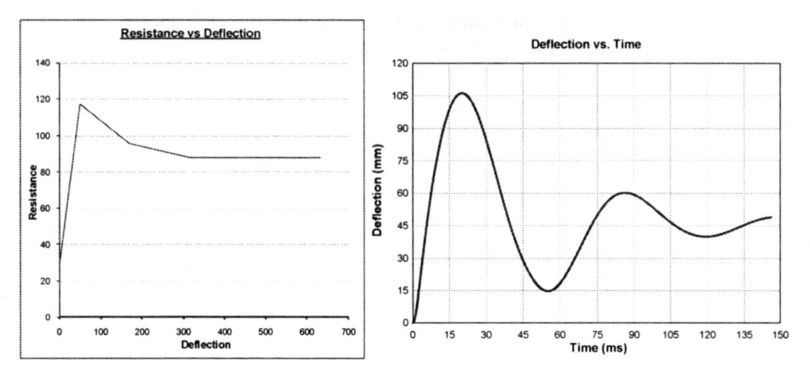

Figure 6: Panel #1 SDOF model input resistance-deflection and output deflection-time graphs.

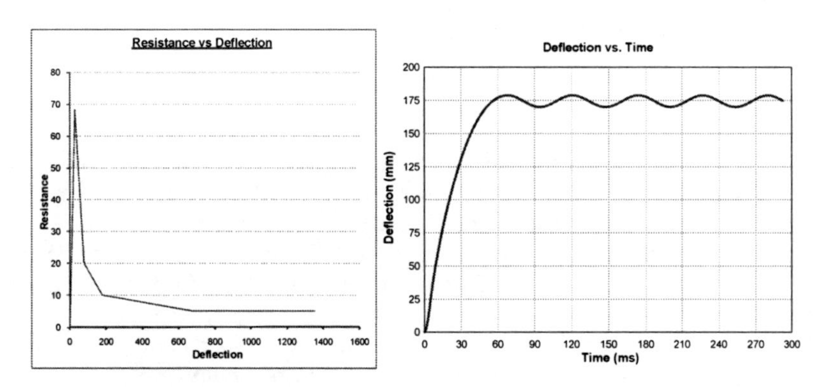

Figure 7: Panel #2 SDOF model input resistance-deflection and output deflection-time graphs.

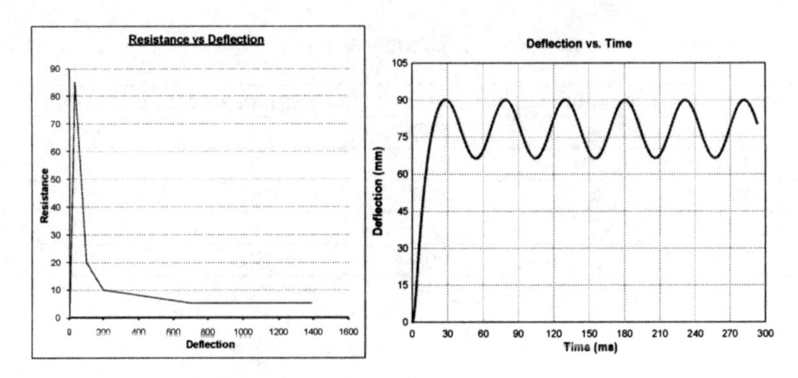

Figure 8: Panel #3 SDOF model input resistance-deflection and output deflection-time graphs.

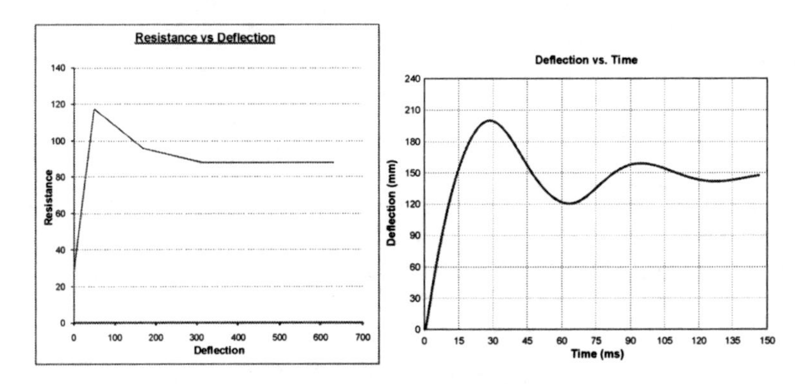

Figure 9: Panel #4 SDOF model input resistance-deflection and output deflection-time graphs.

4 Discussion

It is clear that the panels #1 and #4 containing both steel fibre and additional steel bar reinforcement have significant spring-back which is partially captured by the SDOF model but not to the extent exhibited in the tests. In fact the recording of test panel #1 showed a rebound deflection of the order of 75 mm outwards. However, the peak deflections modelled in the SDOF procedure for panels #1 and #4 correlate much better with the test results than the permanent deflections. Test panels #2 and #3, which contained only steel fibre reinforcement, exhibited characteristic brittle behaviour following cracking and yielding. It would appear that this softening behaviour was captured by the SDOF procedure since the permanent deflections correlate reasonably well. It is interesting to note that the SDOF method predicts a small elastic rebound. It is also noted that the SDOF procedure even with its limitations is capable of bounding the problem. The uplift in strength introduced into the modelling was considered appropriate at this rate of loading. The influence of the additional steel bar reinforcement is clearly significant, the extent of which may only be realised in a detailed finite element modelling procedure. The yield line failure at or near the mid-span of the panel together with no rear face spalling due to the embedded steel fibres was entirely consistent with this type of structure.

5 Conclusions

Four UHPFRC full-size panels were subjected to air blast loading arising from a 100kg TNT equivalent explosive charge. Two of the panels contained both steel bar and steel fibre reinforcement while the other panels contained only steel fibre reinforcement. One panel contained 4% steel fibres by volume while the remaining panels contained 2% steel fibres by volume. Blast loading caused various degrees of permanent damage in the form of a yield line at or near the mid-span but not total structural collapse, by virtue of the stand-off distance from the charge. A consistent approach to modelling the four test panels as a one-way spanning member using an SDOF procedure gave reasonably good correlation and demonstrated the versatility of the SDOF method. A particular feature of the SDOF procedure that contributed to a successful result was the definition of the resistance-deflection relationship to include softening behaviour. The simplified procedure was unable, however, to entirely capture the significant elastic rebound exhibited in the panels with additional steel bar reinforcement. Further investigation is necessary.

Acknowledgements

This study has been carried out with support from the Engineering and Physical Sciences Research Council (EPSRC), under the "Think Crime-4" managed programme. The investigators are grateful for the support of the following industrial collaborators: Centre for the Protection of National Infrastructure (CPNI), Bekeart Ltd, Elkem Ltd, VSL (Australia), Fosroc Ltd, Castle Cement, and the Appleby Group.

References

[1] Richard, P. & Cheyrezy, M., Composition of reactive powder concretes. *Cement Concrete Research*, **25(7)**, pp. 1501-1511, 1995.

[2] Rebentrost, M. & Wight, G., Behaviour and resistance of ultra high performance concrete to blast effects. *Proc. of 2nd Int. Symposium on Ultra High Performance Concrete*, eds. E. Fehling, M. Schmidt, and S. Stürwald, pp. 735-742, 2008.

[3] Barnett, S.J., Millard, S.G., Soutsos, M.N., Schleyer, G.K., & Tyas, A., Flexural Performance of UHPFRC. *Proc. of PROTECT 2007: Structures under Extreme Loading*, 2007.

[4] Millard, S.G., Barnett, S.J., Schleyer, G.K., Tyas, A. & Rebentrost, M., Explosion and Impact Resistance of Ultra High Performance Fibre Reinforced Concrete. *Proc. of PROTECT 2009: Structures under Extreme Loading*, 2009.

[5] Magnusson, J., & Hallgren, M., Reinforced high strength concrete beams subjected to air blast loading. *Proc. of 8th Conf. on Structures under Shock and Impact*, eds. N. Jones & C.A. Brebbia, WIT, pp. 53-62, 2004.

[6] Millard, S.G., Molyneaux, T.C., Barnett, S.J. & Gaob, X., Dynamic enhancement of blast-resistant concrete under flexural and shear loading. Submitted to *Int. Journal of Impact Engineering*, 2010.

[7] Malvar, L.J. & Ross, C.A., Review of strain rate effects for concrete in tension. *ACI Materials Journal*, **95(6)**, pp. 735-739, 1998.

[8] Solomos, G. & Berra, M., Compressive behaviour of high performance concrete at dynamic strain-rates. *Proc. of 6th RILEM Symposium on Fibre-Reinforced Concretes*, eds. M. Di Prisco, R. Felicetti & G.A. Plizzari, RILEM Publications S.A.R.L., pp. 421-430, 2004.

[9] Ulfkjaer, J., Labibes, K., Solomos, G. & Albertini, C., Tensile failure of normal concrete and steel fibre reinforced concrete at high strain rates. *Proc. of 3rd Fracture Mechanics of Concrete Structures*, pp. 585-592, 1998.

[10] TM 5-1300, Structures to resist the effects of accidental explosions, 1990, http://www.ddesb.pentagon.mil/tm51300.htm.

[11] Biggs, J.M., *Introduction to Structural Dynamics*, McGraw Hill, 1964.

[12] Baker, W.E., Cox, P.A., Westine, P.S., Kulesz, J.J. & Strehlow, R.A., *Explosion Hazards and Evaluation*, Elsevier, 1983.

[13] SBEDS v.4.1, Protective Design Centre, U.S. Corps of Engineers, 2009, https://pdc.usace.army.mil/software/sbeds/.

Structural retrofit of glazing systems with polymer materials for blast resistance

K. Marchand[1], C. Davis[1], E. Conrath[1], P. Votruba-Drzal[2], E. Millero[2] & G. Yakulis[2]
[1]*Protection Engineering Consultants, USA*
[2]*PPG Industries, Inc., USA*

Abstract

Protection Engineering Consultants (PEC), in cooperation with PPG Industries, Inc., is investigating the performance of a clear polymer coating material for windows subjected to blast loading. PEC developed a static test fixture and performed static tests to evaluate the retrofit window response. PEC then performed preliminary single-degree-of-freedom (SDOF) analysis, using a resistance function for the clear polymer coated glazing based on the quasi-static test results. Dynamic (full-scale blast) tests were then performed in Yancey, Texas with the assistance of Southwest Research Institute (SwRI). The dynamic test results were then used to determine strain rate effects, resulting in dynamic increase factors for the static resistance functions. The clear polymer coated windows absorbed energy during the blast load through large deflections and allowed very few fragments within the test structure. This paper summarizes the test results and analysis and provides recommendations for further development of this composite glass/polymer system.
Keywords: blast loads, glass fracture, glass hazards, polymer materials.

1 Introduction

Monolithic (non-laminated) glass is a brittle material that shatters upon reaching fracture stress or displacement under wind, impact or explosive loads. Injury studies have demonstrated the hazards associated with skin laceration and shard blunt trauma. Materials such as polyethylene (PET) films have been used for many years to retain shards upon glass lite fracture. To have significant effect on

WIT Transactions on The Built Environment, Vol 113, © 2010 WIT Press
www.witpress.com, ISSN 1743-3509 (on-line)
doi:10.2495/SU100161

hazard reduction, these PET films must not only be adhered to the glass, but must be mechanically anchored to the window supporting structure with screw-attached battens or through "gluing" with materials such as structural silicone. To reduce the labor and to improve the performance of a retrofitted materials designed to reduce glass shard hazards, PPG Industries, Inc. has formulated optically clear polymer materials (coatings) that can be spray-applied to glass. An additional benefit of these materials is that the same spray application can be used to "overspray" the polymer onto supporting frames and mullions, thereby achieving the mechanical anchorage necessary for complete hazard reduction without additional materials (battens) or labor.

PPG provided material properties through dynamic mechanical analysis and Instron tensile testing of four clear polymer samples. The first sample consists of a clear polymer with a low modulus of elasticity in a thin or thick configuration, while the second sample is a clear polymer with a high modulus of elasticity. Both clear polymer formulations were incorporated into the test plan. Adhesion strength was also considered as controlled by the application of adhesion promoters to the glass and frame. Application of the adhesion promoter resulted in a high adhesion specimen with a peel-off strength of more than 25 lb/in. By not applying adhesion promoter to the glass of the low adhesion specimen, the low adhesion polymer separated from the glass during dynamic response while remaining adhered to the frame. The peel-off strength was less than 5 lb/in.

2 Quasi-static tests

Quasi-static tests were performed to generate static load-deflection curves (static resistance functions) and to determine failure mechanisms for clear polymer coated windows. Twenty-three tests were performed. The test matrix, as shown in Table 1, included 23 test specimens and considered six test variables: size of window, type of connection, glazing material, glass thickness, polymer thickness, and type of polymer. Only high adhesion, low modulus of elasticity polymer was used for the static tests.

The two nominal window sizes were 2-ft × 3-ft and 4-ft × 4-ft (width × height), which corresponds to the following actual glazing dimensions: 24.75-in ×34.75-in and 48.75-in × 47.25-in, respectively. The type of connection refers to the connection between the aluminum window frame, glazing, and clear polymer.

All frame pieces were 6061-T6 aluminum. Connections used a 1.25-in, 3M VHB Structural Glazing (grey) Tape G23F on both sides. When polymer was used with a "tape only" connection, the polymer coated glazing was captured within the aluminum frame, where essentially the 3M tape provides the only connection between the polymer-coated glazing and frame. Where "spray 4 sides" is indicated, the polymer was sprayed on the frame, providing a mechanical connection. The remaining variables pertain to the glass and polymer properties.

Table 1: Quasi-static test matrix.

PPG PEC Test Data					
Test #	Nominal Window Size (in)	Type of Connections	Day Light Opening (in)	Actual Glass Thickness (in)	Actual Polymer Thickness (in)
1	24x36x1/4 AN	Tape Only	22.5" x 33.75"	0.225	none
2	24x36x1/8 AN	Tape Only	22.5" x 34.25"	0.115	none
3	24x36x1/4 AN	Tape Only	22.5" x 34.25"	0.225	none
4	24x36x1/8 FT	Tape Only	22.1875" x 34"	0.125	none
5	24x36x1/8 FT	Tape Only	22.1875" x 34"	0.125	none
6	24x36x1/8 AN	Tape Only	22.125" x 34"	0.115	none
7	24x36x1/8 AN	Tape Only	22.125" x 34"	0.115	none
8	24x36x1/8 FT	Tape Only	22.1875" x 33.875"	0.125	none
9	24x36x1/4 AN 0.030 clear	Tape Only	22.1875" x 33.875"	0.225	0.033
10	24x36x1/8 AN 0.030 clear	Tape Only	22.125" x 34"	0.115	0.032
11	24x36x1/4 AN 0.030 clear	Tape Only	22.125" x 33.875"	0.225	0.034
12	24x36x1/8 AN 0.030 clear	Tape Only	22.125" x 34"	0.115	0.033
13	24x36x1/4 AN	Tape Only	22.1875" x 33.875"	0.225	none
14	24x36x1/8 AN 0.030 clear	Spray 4 Sides	22.5" x 34.25"	0.115	0.036
15	24x36x1/8 AN 0.015 clear	Spray 4 Sides	22.5" x 34.25"	0.115	0.0156
16	48x48x1/4 AN	Tape Only	44.5" x 46.5"	0.219	none
17	48x48x1/4 AN 0.060 clear	Spray 4 Sides	44.5" x 46.1875"	0.221	0.062
18	48x48x1/4 AN 0.030 clear	Tape Only	44.5" x 46.3125"	0.221	0.037
19	48x48x1/4 AN 0.125 clear	Spray 4 Sides	44.5" x 46 .5"	0.224	0.119
20	48x48x1/4 AN 0.030 clear	Tape Only	44.625" x 46.375"	0.212	0.036
21	48x48x1/4 AN 0.030 clear	Tape Only	44.25" x 46 .5"	0.225	none
22	48x48x1/4 AN	Spray 4 Sides	44.625" x 46.375"	0.221	0.0318
23	48x48x1/4 AN	Tape Only	44.25" x 46.5"	0.230	none

Two types of glazing materials (annealed and fully tempered) were coated with low modulus of elasticity (E), high adhesion clear polymer. Based on preliminary resistance functions and the test frame capacity, a range of glass and polymer thicknesses were specified to encompass a variety of window responses.

Each window was attached to a load controlled water pressure test tank with a steel mask corresponding to the nominal window size. The side of the window facing the inside of the test tank is denoted as the "blast" face and represents the exterior face of a window in a building. Therefore, the interior side of the window, with polymer coating, was visible. The test tank has a 4-ft by 6-ft opening and is 9-in deep. The mask was bolted to the test tank to decrease the opening to the nominal window size. The non-responding window frame, constructed out of two 4-in by ¼-in aluminum plates that sandwich the glass lite, was bolted to the steel mask. A rubber gasket was placed between the tank/mask and mask/frame interfaces to create a watertight seal. After shimming the frame as required, the bolts were snugly tightened around the frame and tank perimeter. Instrumentation during each test included a combination of linear potentiometers, bi-axial strain gauges, and pressure gauges. The pressure gauge, mounted on the test tank sidewall, measured the tank water pressure, which was assumed equivalent to the applied pressure on the glass.

Comparisons between mode of failure, resistance function, and overall response for windows with and without clear polymer coating were made. Glass without clear polymer coating was tested first to serve as a baseline for comparison to coated glass and to determine the surface flaw parameters required in the glass failure prediction model (GFPM). As expected, the non-coated annealed glass fractured in large, jagged fragments, while the fully-tempered glass fractured in small, smooth shards. Oval fracture patterns were also noted in the majority of tests, as seen in previous research and Figure 1. In most cases for polymer coated glass, the polymer failed by bridging over cracks locally near edges, as shown in Figure 2. Results from the quasi-static tests are summarized in Table 2. The polymer performed better when the glass broke into smaller, evenly distributed fragments because the polymer stress was less localized. PEC recommended a higher strength polymer with a higher modulus of elasticity to improve the polymer response at localized cracks. It was also suggested that, during a dynamic test, a polymer with less adhesion might help prevent tearing at localized cracks, which leads to polymer failure. Anecdotal evidence from Test 17 also supported the theory that less adhesion, through extensive cracking in this case, allowed the polymer to pull free from the glass earlier in time and absorb more energy through large displacements prior to polymer failure. After Test 17, it was possible to pull the polymer off the glass by hand, which was not possible during any other tests.

Table 2: Quasi-static test results.

PPG PEC Test Data			Glass Break					Maximum Polymer failure		
Test #	Nominal Window Size (in)	Type of Connections	Pressure (psi)	midpoint deflection (in)	1/4 point deflection (in)	Strain 1 (μstrain)	Strain 3 (μstrain)	Pressure (psi)	midpoint deflection (in)	1/4 point deflection (in)
1	24x36x1/4 AN	Tape Only	2.84	0.487	0.334					
2	24x36x1/8 AN*	Tape Only	1.78	0.657	0.437					
3	24x36x1/4 AN*	Tape Only	2.33	0.360	0.228					
4	24x36x1/8 FT	Tape Only	8.21	1.370	1.069					
5	24x36x1/8 FT	Tape Only	8.01	1.362	1.058					
6	24x36x1/8 AN	Tape Only	1.59	0.486	did not use					
7	24x36x1/8 AN	Tape Only	1.15	0.487	did not use	501	330			
8	24x36x1/8 FT	Tape Only	7.40	1.300	1.029	833	1202			
9	24x36x1/4 AN 0.030 clear	Tape Only	0.592	0.10	0.06	Glass pre-cracked		1.195	0.50	0.39
10	24x36x1/8 AN 0.030 clear	Tape Only	1.927	0.58	0.38			0.362	2.72	1.91
11	24x36x1/4 AN 0.030 clear	Tape Only	5.066	0.71	0.50	1051	653	0.734	4.28	2.61
12	24x36x1/8 AN 0.030 clear	Tape Only	0.822	0.35	0.20	466	234	0.187	0.90	0.62
13	24x36x1/4 AN	Tape Only	2.71	0.418	0.283	658	294			
14	24x36x1/8 AN 0.030 clear	Spray 4 sides				Glass Pre-Cracked				
15	24x36x1/8 AN 0.015 clear	Spray 4 sides	1.179	0.45	0.29					
16	48x48x1/4 AN	Tape Only	1.794	0.79	0.55					
17	48x48x1/4 AN 0.060 clear	Spray 4 sides	1.284	0.75	0.52			0.234	3.35	2.46
18†	48x48x1/4 AN 0.030 clear	Tape Only	0.634	0.43	0.28	219	247	0.183	1.81	0.70
19	48x48x1/4 AN 0.125 clear	Spray 4 sides	1.715	0.78	0.56			0.210	2.26	1.24
20	48x48x1/4 AN 0.030 clear	Tape Only	1.223	0.72	0.50			0.236	2.78	1.49
21	48x48x1/4 AN 0.030 clear	Spray 4 sides	0.778	0.47	0.32	Glass Pre-Cracked		0.228	2.12	0.92
22	48x48x1/4 AN	Tape Only	1.047	0.54	0.35					
23	48x48x1/4 AN	Tape Only	1.184	0.59	0.39	317	287			

* test weak (tin) side of glass, otherwise test strong side

† glass break while bleed out air

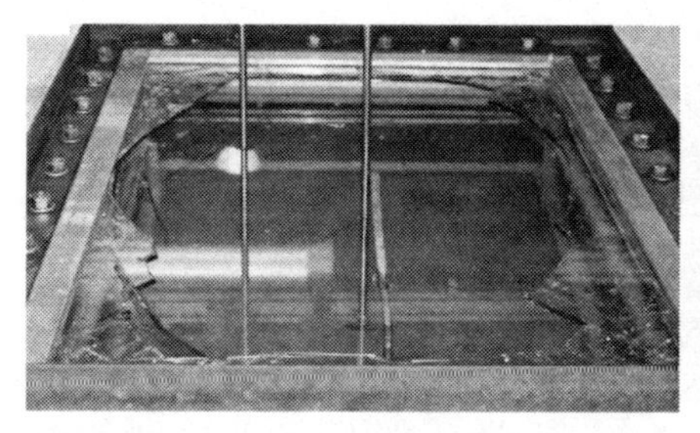

Figure 1: Oval fracture pattern (no poly).

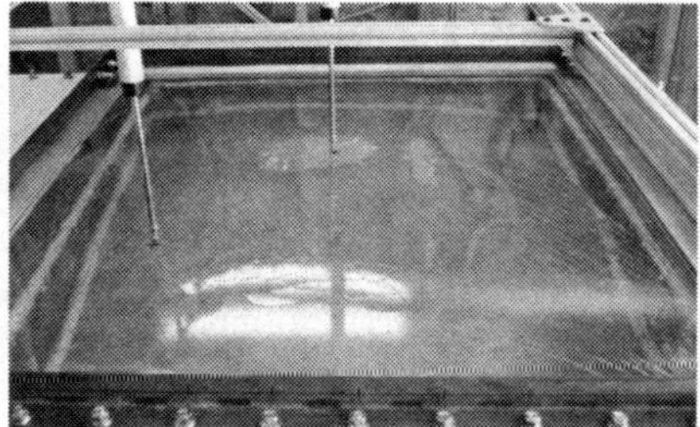

Figure 2: Polymer test with bridging.

3 Analysis and resistance function development

The pressure and displacement histories from each test were used to determine a static resistance function for each window assembly. The measured tank pressure was equivalent to the resistance of the window assembly in this analysis. Figure 3 illustrates the static resistance functions for tests 16-23. The first slope represents the glass resistance up to fracture (AN – annealed). The second positive slope represents the polymer resistance after glass fracture. PEC evaluated the polymer contribution in two regimes: small displacement glass-polymer response and large displacement polymer response only. Comparisons of tests with and without polymer coating show no stiffness enhancement prior to glass fracture for clear polymer coated glass, as shown in Figure 4 (again for tests 16-23). In addition, different polymer connections (tape only vs. spray all sides) performed similarly during static tests. The shim thickness was noted to have a small effect on the stiffness prior to glass failure because the thickness difference changed the amount of fixity at the support. Therefore, the main variables that affected glass break were window size, type of glass, and glass thickness.

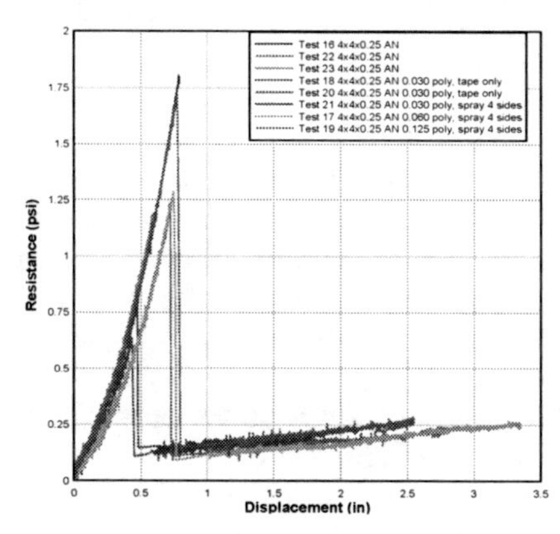

Figure 3: Glass/poly resistance (16-23).

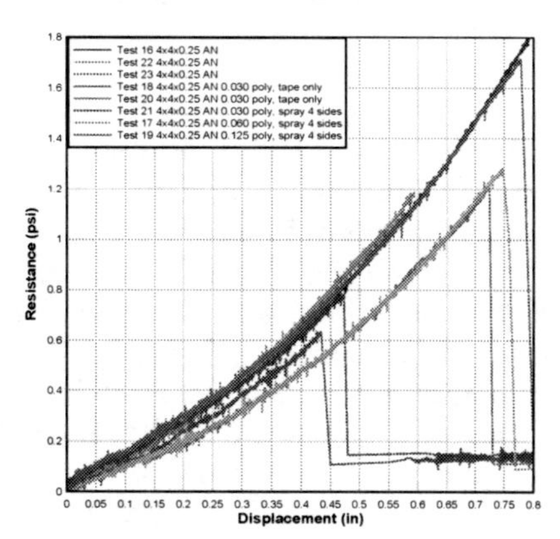

Figure 4: Glass resistance (16-23).

The polymer response after glass fracture thus only varies with the polymer type and polymer thickness. Based on the static test results, a polymer thickness range of 0.030-in to 0.060-in was determined to be most effective. For example, Test 15 had a 0.015-in thick polymer coating and experienced polymer failure concurrent with glass failure; therefore, statically the 0.015-in thickness does not enhance the capacity of the system. Test 19 evaluated a window with 0.125-in thick polymer coating, which performed similar to 0.060-in polymer coated windows at best, illustrating the lack of benefit above a thickness of 0.060-in. The maximum polymer displacement of 4.3-in occurred during Test 11 for a 2x3 AN window with 0.030-in clear polymer. For a 4x4 AN window, a maximum displacement of 3.4-in occurred during Test 17 using 0.060-in thick polymer coating. Overall, the polymer membrane response, while adding significant ductility to the glazing system, was lower than expected based on preliminary dynamic predictions. Therefore, polymer rate effects to be determined through dynamic (blast) testing were deemed essential.

4 Dynamic (explosive) tests

To evaluate the performance of the polymer coated windows under dynamic loads (strain rate effects and failure mechanisms), five full-scale explosive tests were completed at a Southwest Research Institute test site near Yancey, Texas. Each blast test investigated two windows mounted in a single steel reaction structure. The reaction structure was placed at standoffs between 120-ft to 265-ft for charges ranging from 200 to 1200 lbs of ANFO (Ammonium Nitrate and Fuel Oil). The test matrix included eight test specimens and accounted for three test variables: size of window, polymer thickness, and polymer modulus of elasticity. Several parameters were held constant during the dynamic testing including glass thickness, glass type, polymer adhesion, and connection type. PPG performed the clear polymer application at PPG Industries and then shipped the coated glass to PEC for framing similar to that used in the quasi-static tests.

Two nominal window sizes were tested; 2-ft × 3-ft and 4-ft × 4-ft (width × height). The only connection between the aluminum window frame and clear polymer coated glazing was a 1.25-in, 3M VHB Structural Glazing (grey) Tape G23F applied to both sides (tape only). Essentially, the glazing was confined within the aluminum frame for ease of construction and shipment. All frame pieces were 6061-T6 aluminum. In all tests, ¼-in annealed glass was coated with low adhesion clear polymer. An assumed actual thickness of 0.225-in was used in the calculations. Based on preliminary resistance functions and the test frame capacity, a range of polymer thicknesses and modulus of elasticity (E) types were specified to encompass a variety of window lay-ups.

PEC designed a steel reaction structure with two 46-in square openings to accommodate two windows per test, as shown in Figures 5 and 6. The closed box design prevented pressure from wrapping around to the back face of the windows. To decrease the window opening for the 2-ft by 3-ft windows, two steel masks with 22.75-in by 34.5-in openings were bolted to the reaction structure when needed. The test specimens were designed to bolt directly to the

Figure 5: Steel reaction "box".

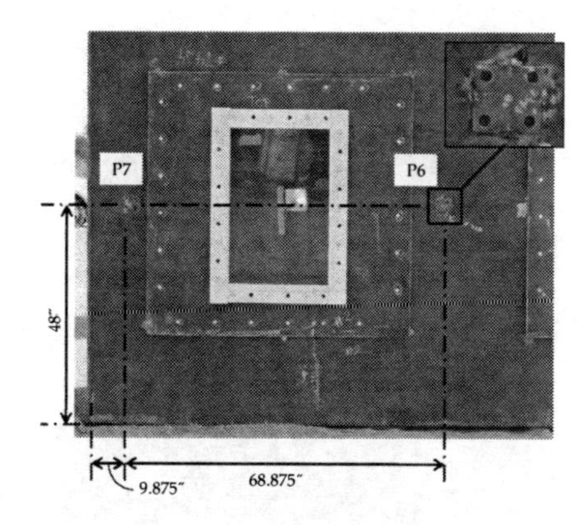

Figure 6: Pressure gage locations.

reaction structure. Openings for instrumentation were also provided on the front and back face for the pressure and scratch gauges, respectively.

The goal of the dynamic clear polymer tests was to evaluate the rate sensitivity and dynamic failure mechanism of clear polymer coated glass. Therefore, charge weights and standoffs were selected to cause membrane response at large post-glass fracture deformations. Six of the eight window specimens (two windows were retested) experienced membrane response. The results from each dynamic test are summarized in Table 3. The displacement was used in combination with the pressure histories and high-speed video

Table 3: Dynamic (explosive) test matrix and results.

			Center Gauge (P6)		Side Gauge (P7)		TNT Equivalency		
Test #	Window Type	Scratch Deflection (in)	Peak Pressure (psi)	Max Impulse (psi-msec)	Peak Pressure (psi)	Max Impulse (psi-msec)	Average for Pressure	Average for Impulse	Test Notes
1A	2x3x1/4 AN 0.030 Low E	0.50	6.10	19.3	5.70	14.9	1.50	0.40	no break
1B	2x3x1/4 AN 0.030 High E	0.50	6.10	19.3	5.70	14.9	1.50	0.40	no break
2A	2x3x1/4 AN 0.030 Low E	N/A	8.80	27.5	8.30	20.5	1.43	0.36	membrane response
2B	2x3x1/4 AN 0.030 High E	6.19	8.80	27.5	8.30	20.5	1.43	0.36	membrane response
3A	2x3x1/4 AN 0.030 Low E	6.13	7.59	27.0	7.39	25.1	1.39	0.40	membrane response
3B	4x4x1/4 AN 0.060 High E	12.38	7.59	27.0	7.39	25.1	1.39	0.40	membrane response
4A	4x4x1/4 AN 0.030 Low E	9.50	5.25	18.49	5.08	27.13	1.26	0.35	membrane response
4B	4x4x1/4 AN 0.030 High E	10.00	5.25	18.49	5.08	27.13	1.26	0.35	membrane response
5A	2x3x1/4 AN 0.015 High E	3.00	8.02	22.91	7.59	30.13	1.52	0.31	tear prior to membrane
5B	2x3x1/4 AN 0.030 High E	1.13	8.02	22.91	7.59	30.13	1.52	0.31	no break

collected. The high-speed video yielded a clear picture of the window response and failure mechanism. In general, the high-speed video shows the significant deformation of glazing specimens during inbound and rebound response.

In general, the clear polymer coated windows performed as predicted by elongating and bridging over cracks after glass failure in the oval yield line pattern noted in the static tests. Tested windows from Test 2 are shown in Figure 7. The membrane is shown to be intact for both configurations, although more glass was lost off of the membrane during rebound in the High E test (b).

a) Test 2A – 2x3x1/4 AN 030 Low E **b) Test 2B – 2x3x1/4 AN 0.030 High E**

Figure 7: Failure mechanism test 2.

The energy absorption through large tensile membrane response indicates that clear polymer is extremely load rate sensitive material. Small tears in the clear polymer were noted in several tests; however, this did not constitute failure and the majority of the glass fragments remained on the outside of the reaction structure. Glass is also a rate sensitive material and produced smaller shards during the dynamic testing. Smaller glass shards combined with less adhesion between the polymer and glass allowed the polymer to release from the glass and distribute the load more uniformly; which provided more places for the polymer to bridge minimizing tears. This results in much larger maximum displacements before failure than observed in the static tests.

A dynamic resistance function for a clear polymer coated window was determined in two parts using the Glass Failure Prediction Model (GFPM) developed by Beason et al. [1] for glass response up to fracture and WinGard [2] equations originally developed by Timoshenko/Urgural for polymer membrane response. The GFPM relates the probability of failure of glass, surface flaw characteristics (m and k), and induced stresses by combining a statistical failure theory for brittle materials (Weibull distribution) with results of geometrically non-linear plate analysis. The surface flaw parameter, m, of 6.79 was determined from the static test results. Measured loads from the blast tests (averages of gages 6 and 7 on the test fixture) were used in the analysis.

Figure 8 illustrates the dynamic resistance function developed for each clear polymer coated window.

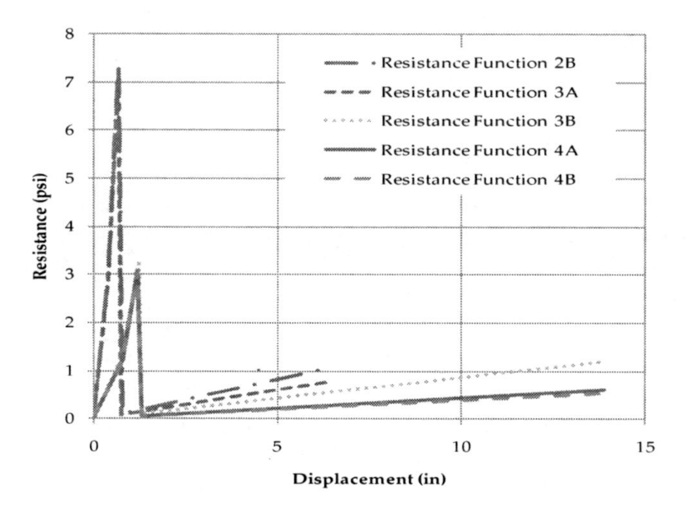

Figure 8: Resistance functions developed from tests and analysis.

The theoretical membrane equations were applied to clear polymer by changing two material properties: Poisson's ratio and modulus of elasticity. A Poisson's ratio of 0.5 was assumed for the clear polymer resistance calculations. An equation for modulus of elasticity in terms of strain across the short span was determined by back calculating a modulus at several deflections for each static clear polymer coated window test. A dynamic failure deflection criterion of 30% of the short span was selected, based on the observed deflection to short span ratio for dynamic tests that experienced significant membrane response. Using the dynamic resistance function and pressure-time history in the SBEDS [3] general SDOF template, the DIF was iterated until the predicted deflection matched the measured deflection. The clear polymer was assumed to have a linear dynamic resistance function defined by the peak dynamic resistance (maximum clear polymer resistance scaled by the DIF) and assumed failure deflection.

5 Conclusions

The blast tests have determined that a minimum thickness of clear polymer of 0.030-in is required to prevent tearing and premature failure of the polymer. Similarly, based on static test results, it is doubtful that a thickness in excess of 0.060-in would be of additional benefit. The low adhesion specimen (glass that is not prepped and primed) seemed to perform well during blast tests, while the modulus of elasticity (high-E vs. low-E) had a very small change in polymer

performance. Both the tape only and spray over all sides connections seem to work adequately.

Comparisons with other "blast resistant" window configurations (security film and laminated glass), as made using range-to-effect (RTE) calculations, show that clear polymer coated glass in the 0.030-in to 0.060-in thickness range perform similarly and in some cases better than 7-mil polyester film and laminated glass (0.030 interlayer).

Acknowledgements

The authors would like to thank Truman Wilt, Mike Mulligan, Howard Senkfor, and Dan Bratys for their support of the development of this technology.

References

[1] Beason, W. Lynn and Morgan, James R., "Glass Failure Prediction Model," ASCE Journal of Structural Engineer, Vol 110, No.2, February 1984.
[2] "Window Glazing Analysis Response and Design, WinGARD," US General services Administration, Public Buildings Service, prepared by Applied Research Associates, August 2005.
[3] "Users Guide for the Single-Degree-of-Freedom Blast Effects Design Spreadsheets (SBEDS)," US Army Corps of Engineers, Omaha District Protective Design Center (PDC), prepared by Protection Engineering Consultants, September 2006.

Comparison of traditional and Herculite® XP glazing subject to blast loads

C. Davis[1], K. Marchand[1], P. Votruba-Drzal[2], E. Conrath[1],
T. Sailock[2] & W. Siskos[2]
[1]*Protection Engineering Consultants, USA*
[2]*PPG Industries, Inc., USA*

Abstract

Protection Engineering Consultants (PEC), in cooperation with PPG Industries, Inc., is investigating the performance of a new glass technology for anti-terrorism/force protection (ATFP) applications. This new PPG glass product is significantly stronger than typical fully tempered glass, which is advantageous in impact, ballistic, and blast resistant window design. This paper compares the performance of this new glass technology, known as Herculite® XP glazing, to that of traditional (annealed and fully-tempered) glazing subjected to blast loads. The performance of the Herculite® XP glazing is based on results from a series of static tests and a large-scale blast test on eight windows. The results of the static and dynamic tests are briefly discussed. A brief discussion is also included on the use of a statistical fracture based glass strength model called the Glass Failure Prediction Model (GFPM) that was used to develop Herculite® XP resistance functions employed in single-degree-of-freedom (SDOF) blast response models.
Keywords: glass technology, blast testing, window design, SDOF analysis.

1 Introduction

Protection Engineering Consultants (PEC) and PPG Industries, Inc. are investigating the performance of Herculite® XP glass, a new glass technology that is stronger than typical fully tempered (FT) glass, for anti-terrorism/force protection (ATFP) applications. To evaluate the performance of Herculite® XP glazing subjected to blast loads, a single large-scale blast test on eight specimens was conducted at the Air Force Research Laboratory (AFRL) at Tyndall Air

WIT Transactions on The Built Environment, Vol 113, © 2010 WIT Press
www.witpress.com, ISSN 1743-3509 (on-line)
doi:10.2495/SU100171

Force Base in Florida. Predictions made with preliminary resistance functions developed using the glass failure prediction model (GFPM) compared well to dynamic test observations. Static tests performed at PEC in Austin, Texas enabled the confirmation of strength values for prediction models. This paper presents a summary of the performance of Herculite® XP glass during each test, an overview of the glass failure prediction model, and comparisons between the performance of Herculite® XP to traditional (annealed and fully-tempered) glazing.

2 Resistance function development

Preliminary resistance functions were developed to perform predictions for the large-scale blast test and to provide an initial comparison between annealed, fully-tempered, and Herculite® XP glazing. The resistance functions utilize a rate dependent model to predict glass break response and a theoretical membrane equation based on equations developed by Timoshenko and Woinowsky-Kreiger [7], adjusted with static and dynamic test data to predict post-crack glazing response. Each model and resulting resistance functions are briefly discussed below.

2.1 Glass failure prediction model overview

The glass failure prediction model (GFPM) by Beason and Morgan [3] is an unrestricted model that predicts glass break strength and deflection. The GFPM is the basis for ASTM E1300, the North American Standard on glass selection for wind loads, and is based on theoretical development as supported by a review of thousands of test data records for glass break. The GFPM (Beason and Morgan [3]) relates the probability of failure of glass, surface flaw characteristics, and induced stresses by combining a statistical failure theory for brittle materials (Weibull distribution) with results of a geometrically nonlinear plate analysis. The model is specifically applicable to the analysis of rectangular glass to resist uniform lateral pressures (not missile impact) and includes all factors that significantly affect the strength of glass including plate geometry, load duration, probability of failure, and glass surface condition (size and distribution of flaws, age of glass, static strength, etc.). The GFPM (Beason et al. [4]) was used to predict glass resistance, while the predicted deflection was determined with a fit by Dalgliesh (CGSB, 1989). The GFPM is a rate-based model that truly accounts for glass being a ceramic-type material where strength is not solely based on the static material properties of the base material, but also on the size and distribution of inherent flaws in the material. In addition, the GFPM allows for analysis of new glass and weathered or aged glass within the same model, as well as different types of glass [annealed (AN), heat strengthened (HS), fully-tempered (FT), and higher strengths].

2.1.1 Load duration

The GFPM is a function of load duration, as shown in Figure 1 (for a 31″x60″x1/2″ [0.219″x0.060″PVBx0.219″] Herculite® XP with a probability of

failure of 0.5). The GFPM predicts larger resistances for shorter duration loads (i.e., 10 ms) than longer duration loads, which is consistent with dynamic increase factors for other materials. In general, most materials experience an increase in strength under rapidly applied loads because the material cannot respond at the same rate at which the load is applied. GFPM predictions with annealed and fully tempered glass show similar trends; small glass sizes and small load durations predict stronger glass.

2.1.2 Probability of failure

Probability of failure (POF) is the number of lites out of 1000 that will fail at the prescribed load. A POF of around 8/1000 (0.008) is typically used for design;

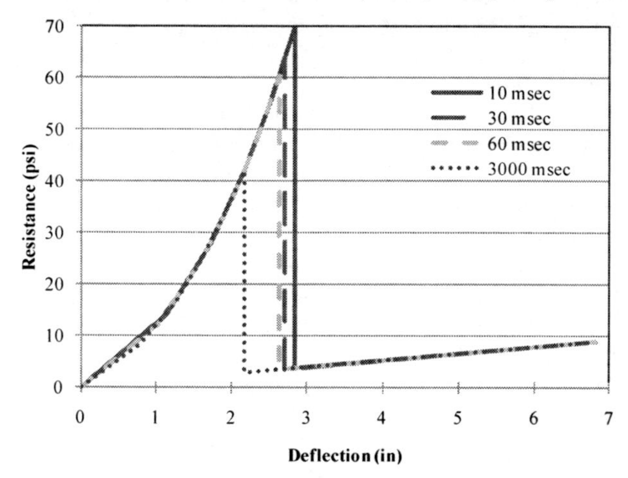

Figure 1: Affect of load duration on GFPM predictions for 31x60x1/2 Herculite® XP glazing (POF = 0.5).

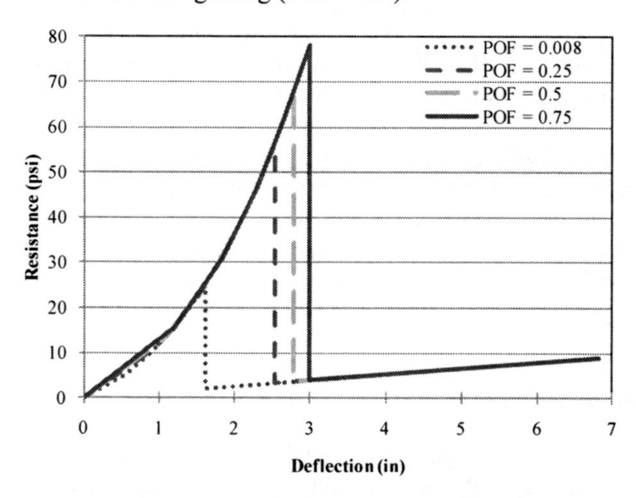

Figure 2: Affect of probability of failure on GFPM predictions for 31x60x1/2 Herculite® XP glazing (15 msec duration).

Table 1: Surface flaw parameters for new glazing.

Glass type	Strength (psi)	m	k
Annealed	3380	6.94	2.86E-53
Fully tempered	13500	6.54	2.86E-53
Herculite XP	34000	6.4	2.86E-53

while a POF of 0.5 was used to best match blast test data generated using "new" (not weathered or "in-service") glass. The glass resistance increases as the POF increases (more glass is allowed to fail which is less conservative), as shown in Figure 2. Analysis of test data was performed to select a set of surface flaw parameters for the GFPM that best match the desired POF of 0.5.

2.1.3 Surface flaw parameters

GFPM is based on the theory that glass plate failure is a result of interaction of glass surface flaws (which result from the manufacturing process and subsequent exposure) and surface tensile stresses from applied loads. The two surface flaw parameters are m and k, which represent the severity and distribution of surface flaws. "Examination of a wide range of theoretically generated glass strength data shows that m is related to the coefficient of variation of the data. As m increases, the coefficient of variation decreases. Further, these examinations show that k is closely related to the mean of the glass strength data." (Beason et al. [4]) After much debate, ASTM E1300 established the following set of surface flaw parameters to be used for annealed (AN) glass design for normal in-service exposures: $m = 7$ and $k = 2.86\times10^{-53}\ N^{-7}m^{12}$. The recommended values do a relatively good job predicting the strength of AN glass when keeping in mind the variations due to glass size and load duration discussed above.

During manufacturing of tempered windows (specifically the heat quench process), residual compressive stresses are induced into the exterior surfaces to increase the nominal glass strength. At the same time, the heat quench process reduces the number and size of flaws present in fully-tempered (FT) glass, which adds to the increase in nominal glass strength. The nominal strength increase due to glass type is accounted for in the GFPM by a strength modifier. However, the strength increase resulting from the reduction in flaws should also be accounted for by the adjustment of the surface flaw parameter, m, for different glass types. Static testing of new annealed, FT and Herculite® XP glazing has allowed the determination of the surface flaw parameter, m, for a probability of failure of 0.5, as summarized in Table 1. The testing has shown that new tempered glass requires a smaller flaw parameter, m, than new annealed glass more flaws, lower strength.

2.2 Membrane model overview

Membrane models predict post-crack glazing response for laminated windows. The laminate interlayer is very thin relative to the glass material and act as membranes after the glass breaks, preventing window failure until they fail.

Membranes displace significantly before rupture and typically provide very significant additional resistance to the window response. PEC is working on development of an updated theoretical membrane model based on static test data and a strain-adjusted modulus of elasticity. The new model was used for all comparisons in this paper, instead of HazL or WinGARD membrane models.

The slope of the PVB (polyvinyl butyral) interlayer resistance is relatively linear, rather than highly non-linear as calculated by theoretical equations. The modulus of the PVB interlayer was considered to be variable, where it initially included a large contribution from the attached glass just after break, and then reduced to a value nearer the static modulus of the PVB as membrane strain increased.

2.3 Resistance functions

The resistance functions utilize the glass failure prediction model to predict glass break and a theoretical membrane equation adjusted with static and dynamic test data to predict post-crack glazing response. The resistance functions were updated after the static and dynamic tests to utilize the observed glass strength and strain rate effects. The modified version of the GFPM was used to evaluate annealed (AN), fully-tempered (FT), and Herculite® XP laminated glazing of varying size and thickness for comparison. Laminated glazing lay-ups included two lites of glass of equal thickness laminated together by a 0.060-in PVB interlayer. A probability of failure of 0.5 and a load duration of 15-msec was used in the GFPM to calculate each resistance function.

Figure 3 illustrates the resistance functions for several annealed, fully-tempered, and Herculite® XP glazings. Note that the slope and peak of the PVB membrane response is constant for similar window geometry (dimensions), while the glass break response highly dependent on the type of glass, window geometry, and thickness. The maximum Herculite® XP glass break resistance is 93% and 58% larger than that of AN or FT, respectively. The glass break deflection for Herculite® XP is also larger than that of AN or FT by 80% and 36%, respectively. The disparities in the point of glass break illustrate the significant increase in strength and energy absorbing capacity (i.e., area under the resistance function) of Herculite® XP glazing over AN and FT. Herculite® XP response up to glass break accounts for 71% of the total area under the resistance function, while FT and AN glass only account for 39% and 4%, respectively, of the total area under the respective resistance functions. Therefore, the response of the PVB is less prominent (however, not insignificant) for higher strength glass. Overall, the contribution of the glass increased by 95% and 45% with the use of Herculite® XP over AN and FT, respectively.

3 Static testing of Herculite® XP

To evaluate the performance of Herculite® XP glazing, a series of large-scale static tests were conducted at the PEC Austin office. The objectives of the static tests were to determine strength values for prediction models (specifically, the

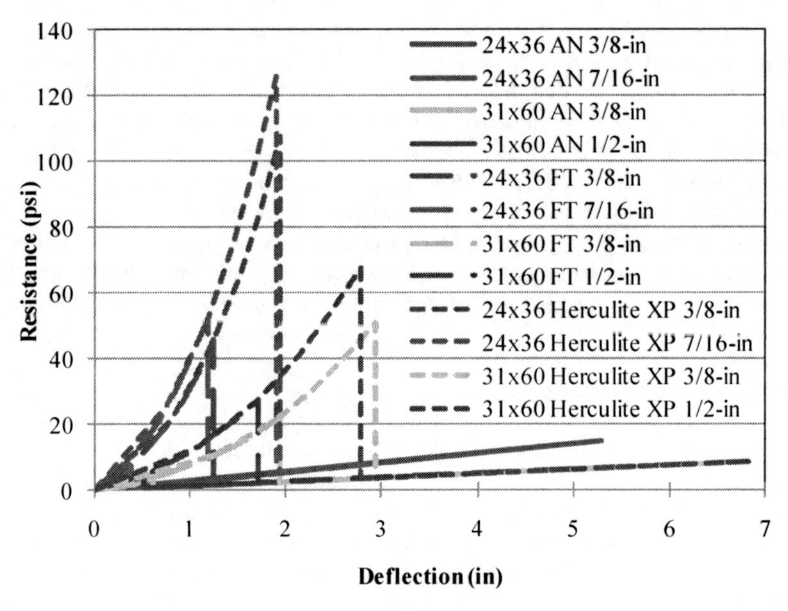

Figure 3: Comparison of resistance functions for various glazing lay-ups.

flaw parameter, *m*) and failure mechanisms of Herculite® XP glazing. The two main test variables include glass thickness and window size. The windows were subjected to static water pressure until failure. Post-test the resistance functions (i.e., GFPM) were updated to provide a better predictive design tool using a single-degree-of-freedom (SDOF) analysis.

3.1 Test set-up

PEC designed a steel test tank, two steel masks, and aluminium window frames for static window testing, while PPG provided all glass specimens. The tank was filled with water, an incompressible medium, by a pump with a pulse-less flow to steadily load the glass (load-controlled test). Air was bled out during the filling process to reduce the amount of air, a compressible medium, in the tank. The lack of air allowed the more accurate capture of post-break PVB response. Neoprene gaskets were used at the tank/mask and mask/specimen interface to create a water tight seal. Glazing construction of test specimens included monolithic lites of Herculite® XP glazing and two lites of Herculite® XP glass of equal thickness laminated together by a 0.060-in PVB interlayer. All glass lites were attached to the frame mounts with 1.25-in wide, 3M VHB Structural Glazing (grey) Tape G23F. The window frame mounts were constructed out of aluminium for ease of construction and installation. To reduce frame rotation during testing, shims were used at bolt locations during installation. The strong side of the glass was tested during each test. Silicone was applied to the water face (weak side) of the test specimen to create a water tight seal and minimize leaks. Instrumentation during the test included two linear potentiometers, a pressure gauge, and video camera. Figure 4 illustrates the static test set-up.

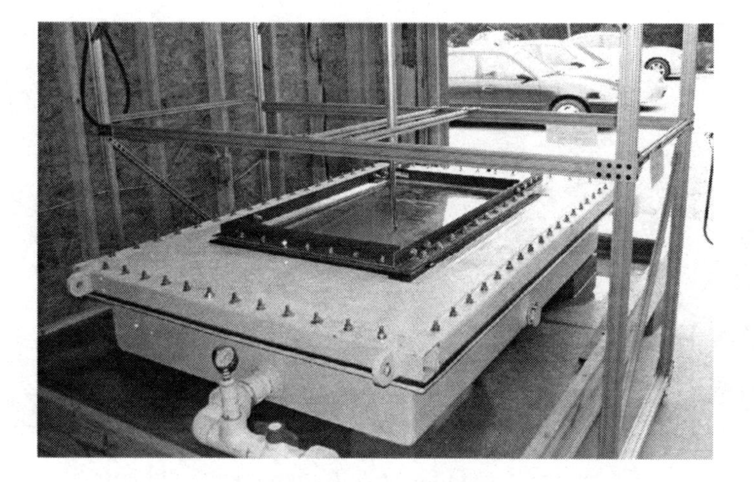

Figure 4: Static test set-up.

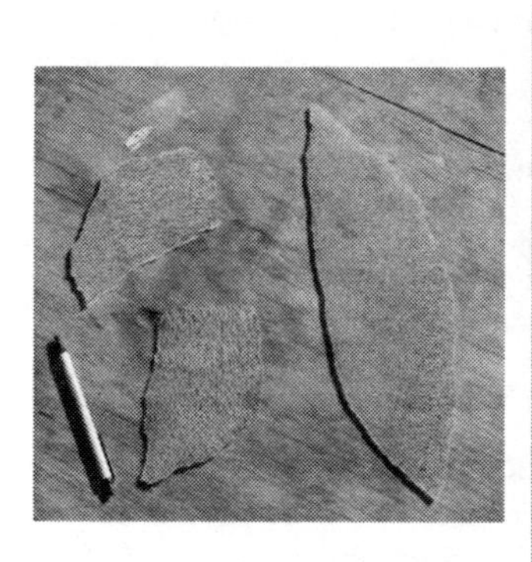

Figure 5: Crack pattern of Herculite® XP glazing.

3.2 Test observations

All windows were tested to failure (i.e. first crack for monolithic specimens and PVB failure for laminated specimens). Test observations include extremely small glass fragments for Herculite® XP glazing, as shown in Figure 5. The Herculite® XP glass fragments were smaller and smoother than both AN and FT glass. Small fragments were beneficial during static testing of laminated Herculite® XP glazing because they reduce the chance of a premature PVB tear and failure; instead allowing significant displacement of the membrane to occur up to membrane rupture, as shown in Figure 6.

Figure 6: Significant deflection prior to failure.

3.3 Test conclusions

An average surface flaw parameter, *m*, of 6.4 was determined from the series of static tests on Herculite® XP glazing. An oval cracking pattern with small fragments was a typical failure mechanism for monolithic Herculite® XP glazing, while significant deflection prior to membrane rupture was noted for laminated Herculite® XP glazing.

4 Dynamic testing of Herculite® XP

To further evaluate the performance of Herculite® XP glazing subjected to blast loads, a single large-scale blast test on eight specimens was conducted at the AFRL at Tyndall Air Force Base in Florida. The objectives of the dynamic test were to evaluate the preliminary resistance function and confirm failure mechanisms. The three main test variables include glass thickness, window size, and window frame material. The windows were subjected to a blast load associated with a predetermined charge weight and standoff. Post-test comparisons between measured deflections and deflections predicted by single-degree-of-freedom (SDOF) analyses illustrated the accuracy of the developed resistance functions.

4.1 Test set-up

AFRL Tyndall provided a reaction structure consisting of four reinforced concrete culverts and 5-ft wide clearing panels around each exterior edge. PEC designed four culvert fronts and eight window frames to modify the existing reaction structure for window testing, while PPG provided all glass specimens. Glazing construction of all test specimens included two lites of Herculite® XP glass of equal thickness laminated together by a 0.060-in PVB interlayer. Nominal glazing thicknesses are shown in Figure 7 along with the dynamic test

set-up. All glass lites were attached to the frame mounts with 1.25-in wide, 3M VHB Structural Glazing (grey) Tape G23F. The window frame mounts were constructed out of steel and aluminium to represent hard and soft mounts, respectively. Instrumentation during the blast test included pressure gauges, strain gauges, deflection gauges, and high-speed video cameras. Eight reflective pressure gauges were installed on the culvert fronts (two per culvert) to determine the applied blast loads. Four interior pressure gauges (one per culvert) were provided, as well as two free-field pressure gauges for explosive yield validation.

4.2 Test observations

The standoff for the charge was selected such that the varying window constructions exhibited a range of responses from no break to near PVB rupture. As predicted, all of the Herculite® XP windows survived the applied blast load, as shown in Figure 8. Four out of the eight windows did not fracture (i.e. no

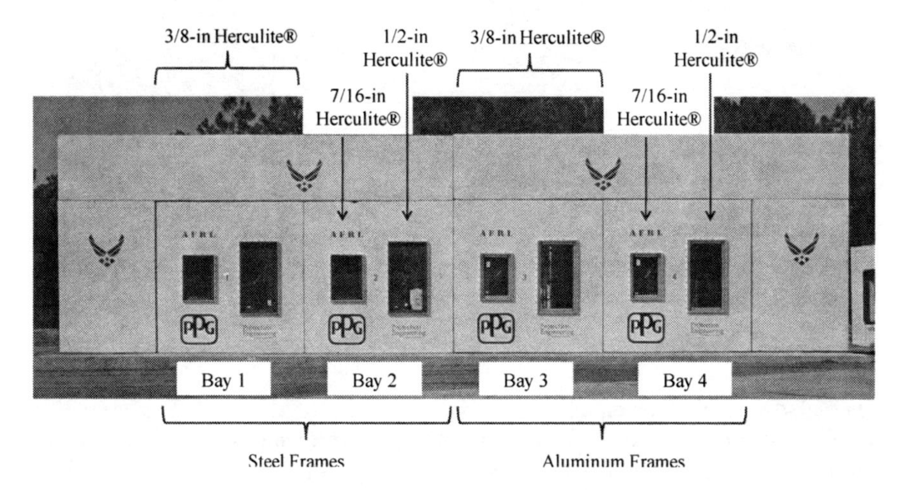

Figure 7: Dynamic test set-up.

Figure 8: Dynamic test observations.

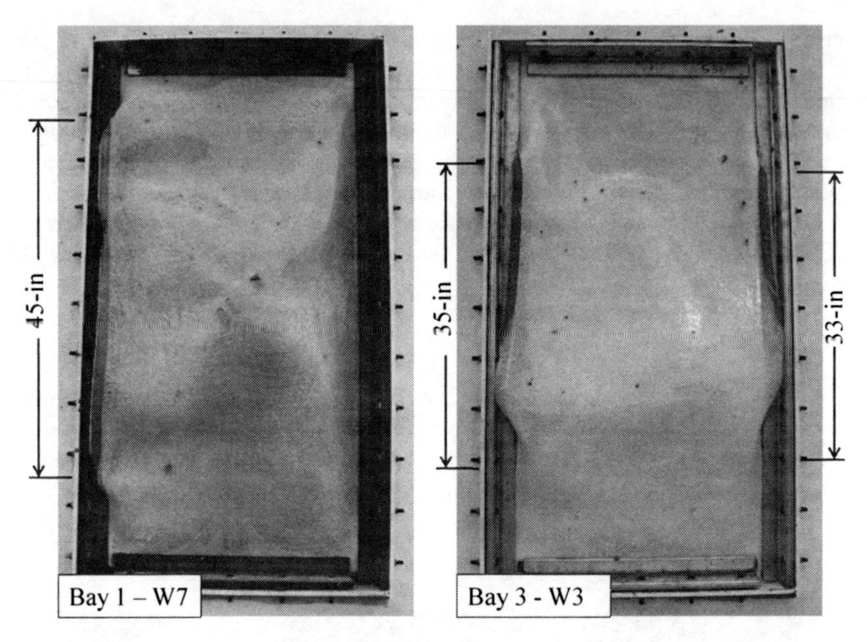

Figure 9: Energy absorption of Herculite® XP laminates.

break). Fracture of the outer lite only was observed for two of the windows (no hazard, according to the UFC 4-010-01 and ASTM 1642-04), while the last two windows experienced fracture of both lites and significant pullout at the frame edges (very low hazard), as shown in Figure 9.

4.3 Test conclusions

Overall, the predictions matched test observations extremely well. The large, thin window predictions included glass fracture and significant exercise of the PVB, but not PVB failure, which was observed. The small, thin window predictions were on the verge of first crack; observations show that one window fractured due to a projectile, while the other did not fracture. Therefore, the large-scale blast test results confirmed response predictions made with the GFPM approach and the validity of the approach for future tests and designs.

5 Comparison of Herculite® XP and fully-tempered glazing

The static and dynamic tests enabled the characterization of Herculite® XP glazing subjected to blast loads and the development of a predictive design tool for this new technology. The tweaked version of the GFPM was used to evaluate fully-tempered and Herculite® XP laminated glazing of varying size and thickness for comparison. Laminated glazing lay-ups included two lites of glass of equal thickness laminated together by a 0.060-in PVB interlayer. A

probability of failure of 0.5 and a load duration of 15-msec was used in the GFPM to calculate each resistance function (see Figure 3).

To make comparisons, the resistance functions developed with the GFPM were employed in single-degree-of-freedom (SDOF) analysis to evaluate window response. Each window was subjected to the Department of Defense (DoD) charge weight I (CWI) load at various standoffs to reach first crack and PVB failure. Figure 10 illustrates the required standoffs for FT and Herculite® XP glazing of various size and thickness. The standoff associated with first crack for Herculite® XP glazing decreased by 32% on average relative to FT glazing. The results also suggest that standoffs required for PVB failure of Herculite® XP glazing decreased by 27% relative to FT glazing standoffs. In addition, the difference in standoff between first crack and PVB failure increases as the glass strength decreases (12% and 5% for FT and Herculite® XP, respectively). Overall, the results suggest that Herculite® XP glazing requires 27% smaller standoffs than FT glazing for PVB failure.

6 Conclusions

The static and dynamic test results confirmed predictions made with the GFPM model which validates the approach for future glazing design. The results also provide data to suggest that standoffs for first crack of fully tempered glass are 32% greater than those for Herculite® XP glazing. Therefore, a stronger glass allows a reduced standoff for the same level of protection which increases the

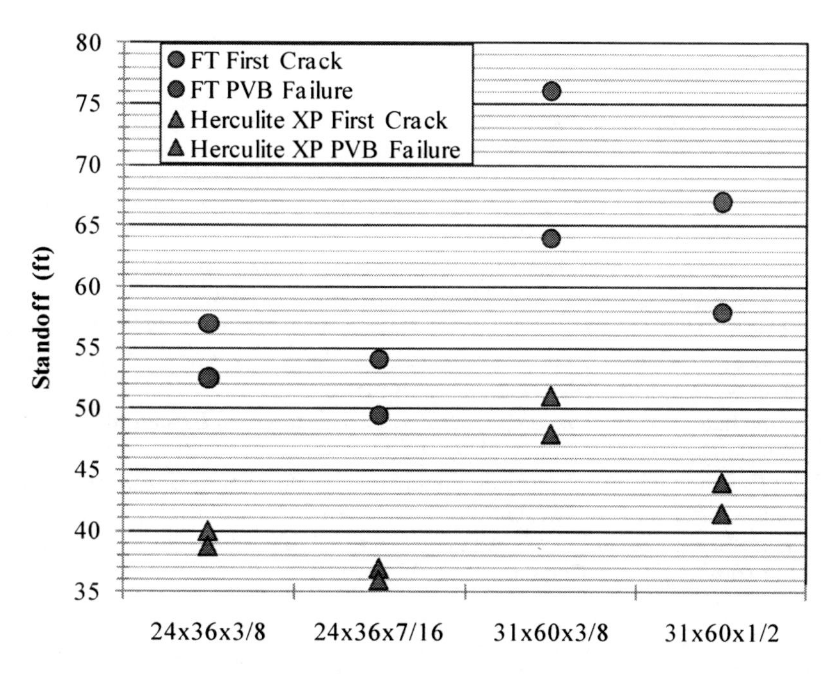

Figure 10: Standoff comparison for glazing lay-ups loaded with DoD CWI.

usable site area and available real estate. If additional standoff is not required, thinner layups with higher strength glass could be used in place of thicker traditional glazing layups to provide a significant cladding weight savings. Extremely small fragmentation was observed during the Herculite® XP testing, which may led to improved PVB membrane response. Overall, laminated glass layups with higher strength glass (i.e., Herculite® XP) provide a good alternative to existing annealed or conventionally tempered layups.

7 Future work

The successful static and dynamic testing of Herculite® XP glazing will be followed by the further development of this new glazing material. The static test data will allow the development of ASTM E 1300 type design guidance for Herculite® XP glazing subjected to wind loads to be used in the GFPM. Impact testing will also be completed in the near future to evaluate the hurricane performance of Herculite® XP glazing. Also, more investigation into the application of Herculite® XP glass in insulated glass units is being pursued.

Acknowledgements

PEC would like to thank Mehran Arbab, Dennis O'Shaughnessy, Lisa Walters, Truman Wilt, and Scott Follett at PPG Industries, Inc. for their support with the development of Herculite® XP glass.

References

[1] ASTM 1642-04, Standard Test Method for Glazing and Glazing Systems Subject to Airblast Loadings, *ASTM International*, February 2004.
[2] ASTM Committee E06, ASTM E 1300: Standard Practice for Determining Load Resistance of Glass in Buildings, *ASTM International*, West Conshohocken, PA, June 2003.
[3] Beason, W.L. and Morgan, J.R., Glass Failure Prediction Model, *Journal of Structural Engineering*, Vol. 110, No. 2, February, 1984.
[4] Beason, W.L, Kohutek, T.L., and Bracci, J.M, Basis for ASTM E 1300 Annealed Glass Thickness Selection Charts, *Journal of Structural Engineering*, Vol. 124, No. 2, February, 1998.
[5] Canadian General Standards Board (CGSB), Structural Design of Glass for Buildings, *CGSB 12.20-M89*, December, 1989.
[6] DoD Minimum Antiterrorism Standards for Buildings, *Unified Facilities Criteria (UFC) 4-010-01*, U.S. Department of Defense, October 2003.
[7] Timoshenko, S. and Woinowsky-Kreiger, S., *Theory of Plates and Shells*, McGraw-Hill Book Company, Inc., 1959.

Shock tube experiments and Fe-simulation of the structural and material non-linear transient response of plates subjected to blast loading

R. Schmidt, M. Stoffel & T. D. Vu
RWTH Aachen University, Germany

Abstract

This paper deals with the experimental investigation, modelling, and finite element simulation of structures exposed to shock-type blast loading conditions taking into account the structural and material non-linear effects. First- and third-order transverse shear deformation theories of plates and shells serve as basis of a finite element algorithm for the simulation of the transient, geometrically non-linear elastic-viscoplastic response. Isoparametric Lagrangian 9-node shell finite elements and the central difference method for the time integration of the non-linear equations of motion are used. Experiments are performed on thin clamped circular aluminium and steel plates in shock tubes. The main advantage of this experimental technique is that the front wave impinging on the structure is plane and yields a uniformly distributed pressure pulse. Consequently, in contrast to other experimental methods reported in literature, the time history of the shock-type loading can be modelled easily during the FE analysis. Comparative numerical simulations using first- or third-order transverse shear deformation plate theory, respectively, show a very good agreement with the experimental results. The best correlation with the experimentally observed transient response and permanent deflection is obtained by the refined, third-order transverse shear deformation model.

Keywords: blast loading, shock tube, first- and third-order transverse shear deformation theory, viscoplasticity.

1 Introduction

The present paper deals with modelling, computational simulation and experimental investigation of the transient large deflection elastic-viscoplastic

WIT Transactions on The Built Environment, Vol 113, © 2010 WIT Press
www.witpress.com, ISSN 1743-3509 (on-line)
doi:10.2495/SU100181

response of plates and shells subjected to impulsive loading. The focus of the paper is on comparative non-linear FE simulation, based on either first- or third order transverse shear deformation plate and shell theory.

The theoretical and numerical developments are based on geometrically non-linear plate and shell theories and associated finite element models available in our earlier papers (Schmidt and Reddy [1], Palmerio *et al.* [2], Vu *et al.* [3]). Experimental investigations of blast loaded plates and comparative numerical simulations of the transient response have been reported already in our papers [4-6], where a very good agreement was achieved by using FE models based on first-order transverse shear deformation theories. Our results showed, however, that the constant transverse shear stress assumption, implicit in first-order transverse shear deformation theories, leads to local over- and underprediction, respectively, of the equivalent plastic strain rates through the thickness that can affect the accuracy of the predicted transient response and permanent deflections. Therefore, in the present paper the third-order transverse shear deformation model is used for comparative FE analysis.

The theoretical predictions of the time history responses and final permanent structural deformations are compared with results of experimental tests performed on thin aluminium and steel plates in shock tubes. By means of this experimental device impulsive surface loads with various intensities and time histories can be applied resulting in geometrically non-linear elastic-viscoplastic vibrations. Experimental investigations of structures subjected to blast loading conditions are frequently performed by shock waves generated by explosives (e.g. gaseous mixtures, charges of plastic explosives) that detonate at some distance of the structure (see Idczak *et al.* [7, 8], Renard and Pennetier [9], Pennetier [10], among others). These experimental techniques lead to a complex non-uniform time-space evolution of the pressure distribution on plane structural elements thus requiring a sophisticated modelling and simulation of the blast loading conditions. The main advantage of the shock tube technique used in this paper is that the shock wave front is plane, thus yielding a uniformly distributed pressure pulse on plates which can be measured easily by only one pressure transducer integrated in the specimen mounting ring. This allows for precise modelling of the transient loading conditions during simulations of the experimentally observed plate response.

The topics addressed in this presentation include (a) the correlation of experimental and simulated transient inelastic response and permanent deflections, (b) the effect of the transverse shear stress distribution (first-order versus third-order transverse shear deformation hypothesis) on the local evolution of the material behaviour and on the global dynamic response, (c) the evolution of deflections, stresses, and plastic zones under blast loading conditions. A good agreement between the computational and experimental results for the elastic-viscoplastic plate response is observed. It turns out that the best correlation is obtained by means of the third-order transverse shear deformation theory. Special emphasis is focused on the evolution of bending moments and equivalent plastic strain rates, where considerable differences can be observed when first- or third-order transverse shear deformation theory, respectively, is applied.

2 Experimental setup

For the experiments shock tubes (diameter 458 mm and 108 mm, respectively) are used in order to subject clamped circular aluminium and steel plates (thickness 2 mm, diameter 553 mm and 138 mm, respectively) to impulsive loadings, see [6]. The high pressure chamber (HPC) is separated from the low pressure chamber (LPC) by a diaphragm. The pressure in the HPC is increased until the diaphragm is ruptured. A shock wave travels through the LPC and impinges on the plate specimen clamped at the end of the tube between two ring flanges leading to a plane high-pressure pulse. The time histories of the centre deflection and of the pressure acting on the plate are recorded by a capacitor and a piezoelectric pressure sensor, respectively.

3 Structural modelling

Consider a plate or shell with volume V, total bounding surface B, mid-surface M and thickness h subjected to blast loading conditions. The principle of virtual displacements of 3-D elastokinetics reads

$$\int_V \left\{ s^{ij}\,\delta E_{ij} - \rho\left(B^i - A^i\right)\delta V_i \right\} dV - \int_B \left({}^{*}s^i + D^i \right)\delta V_i\, dB = 0\,, \qquad (1)$$

where V_i denote the covariant components of the displacement vector in an arbitrary point of the body, s^{ij} and E_{ij} are the components of the second Piola-Kirchhoff stress and Green strain tensor, ρ is the density of the undeformed body, and B^i, A^i are components of the body force and acceleration vectors per unit volume of the undeformed body, respectively. Furthermore, ${}^{*}s^i$ and D^i are components of the prescribed external stress vector and of the damping force vector, respectively, per unit area of the undeformed shell boundary surface B. For viscous damping the damping force vector components may be written as $D^i = -D^{ij}\,\dot{V}_j$ with D^{ij} denoting the components of the damping tensor. Here and throughout the paper the Einstein summation convention is used with Latin indices ranging from 1 to 3 and Greek indices ranging from 1 to 2. The displacement vector $\underset{\sim}{V}$ of any point in the shell space can be referred to the contravariant base vector triad of the reference surface as

$$\underset{\sim}{V} = v_\alpha\, \underset{\sim}{a}^\alpha + v_3\, \underset{\sim}{n}\,. \qquad (2)$$

Refined non-linear first-order transverse shear deformation shell theories (FOSD), which account for small strains (of $O(\eta)$, $\eta \ll 1$), small rotations (of $O(\eta)$) about the normal, and moderate rotations (of $O(\eta^{1/2})$) of the normal, have been derived by Schmidt and Reddy [1]. They are based on the hypothesis of a linearly varying displacement field through the thickness, i.e.

$$v_\alpha = \overset{(0)}{v}_\alpha + \theta^3 \overset{(1)}{v}_\alpha, \qquad v_3 = \overset{(0)}{v}_3, \tag{3}$$

where $\overset{(0)}{v}_\alpha$ and $\overset{(0)}{v}_3$ denote the tangential and normal displacement components at the reference surface, $\overset{(1)}{v}_\alpha$ stands for the rotations of the normal and θ^3 is the distance from the reference surface. Based on the above assumptions the Green strain tensor components can be approximated, to within a negligible small relative error margin (of $O(\eta)$), by the series expansions

$$E_{\alpha\beta} = \overset{(0)}{E}_{\alpha\beta} + \theta^3 \overset{(1)}{E}_{\alpha\beta} + \left(\theta^3\right)^2 \overset{(2)}{E}_{\alpha\beta}, \quad E_{\alpha 3} = \overset{(0)}{E}_{\alpha 3} + \theta^3 \overset{(1)}{E}_{\alpha 3}, \quad E_{33} = \overset{(0)}{E}_{33}. \tag{4}$$

The non-linear moderate rotation FOSD shell strain-displacement relations of Schmidt and Reddy [1] specified for the kinematical hypothesis (3) with all transverse normal strain effects neglected yield the following five-parameter variant of the 2-D strain measures

$$\begin{aligned}
\overset{(0)}{E}_{\alpha\beta} &= \overset{(0)}{\theta}_{\alpha\beta} + \frac{1}{2} \overset{(0)}{\varphi}_\alpha \overset{(0)}{\varphi}_\beta, \\
\overset{(1)}{E}_{\alpha\beta} &= \frac{1}{2}\left(\overset{(1)}{v}_{\alpha|\beta} + \overset{(1)}{v}_{\beta|\alpha} \right) - \frac{1}{2}\left(b_\alpha^\lambda \overset{(0)}{\varphi}_{\lambda\beta} + b_\beta^\lambda \overset{(0)}{\varphi}_{\lambda\alpha} \right) + \frac{1}{2}\left(\overset{(0)}{\varphi}_\alpha b_\beta^\lambda \overset{(1)}{v}_\lambda + \overset{(0)}{\varphi}_\beta b_\alpha^\lambda \overset{(1)}{v}_\lambda \right), \\
\overset{(2)}{E}_{\alpha\beta} &= -\frac{1}{2}\left(b_\alpha^\lambda \overset{(1)}{v}_{\lambda|\beta} + b_\beta^\lambda \overset{(1)}{v}_{\lambda|\alpha} \right) + \frac{1}{2} b_\alpha^\lambda b_\beta^\kappa \overset{(1)}{v}_\lambda \overset{(1)}{v}_\kappa, \\
\overset{(0)}{E}_{\alpha 3} &= \frac{1}{2}\left(\overset{(0)}{\varphi}_\alpha + \overset{(1)}{v}_\alpha \right) + \frac{1}{2} \overset{(1)}{v}{}^\lambda \overset{(0)}{\varphi}_{\lambda\alpha}, \qquad \overset{(1)}{E}_{\alpha 3} = \frac{1}{2} \overset{(1)}{v}{}^\lambda \overset{(1)}{v}_{\lambda|\alpha}, \qquad \overset{(0)}{E}_{33} = 0.
\end{aligned} \tag{5}$$

Here the following abbreviations have been used:

$$\overset{(0)}{\theta}_{\alpha\beta} = \frac{1}{2}\left(\overset{(0)}{v}_{\alpha|\beta} + \overset{(0)}{v}_{\beta|\alpha} \right) - b_{\alpha\beta} \overset{(0)}{v}_3, \quad \overset{(0)}{\varphi}_{\alpha\beta} = \overset{(0)}{v}_{\alpha|\beta} - b_{\alpha\beta} \overset{(0)}{v}_3, \quad \overset{(0)}{\varphi}_\alpha = \overset{(0)}{v}_{3,\alpha} + b_\alpha^\lambda \overset{(0)}{v}_\lambda . \tag{6}$$

The notations $(.)_{,\alpha}$ and $(.)_{|\alpha}$ stand for partial and covariant differentiation on the reference surface and $b_{\alpha\beta}$, b_α^λ denote the covariant and mixed components of the curvature tensor. For zero curvature the equations above yield the von Kármán-type non-linear FOSD plate theory.

The results referring to the third-order shear deformation theory (TOSD) are obtained with a finite plate element developed by Vu *et al.* [3]. The finite element is based on the von Kármán-type non-linearity, too. The TOSD hypothesis for the through-thickness variation of the tangential and normal displacement components reads

$$v_\alpha = \overset{0}{v}_\alpha + \theta^3 \overset{1}{v}_\alpha + \left(\theta^3\right)^2 \overset{2}{v}_\alpha + \left(\theta^3\right)^3 \overset{3}{v}_\alpha, \quad v_3 = \overset{0}{v}_3, \tag{7}$$

where the constant, linear, quadratic and cubic terms are denoted by 0, 1, 2 and 3, respectively. The above assumption for the tangential and normal displacement field yields 9 degrees of freedom per nodal point. In order to reduce the degrees of freedom from 9 to 5, it is assumed that no shear traction is applied on the bottom (θ^3=-$h/2$) and on the top surface (θ^3=$h/2$) of the plate. For plates of isotropic material or laminated of orthotropic layers this is equivalent to

$$E_{\alpha 3}\left(\theta^3 = \frac{h}{2}\right) = 0, \quad E_{\alpha 3}\left(\theta^3 = -\frac{h}{2}\right) = 0\,, \tag{8}$$

which yields

$$\overset{2}{v}_\alpha = 0, \quad \overset{3}{v}_\alpha = -\frac{4}{3h^2}\left(\overset{1}{v}_\alpha + \overset{0}{v}_{3,\alpha}\right). \tag{9}$$

Taking all assumptions into consideration, the Green-Lagrange strain components for the von Kármán-type non-linearity TOSD plate element can be expressed as:

$$E_{\alpha\beta} = \overset{0}{E}_{\alpha\beta} + \Theta^3\,\overset{1}{E}_{\alpha\beta} + \left(\Theta^3\right)^3\overset{3}{E}_{\alpha\beta}, \quad E_{\alpha 3} = \overset{0}{E}_{\alpha 3} + \left(\Theta^3\right)^2\overset{2}{E}_{\alpha 3}, \quad E_{33} = 0\,, \tag{10}$$

where

$$\begin{aligned} &\overset{0}{E}_{\alpha\beta} = \frac{1}{2}\left(\overset{0}{v}_{\alpha|\beta} + \overset{0}{v}_{\beta|\alpha} + \underline{\overset{0}{v}_{3,\alpha}\,\overset{0}{v}_{3,\beta}}\right), \quad \overset{1}{E}_{\alpha\beta} = \frac{1}{2}\left(\overset{1}{v}_{\alpha|\beta} + \overset{1}{v}_{\beta|\alpha}\right), \\ &\overset{3}{E}_{\alpha\beta} = -\frac{2}{3h^2}\left(\overset{1}{v}_{\alpha|\beta} + \overset{1}{v}_{\beta|\alpha} + 2\,\overset{0}{v}_{3|\alpha\beta}\right), \end{aligned} \tag{11}$$

and

$$\overset{0}{E}_{\alpha 3} = \frac{1}{2}\left(\overset{1}{v}_\alpha + \overset{0}{v}_{3,\alpha}\right), \quad \overset{2}{E}_{\alpha 3} = -\frac{2}{h^2}\left(\overset{1}{v}_\alpha + \overset{0}{v}_{3,\alpha}\right). \tag{12}$$

The cubic term of the tangential strains is neglected in the numerical applications. This can be justified for thin-walled structures.

The transformation of the principle of virtual displacements of 3-D elastokinetics Eq. (1) into a 2-D formulation yields

$$\begin{aligned} &\int_M \left\{ \sum_{n=0}^{2} \overset{(n)}{L}{}^{\alpha\beta}\,\delta\overset{(n)}{E}_{\alpha\beta} + 2\sum_{n=0}^{1} \overset{(n)}{L}{}^{\alpha 3}\,\delta\overset{(n)}{E}_{\alpha 3} - \sum_{n=0}^{1}\left[\left(\overset{(n)}{B}{}^{\alpha} - \overset{(n)}{I}{}^{\alpha} + \overset{(n)}{D}{}^{\alpha} + \overset{(n)}{p}{}^{\alpha}\right)\delta\overset{(n)}{v}_{\alpha}\right] - \left(\overset{(0)}{B}{}^{3} - \overset{(0)}{I}{}^{3} + \overset{(0)}{D}{}^{3} + \overset{(0)}{p}{}^{3}\right)\delta\overset{(0)}{v}_{3}\right\} dM \\ &- \int_C \left\{ \sum_{n=0}^{1}\left[{}^{*}\overset{(n)}{L}{}^{\alpha\beta}\,\upsilon_\alpha\,\upsilon_\beta\,\delta\overset{(n)}{v}_{\nu} + {}^{*}\overset{(n)}{L}{}^{\alpha\beta}\,t_\alpha\,\upsilon_\beta\,\delta\overset{(n)}{v}_{t}\right] + {}^{*}\overset{(0)}{L}{}^{3\beta}\,\upsilon_\beta\,\delta\overset{(0)}{v}_{3}\right\} dC = 0\,. \end{aligned} \tag{13}$$

Here, $\overset{(n)}{L}{}^{ij}$ are the gross stress resultants, $\overset{(n)}{B}{}^{i}$, $\overset{(n)}{I}{}^{i}$, and $\overset{(n)}{D}{}^{i}$ denote the gross body, inertia and damping couples, while $\overset{(n)}{p}{}^{i}$, and ${}^{*}\overset{(n)}{L}{}^{ij}$ are the gross surface and boundary load couples, respectively, of the n-th order. They are obtained by integration of the respective physical quantities through all layers and can be found in detail in Librescu and Schmidt [11], Schmidt and Reddy [1]. Furthermore, in Eq. (13) C denotes the boundary curve of the reference surface M, υ_α and t_α are the components of the unit outward normal and tangent vector of C, while indices ν and t denote normal and tangential displacement and rotation components, respectively.

For viscoplastic transient analysis the constitutive equations of Chaboche [12] are employed. Details of the material parameter identification by uniaxial tension tests are available in Stoffel *et al.* [4]. A layered shell model is used that permits to trace the evolution of the material properties in each individual layer. Numerical solutions are obtained by using the isoparametric Lagrangian 9-node shell finite element and the central difference method for the time integration of the non-linear equations of motion (see [2, 13, 14]).

4 Results

Fig. 1 shows the evolution of the pressure and centre deflection, respectively, recorded during a typical shock tube experiment using aluminium plate specimens (thickness 2 mm, diameter 553 mm) as described in Chapter 2. One can observe a very good agreement between the results predicted by FE simulation of the elastic-viscoplastic transient response and the experimental observation. The results obtained by TOSD theory mach the experimental results closer than those obtained by FOSD theory.

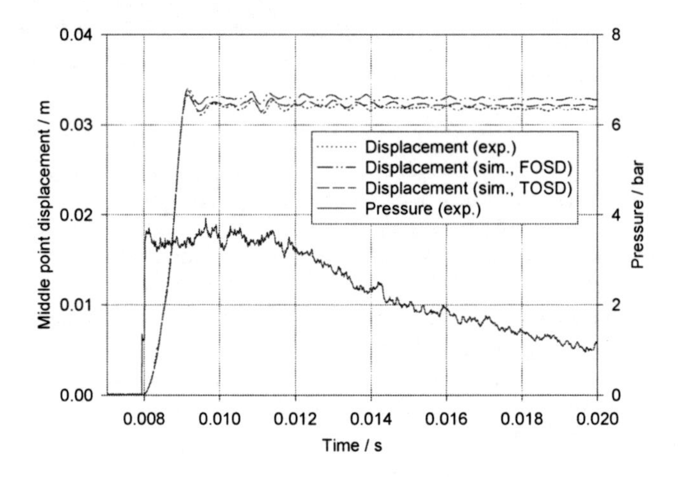

Figure 1: Time history of pressure and centre displacement.

In the above experiment the peak pressure during blast loading was 3.5 bar. Fig. 2 shows FE simulation results for the evolution of the bending moment at the plate centre and at the plate boundary.

It can be observed from Fig. 2 that immediately after the arrival of the shock wave high bending moments occur only close to the boundary. Then the bending action spreads inwards and reaches the plate centre with a delay. Note also that in the plate centre oscillations of the bending moment occur. First small oscillations are visible, than the amplitudes increase, and finally the bending moment tends to zero. This indicates that the movement of the centre during the deformation is much more complicated than it is suggested by Fig. 1. In addition to the bending moment a rapid increase of the membrane force occurs almost simultaneously in all points. This reflects the von Kármán effect and indicates that the blast loading problem turns into a geometrically non-linear one already immediately after the arrival of the shock wave.

The above observations can be explained by flexural waves that originate at the boundary and travel towards the plate centre at different speeds. In the

present case the impulsive transversal loading of the boundary, which occurs as reaction to the pressure impulse on the plate, generates such flexural waves with a wide spectrum of frequencies. According to Doyle [15] the propagation velocity of flexural waves increases with increasing frequency. Furthermore, spectral analysis of the investigated plate vibration indicates that wave amplitudes decrease with increasing frequency. Consequently, those waves with high frequency and small amplitude reach the plate centre first. This results in small amplitude bending oscillations in the plate centre as can be observed in Fig. 2. With a considerable delay the flexural waves with the highest amplitude and lowest frequency reach the plate centre at last what can be seen in Fig. 2, too. At the boundary these oscillations do not occur, because the waves originate there all at the same instant of time.

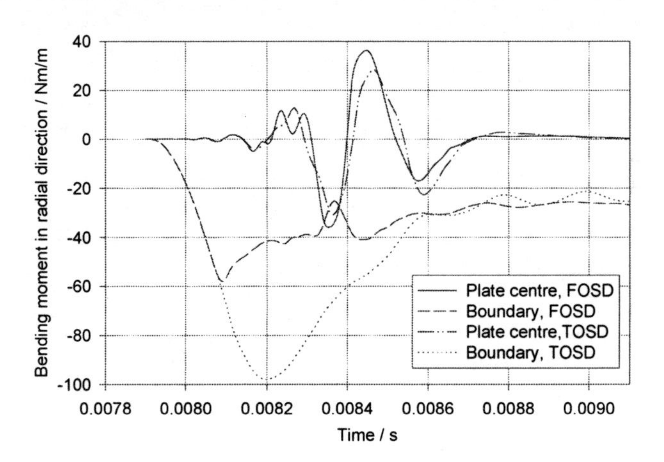

Figure 2: Evolution of the bending moment.

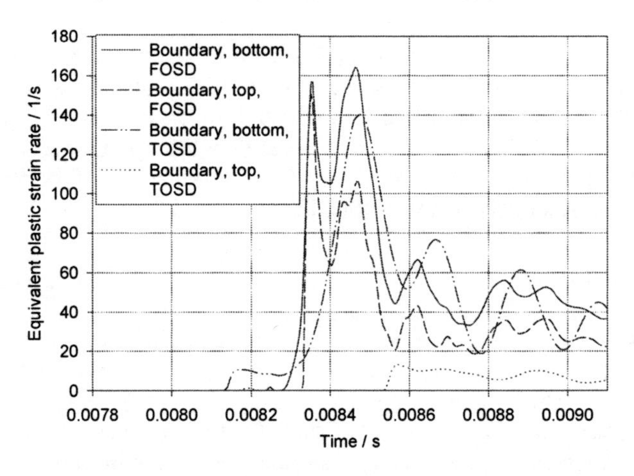

Figure 3: Evolution of the equivalent plastic strain rate at the boundary.

The above observations are also reflected in the evolution of the equivalent plastic strain rate shown in Fig. 3. At the boundary the negative moment causes stretching at the bottom of the plate. Together with the action of the tensile membrane forces and shear forces plastic zones develop first at the boundary bottom layer. It is interesting to note that for both the top and bottom layer the FOSD theory predicts higher equivalent plastic strain rates than the TOSD theory, in time intervals where the predicted bending moments are almost equal. This is due to the fact that according to the FOSD hypothesis the transverse shear stress distribution is constant through the thickness, while in the TOSD theory it is parabolic and tends to zero (see Eq. (8)) at the bounding surfaces.

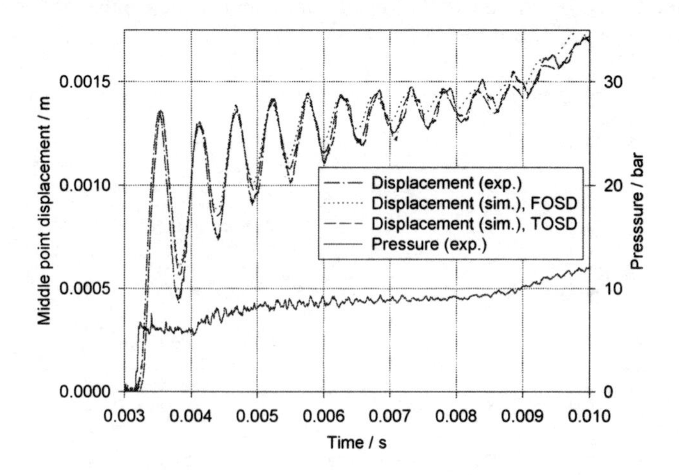

Figure 4: Time history of pressure and centre displacement.

Fig. 4 shows the evolution of the pressure and centre deflection, respectively, recorded during a typical shock tube experiment using steel plate specimens (thickness 2 mm, diameter 138 mm) as described in Chapter 2. Also here one can observe a very good agreement between the results predicted by FE simulation of the elastic-viscoplastic transient response and the experiment. Again it can be noticed that the TOSD theory performs better than the FOSD theory, especially for increasing time. Hence, also the permanent deformations of the structure are predicted more accurately by the TOSD theory.

5 Conclusions

The results presented in this paper demonstrate that simulation of blast loaded structures cannot be based on bending or non-linear membrane theories but require models which take into account the complicated membrane-bending interaction already during the impulse duration. Furthermore, the results of the present investigations show clearly that the constant transverse shear stress assumption, implicit in first-order transverse shear deformation theories, leads to local over- and underprediction, respectively, of the equivalent plastic strain rates

through the thickness, depending on the distance of the point under consideration from the mid-surface. This can affect the accuracy of the predicted transient response and permanent deflections. Therefore, in the present paper the results of simulations based on third-order transverse shear deformation theory showed a better correlation with the experimental results.

References

[1] Schmidt, R., Reddy, J.N. : A refined small strain moderate rotation theory of elastic anisotropic shells, ASME J. Appl. Mech. 55 (1988), 611-617.
[2] Palmerio, A.F., Reddy, J.N., Schmidt, R. : On a moderate rotation theory of laminated anisotropic shells, Part 1: Theory, Part 2: Finite element analysis, Int. J. Non-Linear Mech. 25 (1990), 687-714.
[3] Vu, T.D., Lentzen, S., Schmidt, R. : Geometrically nonlinear FE-analysis of piezolaminated plates based on first- and third-order shear deformation theory, Proceedings of the 8th International Conference on Mechatronics Technology, ICMT 2004, Hanoi, Vietnam, 8 – 12 November 2004, eds. Nguyen Khoa Son, Pham Thuong Cat, Pham Anh Tuan, 267-272, Vietnam National University Publisher, Hanoi 2004.
[4] Stoffel, M., Schmidt, R., Weichert, D. : Shock wave-loaded plates, Int. J. Sol. Struc. 38 (2001), 7659-7680.
[5] Stoffel, M., Schmidt, R., Weichert, D. : Simulation and experimental validation of the dynamic response of viscoplastic plates, in "Computational Methods and Experimental Measurements X", 505-514, WIT Press, Southampton 2001.
[6] Schmidt, R., Stoffel, M. : Modelling, Simulation and experimental investigation of plates subjected to blast loading conditions, in: "Computational Ballistics II", eds. V. Sanchez-Galvez, C.A. Brebbia, A.A. Motta, G.E. Anderson, 249-258, WIT Press, Southampton-Boston 2005.
[7] Idczak, W., Rymarz, Cz., Spychala, A. : Large deflection of a rigid-viscoplastic impulsively loaded circular plate, J. Tech. Phys. 21 (1980), 473-487.
[8] Idczak, W., Rymarz, Cz., Spychala, A. : Studies on shock-wave loaded, clamped circular plates, J. Tech. Phys. 22 (1981), 175-184.
[9] Renard, J., Pennetier, O. Nonlinear dynamic response of plates submitted to an explosion – Numerical and experimental study, Structural Dynamics – Eurodyn`96, 689-694, Balkema, Rotterdam (1996).
[10] Pennetier, O. : Interaction Structures-Detonations, Thesis, Université d'Orleans 1998.
[11] Librescu, L., Schmidt, R. : Refined theories of elastic anisotropic shells accounting for small strains and moderate rotations, Int. J. Non-Linear Mech. 23 (1988), 217-229.
[12] Chaboche, J.L. : Constitutive equations for cyclic plasticity and cyclic viscoplasticity, Int. J. of Plast. 5 (1989), 247-302.
[13] Klosowski, P., Schmidt, R. : Geometrically and physically nonlinear transient analysis of structures, in "Dynamical Problems in Mechanical

Systems IV", 171-182, eds. R. Bogacz, G.P. Ostermeyer, K. Popp, Polish Academy of Sciences, Warsaw 1996.
[14] Kreja, I., Schmidt, R., Reddy, J.N. : Finite elements based on a first-order shear deformation moderate rotation shell theory with application to the analysis of composite structures, Int. J. Non-Linear Mech. 32 (1997), 1123-1142.
[15] Doyle, J.F. : Wave propagation in structures, Springer Verlag, Berlin-Heidelbeg-New York, 1997.

Section 6
Structural crashworthiness

Finite element modelling of cable median barriers under vehicular impacts

J. Bi[1], H. Fang[1] & D. C. Weggel[2]
[1]*Department of Mechanical Engineering and Engineering Science, The University of North Carolina at Charlotte, USA*
[2]*Department of Civil and Environmental Engineering, The University of North Carolina at Charlotte, USA*

Abstract

Cable median barriers (CMBs) are safety devices installed on highway medians to prevent cross-median crashes. Although fatality rates decrease notably after installation of CMBs, a small percentage of vehicles are found to under-ride the CMBs and enter the opposing traffic lanes. Retrofitting the current CMBs could reduce the number of these events and further increase the reliability of these barrier systems. Owing to the high cost and restrictions of full-scale physical crash tests, detailed analysis and parametric studies of CMB designs have been recently oriented towards numerical modelling of vehicle-barrier impacts. In this study, a three-strand CMB system installed on a sloped median and impacted by a passenger vehicle is modelled and evaluated using the nonlinear finite element (FE) method. Detailed FE models of the CMB and the vehicle are presented. Various modelling issues involved in the nonlinear FE analysis, such as contact modelling and numerical instabilities, are discussed. Based on an analysis of vehicle-barrier interactions using FE simulations, a new CMB design is derived and its performance is compared to the current design.

Keywords: finite element modelling, cable median barrier, sloped median, roadside safety, cross-median collisions.

1 Introduction

Highway safety has been a public concern for decades. Over the years, researchers have developed several types of roadside safety devices including median barriers for preventing cross-median collisions. Depending on site

www.witpress.com, ISSN 1743-3509 (on-line)
doi:10.2495/SU100191

conditions, rigid, semi-rigid and flexible barrier systems are considered and installed on U.S. highways; they must conform to guidelines provided by the American Association of State Highway and Transportation Officials (AASHTO [1-3]) and are obliged to satisfy the safety requirements specified by the NCHRP Report 350 (Ross et al. [4]).

As a flexible barrier system, cable median barriers (CMBs) are normally adopted for medians with a width of 7.3 m, the minimum width to accommodate lateral cable deflections. Fatality rates decrease notably after installations of CMBs (Sposito and Johnston [5], McClanahan et al. [6], WSDOT [7], Ray et al. [8]), with an estimated reduction of 90 percent of the cross-median collisions. However, under certain site and impact conditions, a small percentage of the vehicles were found to under-ride CMBs and crash into vehicles in opposing traffic, causing fatal injury (Johnson et al. [9], MacDonald and Batiste [10]). These incidents require a more comprehensive evaluation of cable-vehicle interactions to increase the reliability of CMBs for small-vehicle impacts.

The design and evaluation of roadside safety devices were largely based on engineering experience and standardized physical tests (Bligh and Mak [11]). Given the high cost and restrictions of full-scale physical crash tests, attention was recently directed to numerical simulations of vehicle-barrier impacts that were made possible by the marked improvement in computer hardware and parallel computing (Ray [12], Ren and Vesenjak [13], Marzougui et al. [14], Marzougui and Opiela [15], Ulker and Rahman [16]).

In this paper, the performance of an in-service three-strand CMB under vehicular impacts is evaluated using numerical simulations. The nonlinear finite element (FE) models are created to simulate a small passenger vehicle impacting a CMB installed on a sloped median. Several modelling issues involved in the nonlinear FE analysis, such as contact modelling and numerical instabilities, are discussed. Based on an analysis of vehicle-barrier interactions using FE simulations, a new CMB design is derived and evaluated. The new design is shown to have better performance than the current design in preventing vehicles from under-riding the CMB.

2 Problem description

Vehicle under-riding was found to dominate reported cases of cross-median crashes involving CMBs. In CMB under-riding, the cable heights in relation to the vehicle's bumper height are critical at the instant of impact. The interaction with cables affects the dynamic response of the vehicle after impact; this is the primary phenomenon studied in this paper. The North Carolina three-strand low-tension CMB is assessed in this study. The heights of the three cables – top, middle and bottom as shown in Fig. 1 – are defined as the vertical distance from the cable to the middle point of the post on the ground line.

The suspension of a vehicle is compressed as it crosses the bottom of the sloped median (see Fig. 2), resulting in a lowered bumper relative to the ground. Suspension compression is affected by the vehicle's speed and impact angle. To address the issue of vehicle under-riding, some of the cables in the current design

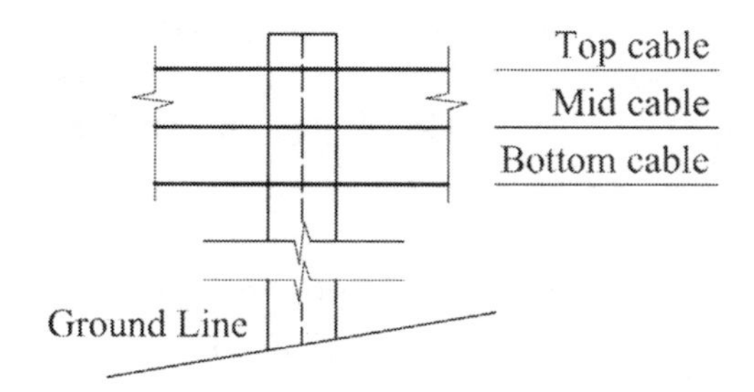

Figure 1: The three cables of a low-tension CMB.

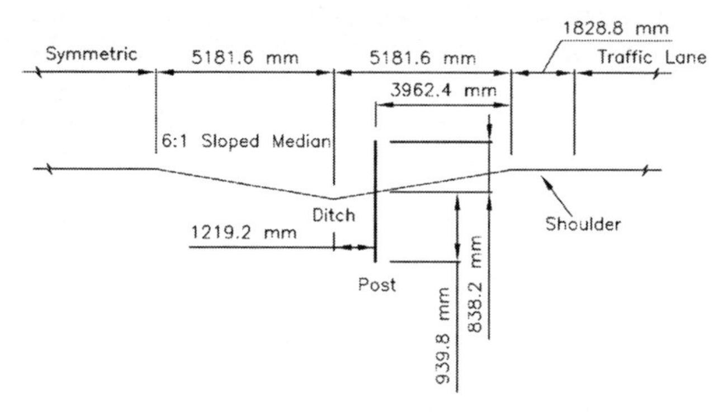

Figure 2: A North Carolina CMB installed on a 6:1 sloped median.

may need to be lowered to improve vehicle-cable interactions. Table 1 gives several design options that have different combinations of cable heights. The height of the top cable remains unchanged so that large vehicles can also be captured. For the bottom and middle cables, different height combinations are investigated to identify the design with better containment of small vehicles.

Table 1: CMB design options.

Design Option	Cable Height (mm)		
	Bottom Cable	**Middle Cable**	**Top Cable**
Current Design	520.7	641.4	762.0
Design 1	330.2	520.7	
Design 2	355.6	558.8	
Design 3	381.0	520.7	
Design 4	381.0	571.5	
Design 5	431.8	596.9	

Table 2 provides the impact conditions of the FE simulations used for all design options. The CMB is located 1.22 m from the ditch centerline, and impacts from both front-side (the side without ditch) and back-side (the side with

Table 2: Impact conditions for the FE simulations.

Impact Side	Impact Angle	Impact Speed (km/hr)			
Front-side	$20^\circ, 30^\circ, 40^\circ$	88.5	104.6	112.7	120.7
Back-side					

ditch) are analyzed. For each side, a total of 12 conditions are simulated, three impact angles over four impact speeds.

The main focus of this study is to evaluate the vehicle's response after impacting the CMB to determine the best cable heights for a retrofit design. Vehicle responses are placed into one of four categories, as shown in Fig. 3, based on the effectiveness of the CMB in redirecting the vehicle. In Case A, the vehicle is redirected within the sloped median and its lateral velocity reaches zero. Cases B and C are similar to Case A, except that in Case B the vehicle is redirected on the shoulder and in Case C the vehicle is redirected in the opposing traffic lane. Case D refers to the scenario where the vehicle enters the opposing traffic lane and has a non-zero lateral velocity (i.e. the vehicle is not redirected). The CMB is considered effective for Cases A and B and to have failed for Cases C and D.

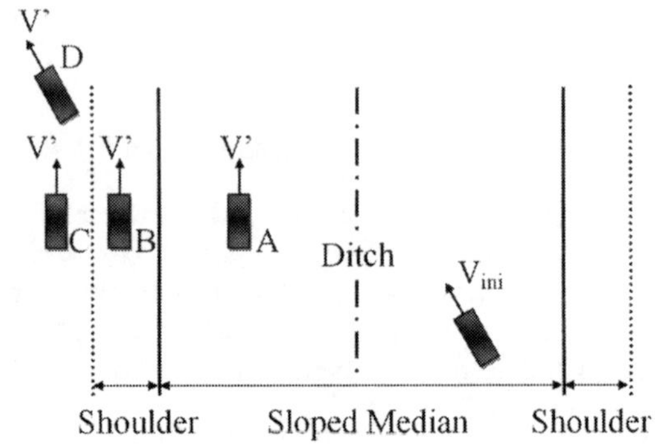

Figure 3: Vehicular responses after impact. A. redirection within sloped median; B. redirection on shoulder; C. redirection in opposing traffic lane; and D. penetration.

3 Finite element modelling

The FE models for the simulations in this paper include a vehicle, a CMB, and a sloped median. The CMB model was obtained from the National Crash Analysis Centre (NCAC) and modified according to NCDOT design specifications (NCDOT [17]). The vehicle model was also obtained from NCAC and modified to correct a number of modelling issues. These models were created for the nonlinear FE code, LS-DYNA (LSTC [18]), which is specifically formulated for dynamic transient analysis. In all simulations, the median slope was 6:1

(horizontal:vertical) and the vehicle was assumed to be fully landed on the sloped median.

3.1 Vehicle model

The vehicle model contains a total of 336 parts, 283859 nodes and 270768 elements (2852 solids, 122 beams, 267786 shells and 8 other elements). 339 constitutive (material) models were used, including the piecewise linear plasticity model defined for most steel components, the elastic spring model for the shock absorber, the crushable foam model for the bumper, and the honeycomb model for the radiator core. Constraints within the vehicle are mostly modelled by rigid links and spot welds. Hourglass control was used on various components that could potentially experience large deformations.

Some shell elements were found to be disconnected and caused instabilities that terminated the simulations. Some shell elements were found to have small Jacobians (less than 0.6) and were unstable under large deformations. Constraints between tires and rims were missing and resulted in unrealistic tilting of the tires. All of these issues, along with a number of initial penetrations, were corrected to ensure simulation stability.

3.2 CMB model

A typical CMB section consists of cables, steel posts, hook-bolts, and reinforced concrete anchor blocks. The post and cable models are shown in Fig. 4. In the NCDOT design specifications (NCDOT [17]), a CMB section has approximately 120-m of effective length (between the third posts from both ends). At a post spacing of 4.88 m, 28 posts are required for each section.

The CMB model of this study was taken from a three-strand low-tension cable system used by WSDOT. The WSDOT model was validated with a full-scale crash test on a flat terrain; it was modified according to NCDOT design specifications (NCDOT [17]) and placed on a 6:1 sloped median in this study.

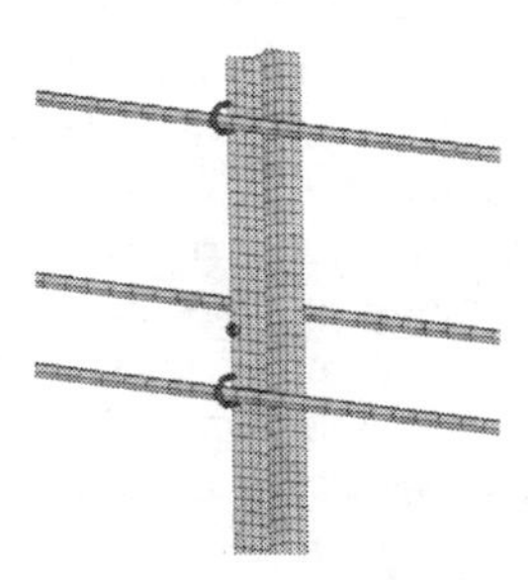

Figure 4: Modified post and cable model.

The modified CMB model includes a total of 555 parts, 148978 nodes and 143643 elements (580 solids, 13085 beams, 129880 shells and 98 other elements). The posts contain 91168 nodes and 87444 elements. The cables contain 48122 nodes and 46765 elements. The anchor blocks contain 9678 nodes and 9430 elements. The median contains 10 nodes and 4 elements. Eighteen material models were used including the piecewise linear plasticity, elastic, elastic spring, rigid, and null material models.

The cables were modelled with beam elements surrounded by six shells rigidly linked to both ends of each beam element. The six shell elements form a hexagonal cross-section that approximated the circular cross-sections of the cables. These shells were used only for contact purposes and were assigned with null materials so that they would not contribute to the mass or stiffness of the cables. The hook-bolts were modelled in the same way as the cables. However, the material properties of the beam elements of the hook-bolts were different from those in the cables. The compacted soil around the posts was modelled with solid elements in the original CMB model. Given the small deformations of the underground portion of the posts as observed in test simulations, the soil elements were removed and the posts were fixed below the ground surface; this approach also significantly reduced the computational cost.

The median (sloped region and both shoulders) was simply modelled using four shell elements. All degrees of freedom of all nodes were fixed to create rigid ground surfaces.

3.3 Contact modelling

Vehicle-cable interactions caused severe deformations and a large number of edge contacts that were difficult to handle in the FE simulations. Undetected contacts could cause unrealistic penetrations and result in numerical instabilities. The penalty method in LS-DYNA was used for contact force calculation in which the contact stiffness is determined by the material properties and size of the contacting surface. LS-DYNA supports several contact checking algorithms including nodes-to-surface, surface-to-surface and single-surface contacts. For nodes-to-surface contact, a slave and a master segment need to be defined. Each slave node is checked for penetration through the master segment. With detected penetrations, contact forces are calculated and applied to separate the two parts. For surface-to-surface contact, penetrations are checked both ways between the slave and master surfaces. For single-surface contact, only the slave segments are defined and penetration is checked within each and among all segments. Single-surface contacts are commonly used in vehicle crash analysis due to the large number of potential contacts among components.

The simulations in this study involved contacts between cables and a large number of components of the vehicle that had complex geometries and could suffer large deformations. To handle this complicated situation, the automatic–general- interior contact algorithm in LS-DYNA was adopted to account for contacts, including both internal and external edges of shell elements.

The connections between the post flange and hook-bolts were originally modelled by a contact tie between the two. Since failure initiated from plastic yielding of either the hook-bolt or the post, the connection between them never failed during an impact. Therefore, these contact ties were replaced with rigid-link connections between hook-bolts and posts. Contacts between cables and posts were handled by the nodes-to-surface algorithm, which was also used for the contacts between the tires of the vehicle and the median surface.

4 Simulation results and discussion

The crash simulations were performed using LS-DYNA 971 running on eight nodes in parallel on a Linux cluster. The simulation results showed that the deformations of the vehicle during an impact were mostly located at the left frontal corner of the vehicle, including the bumper cover, fender, hood and several parts under the hood. Upon impacting the fender, the cables could fold the fender and engage with the internal parts. The cables could also slide up or slide down the vehicle without engagement, depending on the points of impact and the geometric features of the bumper, fender and hood. The simulation results confirmed the sensitivity of cable-vehicle engagement to impact location. It should be noted that some cables initially engaged with the vehicle for a short duration but eventually slid up or down the vehicle. In this case, these cables contributed to the retention of the vehicle but were less effective than those that engaged with the vehicle throughout the entire impact duration.

Figure 5 shows a typical scenario of cable engagements during impact. The top cable slides up the vehicle along the outer frame with no engagement. The middle cable deforms the fender inward and engages with parts in the engine compartment. The bottom cable engages with the bumper cover and the tire for a short duration and then slides under the vehicle. Since the kinetic energy of the impacting vehicle is generally absorbed by the deflections in cables and the resulting plastic deformations of hook-bolts and posts, more cable engagements would utilize more of the CMB's capacity to redirect the vehicle.

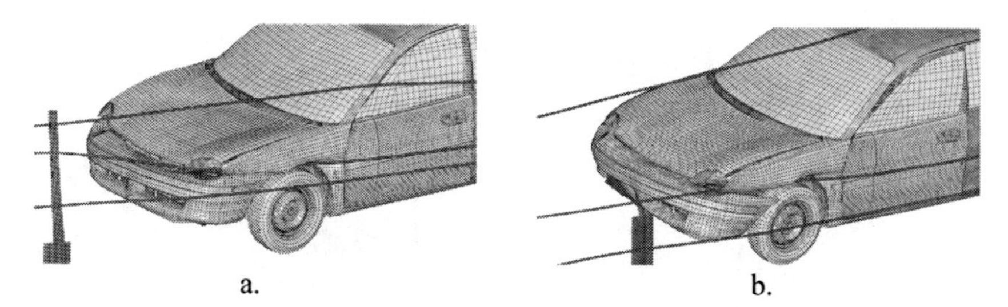

Figure 5: Typical cable-vehicle engagement. a. t=0.15 sec; and b. t=0.42 sec.

The current design and the five retrofit designs were run using the impact conditions given in Table 2. The current CMB design performed well in front-side impacts; the vehicle was redirected within the sloped median under all impact conditions. In the 20° impacts, only the bottom cable engaged with the vehicle. In the 30° and 40° impacts, both the middle and bottom cables engaged with the vehicle providing more retention than that of the 20° impacts. Figure 6 shows four snap shots of the front-side impact at 40° and 120.7 km/hr. It can be seen that the front part of the vehicle (e.g., the hood and bumper cover) underwent severe deformations due to cable engagement.

For back-side impacts, there was 2.44 m less median distance to redirect the vehicle. In all back-side impacts, only the bottom cable engaged with the vehicle due to the low vehicle profile from suspension compression. Consequently, the CMB failed to safely redirect the vehicle for 30° and 40° impacts. Moreover, yielding of the hook-bolts allows the cable to "unzip" from the post, further decreasing the stiffness of the system under these impact scenarios. Figure 7

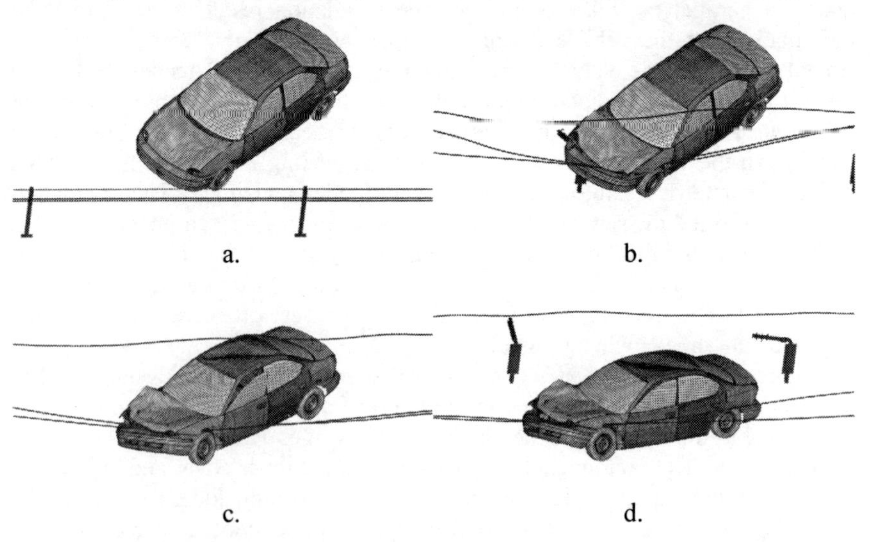

Figure 6: The current design under front-side impact at 40° and 120.7 km/hr. a. t = 0.00 sec; b. t = 0.10 sec; c. t = 0.20 sec; and d. t = 0.30 sec.

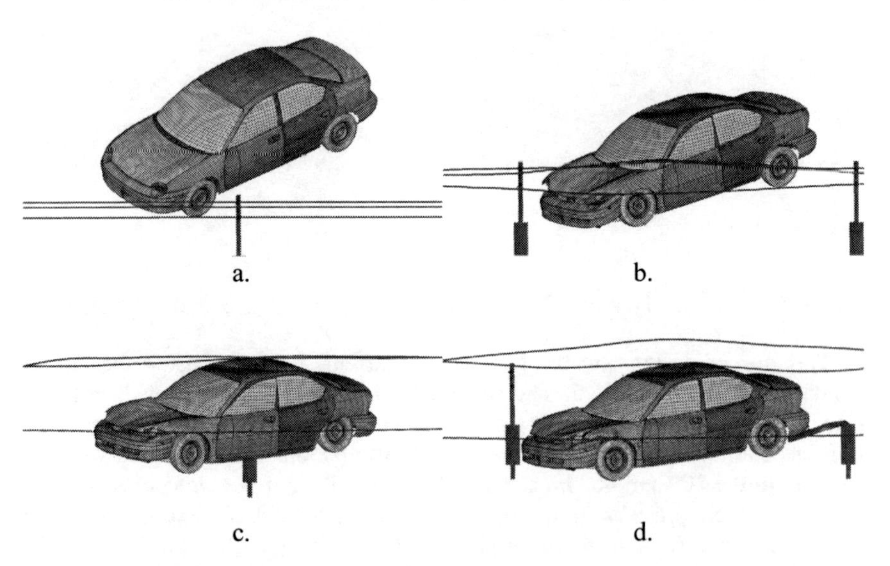

Figure 7: The current design under back-side impact at 40° and 120.7 km/hr. a. t = 0.00 sec; b. t = 0.10 sec; c. t = 0.20 sec; and d. t = 0.30 sec.

shows four snap shots of the back-side impact at 40° and 120.7 km/hr. Severe deformations of the hood, fender, bumper cover, and the roof were observed. Based on the above analysis, a successful retrofit design should aim at increasing the system performance under back-side impacts, while maintaining good performance under front-side impacts.

Retrofit design 1 performed slightly better for back-side impacts. However, for front-side, 40° impacts, the system became ineffective since the bottom cable did not engage with the vehicle. For the same reason, retrofit design 3 under 40°, front-side impacts performed worse than the current design. Since the bottom cable always slid under the vehicle in front-side impacts for retrofit designs 1 and 3, it was concluded that a cable height lower than 381 mm would be too low for effective engagement with the vehicle (for front-side impacts). It was further observed that the top cable (at 762 mm) always slid over the vehicle, leaving only the middle cable to engage with the vehicle. However, as seen from simulation results of design 1, engagement with the middle cable alone was not sufficient to redirect the vehicle for 30° and 40° back-side impacts. Therefore, retrofit designs 1 through 4 were considered ineffective and thus abandoned. Despite this, simulation results confirmed the strong correlation between cable heights and vehicle-cable engagement.

An analysis of the simulation results showed that the cables were required to have a minimum height of 431.8 mm for front-side impacts. Cables lower than this height would be ineffective in engaging with small passenger vehicles. Cables at a height of 520.7 mm above grade were found to perform well for both front-side and backside impacts. Effective cable heights for backside impacts ranged from 431.8 to 571.5 mm for 20° impacts, 355.6 to 520.7 mm for 30° impacts, and 330.2 to 520.7 mm for 40° impacts. Therefore, cable heights of 431.8 to 520.7 mm were considered the most effective for all the back-side impacts and were further investigated for the improved design in this study.

The cable-mounting options on the posts affect the resistance of the CMB. This is because more resistance can be provided by pushing the cable *into* the post and yielding it, than by just yielding a hook-bolt without yielding the post. It was therefore proposed in the improved design that the current cable installation sides be reversed (see Fig. 8). Once the bottom cable engaged with the vehicle impacting from the back-side, it would need to yield the post before being released and thus would provide more resistance to back-side impacts.

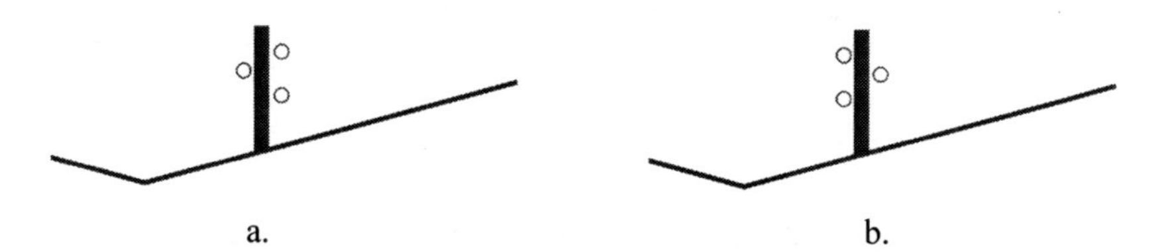

Figure 8: Cable mounting options. a. current designs; and b. improved design.

Based on the correlation of cable heights with cable-vehicle interactions, an improved design was suggested by mounting the three cables as shown in Fig. 8b at 431.8, 520.7, and 762.0 mm above grade. The simulation results of the improved design showed that the vehicle was successfully redirected for all back-side and front-side impacts. The redirection for back-side impacts was improved remarkably with full engagement of both the bottom and middle cables in all scenarios. Figure 9 shows a plan view of several snap shots of the simulation results for the current and improved designs under back-side impacts at 40° and 120.7 km/hr. The current design showed no sign of redirection through 0.4 sec after the impact; the vehicle's travel direction was nearly unchanged. In the improved design, the vehicle's travel direction changed significantly at 0.4 sec, and the vehicle was completely redirected by 0.6 sec.

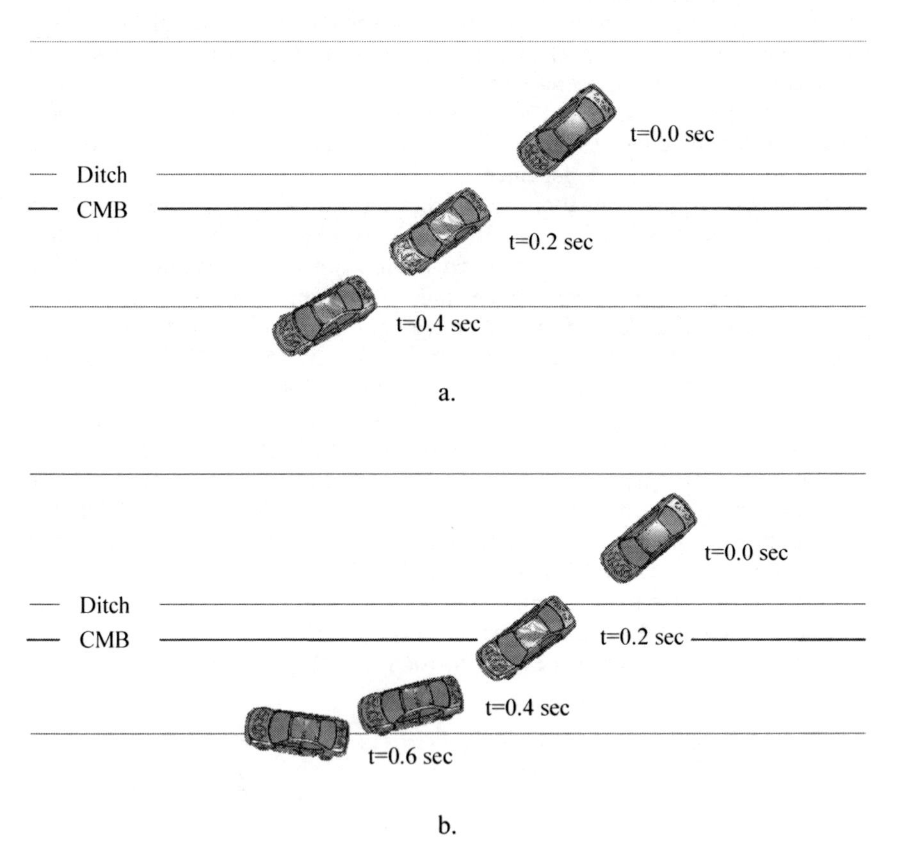

Figure 9: Vehicle redirection under back-side impacts at 40° and 120.7 km/hr. a. current design; and b. improved design.

5 Concluding remarks

In this paper finite element simulations were performed to evaluate the safety performance of a CMB installed on a 6:1 sloped median and impacted by a small passenger vehicle. The vehicle model was obtained from NCAC and was modified to accommodate sloped median crash simulations. The WSDOT CMB model was obtained and modified according to NCDOT design specifications (NCDOT [17]). Contact modelling, boundary conditions and numerical instabilities were discussed.

The in-service CMB design and five retrofit designs were evaluated over a range of impact conditions. Simulation results showed that cable heights were critical to cable-vehicle interactions. Engagement of more cables with the vehicle produces a higher effective resistance of the barrier and reduces CMB lateral deflections. It was shown that cables lower than 431.8 mm above grade are ineffective for front-side impacts. Cables 431.8 to 520.7 mm above grade are effective for back-side impacts; top cables 762.0 mm above grade always slide over the small-sized vehicle. Based on the analysis of cable heights, the improved design is proposed with bottom and middle cable heights at 431.8 and 520.7 mm above grade and the top cable height remains unchanged for the redirection of large-sized vehicles.

The cable mounting options on the posts also affect the resistance of the CMB. This is because more resistance can be provided by pushing the cable *into* the post, and thus yielding the post, than just yielding a hook-bolt without yielding the post. The reversed cable mounting option was adopted in the improved design to maximize effective resistance of the CMB for back-side impacts. The improved design was proved capable of redirecting small vehicles under all the investigated impacting conditions.

Acknowledgement

This study was funded by the North Carolina Department of Transportation (NCDOT) under project No. 2008-10.

References

[1] AASHTO, 2002 Roadside Design Guide. American Association of State Highway and Transportation Officials, Washington, D.C., 2002

[2] AASHTO, Revised Chapter 6: Median Barriers, AASHTO Roadside Design Guide (Ballot Draft). American Association of State Highway and Transportation Officials, Washington, D.C., 2005

[3] AASHTO, Roadside Design Guide (3rd Edition) 2006, with Updated Chapter 6. American Association of State Highway and Transportation Officials, Washington, D.C., 2006

[4] Ross, H. E., Jr., Sicking, D. L., Zimmer, R. A., and Michie, J. D., Recommended Procedures for the Safety Performance Evaluation of

Highway Features. Transportation Research Board, National Research Council, Washington D.C. (NCHRP Report 350), 1993

[5] Sposito, B., and Johnston, S., Three-Cable Median Barrier Final Report. Oregon Department of Transportation, Salem, 1998

[6] McClanahan, D., Albin, R. B., and Milton, J. C., Washington State Cable Median Barrier In-Service Study. 83rd Annual Meeting of the National Transportation Research Board, Washington D.C., 2004

[7] WSDOT, I-5 Marysville Cable Median Barrier. In: WSDOT Report, Washington State Department of Transportation, Olympia, WA, 2006

[8] Ray, M.H., Silvestri, C., Conron, C.E., and Mongiardini, M., Experience with Cable Median Barriers in the United States: Design Standards, Policies, and Performance. Journal of Transportation Engineering-ASCE, 135(10), 711-720, 2009

[9] Johnson, M., Zukowski, R., Andraschko, M., Austin, T., VerGowe, J., and Meyers, D., Collision Analysis and Reconstruction Report - Incident Report # 2007-327-SE, 2007

[10] MacDonald, D. B., and Batiste, J. R., Cable Median Barrier - Reassessment and Recommendations. WSDOT Report, Washington State Department of Transportation, Olympia, WA, 2007

[11] Bligh, R. P., and Mak, K. K., Crashworthiness of Roadside Features across Vehicle Platforms. Transportation Research Record (1690), 68-77, 1999

[12] Ray, M. H., Use of Finite Element Analysis in Roadside Hardware Design. Transportation Research Board (453), 61-71, 1996

[13] Ren, Z., and Vesenjak, M., Computational and Experimental Crash Analysis of the Road Safety Barrier. Engineering Failure Analysis, 12(6), 963-973, 2005

[14] Marzougui, K., Mohan, P., Kan, C. D., and Opiela, K., Performance Evaluation of Low-Tension Three-Strand Cable Median Barriers. Transportation Research Record (2005), 34-44, 2007

[15] Marzougui, D., and Opiela, K., Evaluation of Rail Height Effects on the Safety Performance of W-Beam Barriers. In: 6th European LS-DYNA User's Conference, Gothenburg, Sweden, 2007

[16] Ulker, M. B. C., and Rahman, M. S., Traffic Barriers under Vehicular Impact: From Computer Simulation to Design Guidelines. Computer-Aided Civil and Infrastructure Engineering, 23(6), 465-480, 2008

[17] NCDOT, North Carolina Cable Guardrail Standard Drawings. http://tig.transportation.org/?siteid=57&pageid=2041, 2006

[18] LSTC, LS-DYNA Keyword User's Manual. Livermore Software Technology Corporation, Livermore, CA, Version 970, 2003

Real size experiments of a car crash against a building column

B. Ferrer[1], S. Ivorra[1], R. Irles[1] & D. Mas[2]
[1]*Departamento de ingeniería de la Construcción, Universidad de Alicante, Spain*
[2]*Instituto Universitario de Física Aplicada a las Ciencias y a las Tecnologías, Universidad de Alicante, Spain*

Abstract

Several real size car crash experiments have been carried out. In these experiments low velocity impact against a building column has been tested. The experiments have been carried out on a reinforced concrete column with a rectangular section. The test was developed in an external area; therefore, both the compression load and the boundary conditions in the column had to be applied with a special device. However, the use of a real car forced one to adopt some design conditions in this device. Furthermore, to avoid the need of a driver in the car, an inclined plane and a mechanical guiding system was used.

In this paper the device, procedures and instrumentation used in these experiments are described. Five experiments were done with this system and their results are presented here.

Keywords: real size, car crash, experiment, low velocity, impact, concrete column.

1 Motivation

The accidental actions due to the horizontal impact of a car against a building have been included in the most recent building codes, such as EUROCODE 1 [1–3], "Minimum design loads for building and other structures" [4], the "International Building Code" [5] or "DIN 1055-9: Actions on structures. Accidental actions" [6]. All of these codes use an equivalent static load (ESL) to represent the effects of an impact.

WIT Transactions on The Built Environment, Vol 113, © 2010 WIT Press
www.witpress.com, ISSN 1743-3509 (on-line)
doi:10.2495/SU100201

However, the indications given in these codes are not consistent, even between different parts of the same code. This can be checked concerning vehicle barriers and studying a particular case, i.e. according to Annex B of part 1.1 of EUROCODE 1 [1], an impact with a vehicle mass of 3000 kg and an initial velocity of 20 km/h provides an ESL of 308 kN, but according to “Minimum design loads for building and other structures” [4], the ESL is 26.7 kN.

These discrepancies are also evident for building structures. For these structures and the same impact considered before, the ESL is 166 kN according to Annex C of Part 1.7 and Annex A of Part 2.7, both from EUROCODE 1 [2, 3], but 50 kN according to Part 1.7 of the same EUROCODE 1 [2]. Therefore, the dispersion between the codes is noticeable. A deeper analysis of the related codes and its dispersion can be found in [7].

The horizontal impact against a building structure can happen in a car park inside of the building or the impact can come from a car driving along a street outside of the building. In the first case, the configuration of the car park is related with the value of the velocity in the moment of the impact. For car parks with long straight ways, the car can reach a velocity of 20 km/h or even higher, not perceiving the driver a potential risk. In the second case, with a building, in an urban area, the velocity of the car can be around 50 km/h. Fortunately, the usual driver reaction before a crash is to brake, and the actual speed at impact is much lower than the circulation speed. Therefore, in the study of the impacts consequences caused by vehicles on structures, it is worth to analyze the low-speed impacts, i.e. below 30 km/h.

Experiments typically found in the bibliography concern to high velocity crashes [8] or impacts with rigid projectile [9, 10]. In this sense, one must take into account that, during the impact, the dissipated energy by the projectile significantly affects the load transmitted to the structure. Then, the use of real cars instead of rigid projectile is essential in the study of the load transmitted to the structure and consequently in its behaviour during and after the impact.

In order to understand the crash process and to analyze its effects on the structure, a series of crash experiments have been performed with real size cars and columns. For these experiments, a column of a park car located under a street was selected. The column was compressed to simulate the load of the structure above, this load being estimated in 70 t. All the experiments were done with an impact speed of 20 km/h. During the impact, the acceleration, displacement and strain of several points in the column were registered and analyzed.

2 Experimental setup

2.1 Test frame

To apply the needed compression in the column, the frame shown in figure 1 was built. It consists on a foundation made with reinforced concrete that has 4 steel anchors and a steel upper part (figure 2). The compression in the column was

Figure 1: Test frame.

(a) (b)

Figure 2: Foundation with four anchors during its construction (a) and upper part of the test frame (b).

(a) (b)

Figure 3: Ramp with metallic guide (a) and metallic wheel to maintain the direction of the car (b).

applied by 4 bar fixed to the 4 anchors and the upper part. The anchors in the foundation were set far from the bottom of the column, to allow the car to crash against the column. An anchor plate was placed in the middle of the foundation to allow the change of the column with a simple procedure.

To provide the needed crash velocity in the cars without driver, a ramp was built. The slope and length of ramp was calculated to give a final velocity of 20 km/h starting from 0 at the top of the ramp. The ramp is 30 m long and has a slope of 8.3%. The fact that no driver is controlling the vehicle makes necessary to design a specific rail to maintain the direction. In figure 3(b) one can see how a metallic wheel is attached to the car in order to maintain it in the centre of the ramp (fig. 3).

2.2 Vehicles

Five different cars were used in the experiments. Their main characteristics are shown in table 1. The cars came from a junkyard but still operational. This feature was necessary to properly position the vehicle on the ramp.

2.3 Column

The column used in all the experiments was made with reinforced concrete. This column is 2.75 high and it has rectangular section of 25x35 cm. The concrete was a 25 MPa one and the reinforcement is shown in figure 4. In the bottom part, an anchor plate was placed to allow an easy replacement of the column in case of damage.

Once placed in its position, a compressive load of 70 t was applied in the column by means of the test frame. To do that, a tensile load of 19.62 t was applied to each bar of the test frame.

Table 1: Main characteristics of cars used in the experiment.

Test	Vehicle	Year	Mass (kg)
1	Seat Ibiza	1979	800
2	Renault 19 Chamade	1994	1085
3	Seat Málaga	1989	975
4	Ford Escort	1989	977
5	Fiat Punto	1994	830

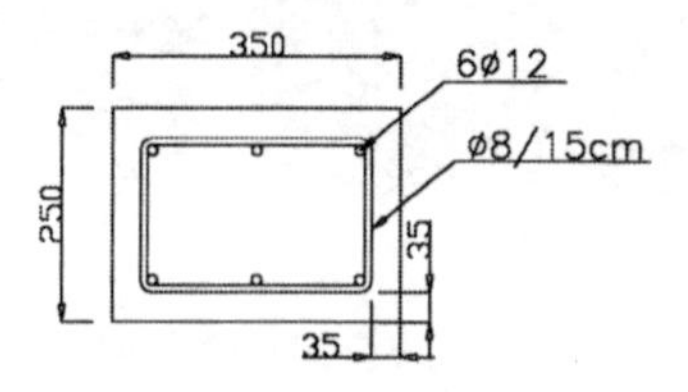

Figure 4: Column section: size (mm) and reinforcement.

2.4 Instrumentation

Acceleration, displacement and strain in the column were registered. Before the first experiments were done, the expected acceleration peak to register was unknown. For this reason four different kinds of accelerometers were used with different measurement range: shock (350B03), triaxial (356A16), structural (333B50), seismic (393A03), all from PCB Piezotronics. The measurement range of these accelerometers is ±10000g, ±50g, ±5g and ±5g, respectively; and the frequency range is from 0.4 to 10000 Hz, 0.5 to 5000 Hz, 0.5 to 3000 Hz and 0.5 to 2000 Hz, respectively. After the first experiment, the registered acceleration was lower than 5 g in all the accelerometers, so shock accelerometers were discarded.

To register the displacement, two different systems were used. A linear displacement sensor (LDS) based on potentiometer transducer (SLS 130 Penny+Giles) with a 150 mm stroke length was attached to an external structure. The frequency range of the LDS is from 0 to 2000 Hz. Additionally, a high speed camera (HSC) with 1000 frames per second and spatial resolution of 800x560 pixels was placed 2 m far from the column. Through this high speed camera, the movement of a target placed in the surface of the concrete was registered by means of a new image processing technique specially developed for these experiments [11]. With the camera, the measurement from linear displacement sensor can be checked to be sure that the external structure did not move at all during the impact.

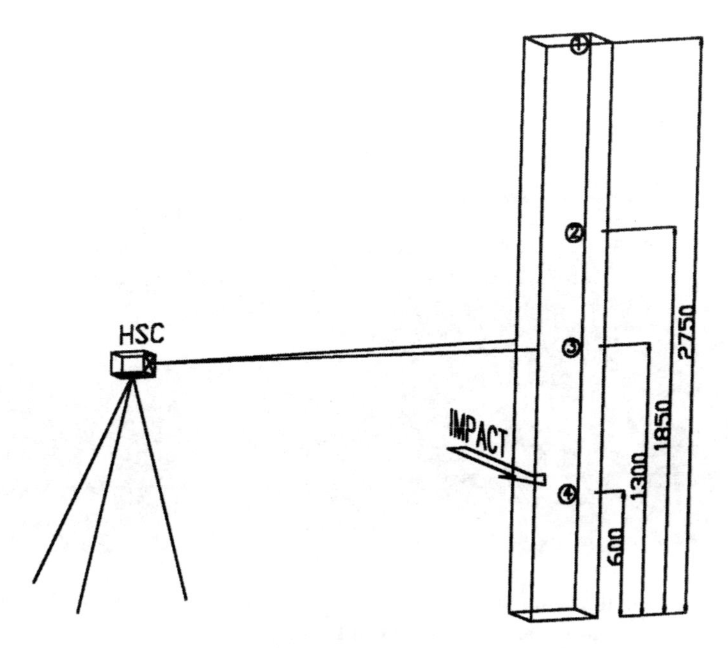

Figure 5: Location of points monitored (mm).

The velocity of the car in the moment of the impact was analyzed by means of a video camera with 30 frames per second, focusing at the end of the ramp. Several marks in the car allowed to measure the distance travelled between the last two frames before the impact and thus obtain the crash velocity.

Additional devices used are: Signal conditioner (482A22 PCB Piezotronics) with 4 channels, sensor interface for accelerometers (PCD-300A Kyowa) with 4 channels, sensor interface for linear displacement sensor and strain gauges (Spider8 HBM). Four points in the column were selected to be analyzed with these sensors. These points are located in the opposite face of the impact at 0.6 m (impact), 1.3 m (maximum displacement expected, 1.85 m and 2.75 m (upper part) high (figure 5).

3 Experimental results and analysis

3.1 Final car condition

One interesting result is the final condition of the car. The occurrence of plastic deformation in the car determines the importance of using of a real car in the experiments instead of a rigid projectile. For crash tests at high velocities large plastic deformations in the car are predictable but this is not so obvious for low velocity impacts.

All the five cars used in the experiments show noticeable plastic deformation (fig. 6). The maximum displacement of the frontal part in each car was measured (fig. 7) and the values are shown in table 2.

Figure 6: Final car condition after the experiments: Seat Ibiza, Renault 19 Chamade, Seat Málaga, Ford Escort, Fiat Punto.

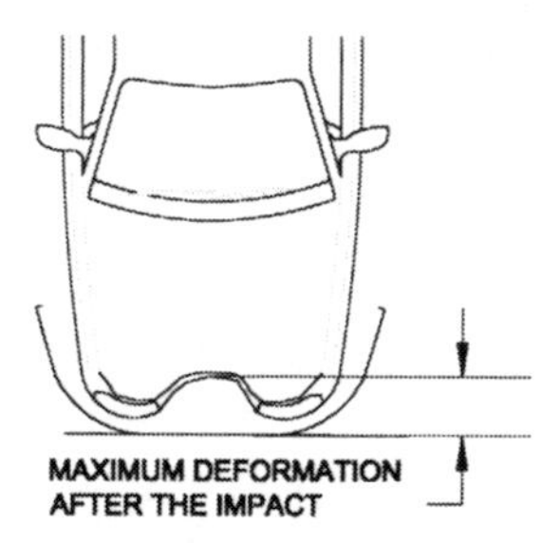

Figure 7: Measured distance in the car after the impact.

Table 2: Maximum plastic deformation in the car, according to fig. 7.

Test	Vehicle	Maximum deformation (mm)
1	Seat Ibiza	250
2	Renault 19 Chamade	400
3	Seat Málaga	300
4	Ford Escort	240
5	Fiat Punto	280

3.2 Column behaviour

An example of registered acceleration is shown in fig. 8. The graph shows the data from the accelerometer located at point 3rd in 4th impact and its Fourier transform. For a more accurate visualization of the behaviour during the impact, an enlarged view of the first time period of the impact is shown in fig. 9. From this figure it is clear that there are two different behaviours. During the first 120 ms of recording, the signal is very irregular and no pattern can be observed. However, from time 120 ms, a harmonic wave can be matched to the acceleration. The first part may correspond with the time during the car is pushing against the column while the second part corresponds to a free vibration of the column after car rebounds. This behaviour was observed in all the experiments and can be also seen with the HSC as we show in fig. 10. The duration of the irregular part coincides in both the accelerometer and the camera registers so the duration of loading in the column can be set in 120 ms approximately (figs. 9, 10).

The peak acceleration was obtained from each accelerometer, as it is shown in table 3. It is noticeable that from 3rd impact, the measured acceleration in point 1 (2.75 m from bottom) is higher than before. This can be due to the lost of load by the bars of the test frame. One must to take into account that the 3 latest experiments were done in the same day, 3 months after the loading of the bars and also 2 experiments after. In fact, during the change of the column done after the 5th impact, the load in the bars could be checked and a value of 15.1 t per bar was found, instead of 19.62 t originally.

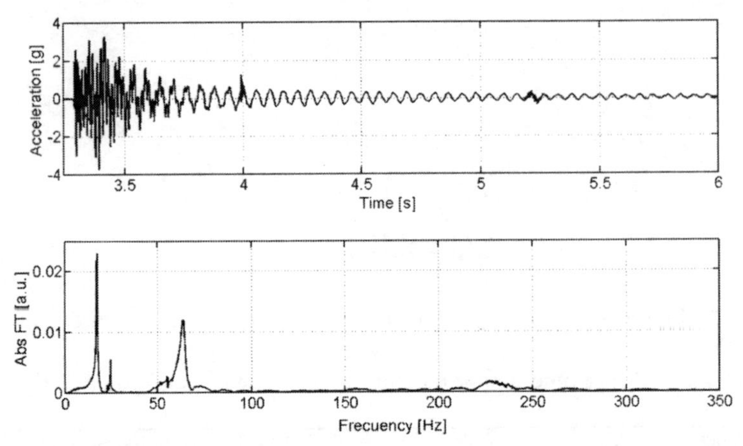

Figure 8: Recorded acceleration in the fourth experiment and Fourier transform.

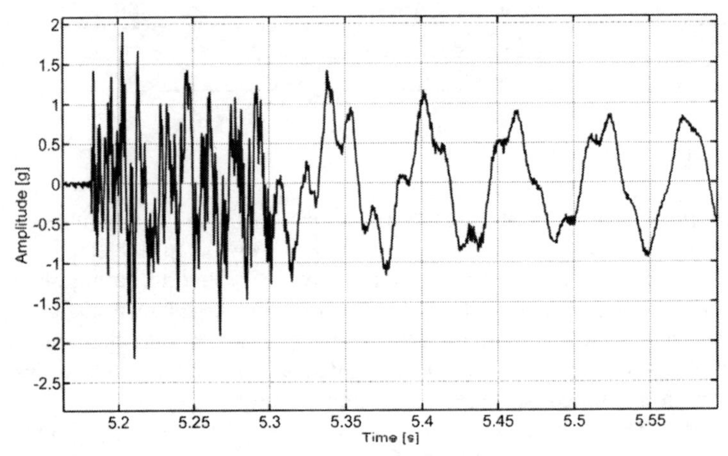

Figure 9: Detail of the recorded acceleration shown in fig. 8 – signal during the first time of the impact.

In fig. 8 one can see that the frequencies of the pillar are 17.3 Hz and 64 Hz. A small peak at 17.4 can also be observed. These frequencies agree with the ones obtained from the HSC, as we show in fig 11. This feature was observed not only for this experiment but also for all of them (table 4). The camera show also very noisy data from 100 Hz, but this is done to flickering in the image and no reliable data can be obtained in this region for the presented setup.

Although the data were also taken with LDS, the obtained frequencies do not always agree with those from the accelerometers. Thus, data from these devices were not considered and camera results were taken as real displacement data. We would also like to emphasize that the observed frequencies are decreasing with successive experiments, what agree with the fact that the bars lose load with the time and the crashes.

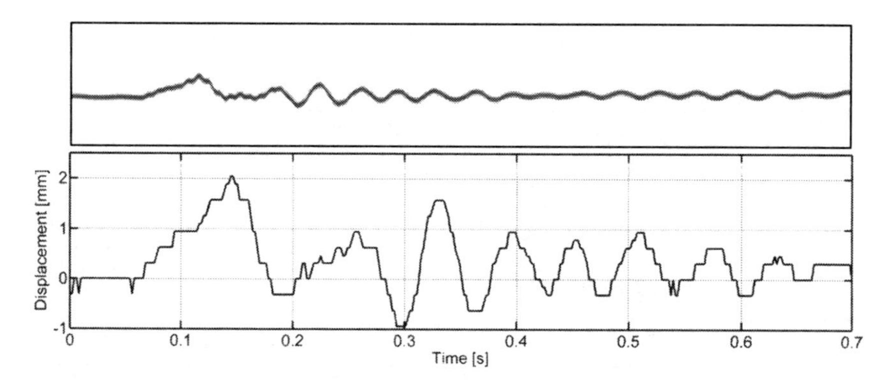

Figure 10: Results obtained from HSC in fourth experiment (top) and numerical reconstruction of this signal (bottom).

Table 3: Peak acceleration for each experiment in all the monitored points.

	Peak acceleration				
POINT	1st impact	2nd impact	3rd impact	4th impact	5th impact
1	3	1.8	5	4.3	4.4
2	3.2	3.5	2.2	3.7	1.9
3	4.5	4.5	---	---	---
4	3	3	2.8	3.8	2.9

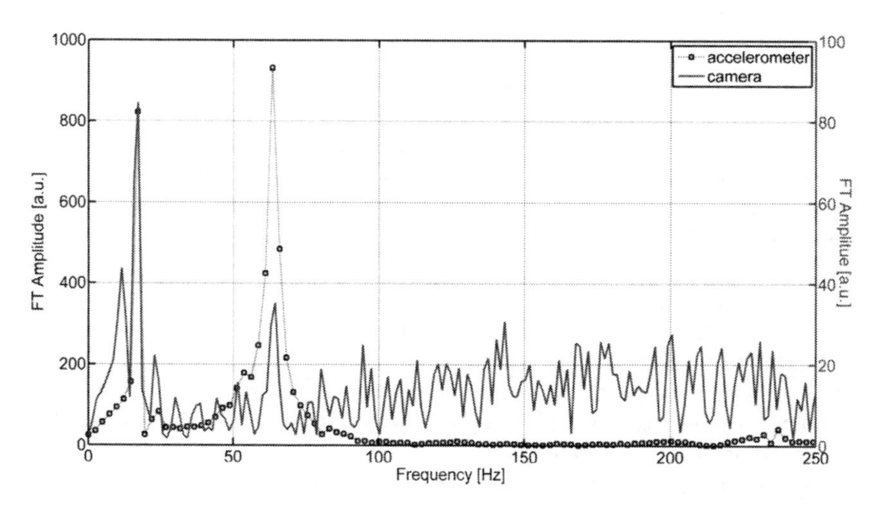

Figure 11: Fourier transform of LDS and HSC signal from the second impact.

Table 4: Frequencies in each experiment and in each point.

Point	Sensor	Frequency (Hz)				
		1st imp.	2nd imp.	3rd imp.	4th imp.	5th imp.
1	ACC	19-69.7	18.5-67.6	17.2-63.6	17.3-63.8	17.5-64.4
	LDS	19.04	18.5-44.2	17.2	24.3	21.3
2	ACC	19-69.8	18.5-67.6	17.5-64.8	17.3-64	17.5-64.4
	LDS	19	---	17.19	24.3	21.3
3	ACC	19-69.8	18.5-67.6	---	---	---
	LDS	19	44.2-56.9	17.16	24.13	21.7
	HSC	---	18-67	17.3-64	17.3-64	17.5-64
4	ACC	19-70	18.5-67.6	17.4-64.4	62.78	17.4-20.7
	LDS	---	44.2-56.9	---	---	---

Table 5: Maximum displacement registered in the column in each impact at each point.

	MAXIMUM REGISTERED DISPLACEMENT (mm)				
POINT	1st impact	2nd impact	3rd impact	4th impact	5th impact
1 (LDS)	1.2	2.1	1.2	2.7	1.8
2 (LDS)	2.7	---	4.4	5.7	3.8
3 (LDS)	1.2	3.8	1.8	2.6	1.7
3 (HSC)	1.7	1.8	2.1	2	1.8

The maximum displacement (table 5) also gives great differences between the values from LDS and those from HSC at point 3. The rest of values for LDS in other points have been included to show the lack of agreement between those values and data from the camera.

4 Conclusions and ongoing work

A new design was developed to perform real size car impact experiments against columns under compression. After some experiments, the movement of the higher part of the column is shown. Nevertheless, we believe that, in combination with a finite element model, the test frame can be useful to calibrate the concrete of the column in the model and, after that, to simulate different boundary conditions in the column.

After 5 experiments it is proved that, at a velocity of 20 km/h, the car has important plastic deformations and therefore, the use of real car in the study of low velocity impact is definitely necessary. In addition, the performed experiments show that, for an impact of 20 km/h against a column, the duration of the loading is around 120 ms.

The use of HSC in the determination of displacement during the impact was more reliable than the use of LDS. By means of HSC was obtained a maximum displacement of the column around 2 mm in all the experiments.

Following this work, the next step will be the numerical modelization by means of finite element models to obtain a complete model that allows studying this problem without expensive experiments.

Another interesting study is the integration of the accelerometers signal to obtain displacement in the rest of measured points. To check this calculation, the values obtained from HSC can be used.

References

[1] "Eurocode 1: Actions on structures – Part 1-1: General actions – Densities, self-weight, imposed loads for buildings" Final Draft prEN 1991-1-1, European Committee for Standardization, 2001

[2] "Eurocode 1: Actions on structures – Part 1-7: General actions – Accidental actions" Final Project Team Draft (Stage 34), Draft prEN 1991-1-7, European Committee for Standardization, 2003

[3] "Eurocode 1: Actions on structures – Part 2-7: Accidental actions due to impact and explosions" ENV 1991-2-7, European Committee for Standardization, June 1998

[4] "Minimum Design Loads for Buildings and Other Structures", ASCE STANDARD, ASCE/SEI 7-05, American Society of Civil Engineers, 2006

[5] "2009 International Building Code", International Code Council, See http://www2.iccsafe.org/states/2009ICodes/Building/Building_Frameset.html for further details. Accessed at 01/09/2009.

[6] "Actions on structures - Part 9: Accidental actions", DIN Deutsches Institut für Normung eV., August 2003

[7] Ferrer B., Ivorra S., Irles R., Low velocity vehicle impact: an outline of relevant codes, Proceedings-ICE: Structures & Buildings, Submitted (2010)

[8] European New Car Assessment Programme (Euro NCAP). Frontal Impact Testing Protocol. Version 5.0. October 2009.

[9] Koh C. G., Liu Z. J., Quek S. T., Numerical and experimental studies of concrete damage under impact. Magazine of Concrete Research, 2001, 53, Nº 6, December, 417-427

[10] Eibl J., Design of concrete structures to resist accidental impact. The Structural Engineer, Volume 65A, nº 1, January 1987

[11] Ferrer B., Espinosa J., Pérez J., Ivorra S., Mas D., Optical scanning for structural vibration measurement on a steel column, Submitted (2010)

Section 7
Seismic behaviour

Simulating seismically isolated buildings under earthquake-induced pounding incidences

P. C. Polycarpou & P. Komodromos
Department of Civil and Environmental Engineering, University of Cyprus, Nicosia, Cyprus

Abstract

Seismically isolated buildings usually experience large horizontal relative displacements during strong earthquakes due to the flexibility that is incorporated, through seismic bearings, at their bases. If the available clearance around a seismically isolated building is, for any reason, limited, then there is a possibility of the building pounding against adjacent structures. This paper, presents a methodology for simulating this problem using numerical methods, in order to investigate the effects of potential pounding on the overall seismic response of seismically isolated buildings.
Keywords: seismic isolation, earthquake, pounding, impacts.

1 Introduction

Seismic isolation is usually based on the incorporation of flexible elastomeric bearings, typically at the base of a building, in order to shift its fundamental period outside the dangerous for resonance range of periods and avoid resonance with the induced earthquake excitation. However, strong seismic actions cause large horizontal relative displacements at the isolation level of seismically isolated structures due to the excessive flexibility that is provided through the seismic bearings. Therefore, a wide clearance, known as "seismic gap", must be provided around a seismically isolated building in order to accommodate the expected large horizontal displacements during a strong earthquake. Nevertheless, the width of the provided seismic gap cannot be unlimited due to practical constraints, especially in cases of retrofitting existing structures. In addition, it is widely accepted that there are several uncertainties about the characteristics of the expected earthquake and the methods of estimating the

WIT Transactions on The Built Environment, Vol 113, © 2010 WIT Press
www.witpress.com, ISSN 1743-3509 (on-line)
doi:10.2495/SU100211

induced relative displacements of the building. Thus, a reasonable concern is the possibility of poundings of a seismically isolated building against either the surrounding moat wall or adjacent buildings during a very strong earthquake.

Earthquake induced pounding incidences between fixed-supported buildings, motivated relevant research in the past [1-5]. However, very limited research work has been carried out for poundings of multi-storey seismically isolated buildings [5-8], which exhibit quite different dynamic characteristics from fixed-supported buildings. Specifically, poundings of a seismically isolated building occur primarily as a result of the large relative displacements at the isolation level (Figure 1), while in the case of conventionally fixed-supported buildings, poundings occur due to the deformations of the superstructure, usually at the building tops. Moreover, it is more likely to have more demanding performance requirements and higher expectations for buildings that utilize an innovative earthquake-resistant design, such as seismic isolation, than for conventionally fixed-supported buildings.

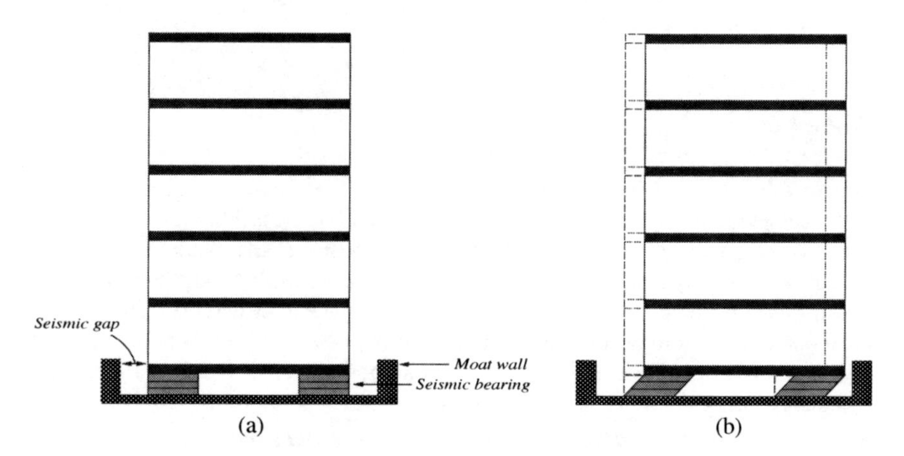

Figure 1: (a) Configuration of a seismically isolated building; (b) Mode of deformation during an earthquake.

This paper presents a simple methodology for the numerical simulation of seismically isolated buildings that undergo earthquake excitations with the possibility of impact occurrences.

2 Methodology

The modelling of the simulated structures is performed in two dimensions (2D), while the multi-storey buildings are modelled as multi-degree of freedom (MDOF) systems, with shear-beam behaviour and the masses lumped at the floor levels, assuming linear elastic behaviour during earthquake excitations. A bilinear model is used for the simulation of the isolation system's behaviour, which corresponds to the use of Lead Rubber Bearings (LRB). In particular, prior to the yielding of the lead core, the isolation system has an initial stiffness

K_1, which is much higher than the post-yield stiffness K_2 that corresponds solely to the stiffness of the rubber.

2.1 Impact modelling

The numerical modelling of impact and the estimation of the impact forces acting on the colliding bodies is an essential topic, not only for the cases of structural poundings, but also for other problems involving numerical simulation of impact. Usually, in numerically simulated dynamic systems, such as multi-storey buildings under earthquake excitations, structural impact is considered using force-based methods, also known as "penalty" methods. These methods allow small interpenetration between the colliding structures, which is justified by their deformability at the vicinity of the impact. Contact springs are automatically formed when an impact is detected, kept as long as the colliding bodies remain in contact and removed as soon as the bodies are detached from each other. The interpenetration depth is used together with the stiffness of the contact spring to estimate, according to the impact model, the contact forces that are applied to the structures, pushing them apart. There are either linear or non-linear impact models, depending on whether the impact force is increasing linearly or exponentially with the indentation.

In the current study, the modified linear viscoelastic impact model [7] is used, assuming an impact spring and an impact dashpot exerting, in parallel, impact forces to the colliding structures whenever their separation distances are exceeded. Actually, it is a small variation of the classical Kelvin-Voigt impact model, in which the tensile forces arisen at the end of the restitution period are omitted and a small plastic deformation is introduced, which increases the available clearance (Figure 2). In particular, when a contact is detected, the impact force is estimated at each time-step using the following formulas:

$$F_{imp}(t+\Delta t) = \begin{cases} k_{imp} \cdot \delta(t) + c_{imp} \cdot \dot{\delta}(t) & \text{when} \quad F_{imp}(t) > 0 \\ 0 & \text{when} \quad F_{imp}(t) \leq 0 \end{cases} \tag{1}$$

where $\delta(t)$, is the interpenetration depth, $\dot{\delta}(t)$ is the relative velocity between the colliding bodies, k_{imp} is the impact spring's stiffness and c_{imp} is the impact damping coefficient. The later is computed according to the following formulas, provided by Anagnostopoulos [1], and based on the conservation of energy before and after impact:

$$c_{imp} = 2 \cdot \xi_{imp} \sqrt{k_{imp} \frac{m_1 \cdot m_2}{m_1 + m_2}} \tag{2}$$

$$\xi_{imp} = -\frac{ln(COR)}{\sqrt{\pi^2 + (ln(COR))^2}} \tag{3}$$

In the above formulas, m_1, m_2 are the masses of the two colliding bodies and COR represents the coefficient of restitution which is defined as the ratio of relative velocities after and before impact ($0<COR\leq 1$). The exact value of the impact stiffness term (k_{imp}) is practically unknown, since its physical meaning is not clearly determined. However, it seems that its value depends on the mechanical properties of the material and the geometry of the contact surface of the colliding bodies. For the simulations in the current study the value of 2500 kN/mm has been chosen for the impact stiffness, while the COR is assumed to be equal to 0.6.

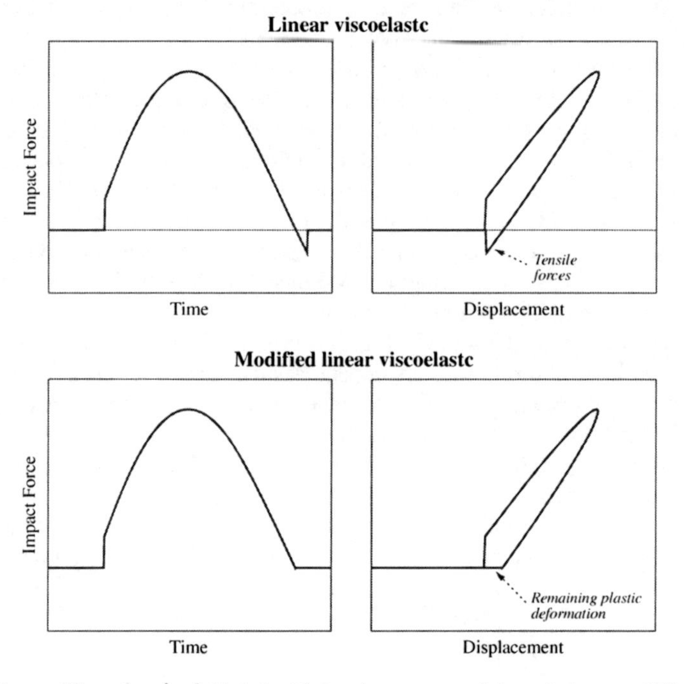

Figure 2: The classical Kelvin-Voigt impact model and the modified linear viscoelastic impact model.

2.2 Equations of motion

The differential equations of motion for a seismically isolated building, which is modeled as a MDOF system, are expressed in the following matrix form:

$$\underline{F}^{I}(t)+\underline{F}^{D}(t)+\underline{F}^{E}(t)=\underline{0} \tag{4}$$

where $\underline{F}^{I}$, $\underline{F}^{D}$ and $\underline{F}^{E}$ are the inertia, damping and elastic forces, respectively, acting on the structure at time t. In the case of a ground excitation with an acceleration time-history $\ddot{U}_g(t)$, the inertia forces are expressed as:

$$\underline{F}^{I}(t)=\underline{M}\cdot\underline{\ddot{U}}(t)+\underline{M}\cdot\underline{\iota}\cdot\ddot{U}_g(t) \tag{5}$$

where $\underline{\iota} = [1 \;\; 1 \;\; \cdots \;\; 1]^T$. The damping forces are expressed in terms of the floor velocities and the damping matrix of the MDOF system:

$$\underline{F}^D(t) = \underline{C} \cdot \underline{\dot{U}}(t) \tag{6}$$

The elastic forces ($\underline{f_s}$) of the superstructure are computed based on the stiffness matrix and the corresponding displacements at time t, while for the seismic isolation system the elastic forces are calculated according to the bilinear model considering the displacement time-history $u(t)$ and the velocity sign at the isolation level at time t:

$$\underline{F}^E(t) = \begin{cases} \underline{f_s}\left(u(t), \dot{u}(t)\right) & \text{no impact} \\ \underline{f_s}\left(u(t), \dot{u}(t)\right) + \underline{e} \cdot F_{imp}^E(t) & \text{during impact} \end{cases} \tag{2.1}$$

$F_{imp}^E(t)$ and $F_{imp}^D(t)$ are the elastic and damping contact forces during impact, respectively, which are calculated according to the corresponding impact model. The impact forces are non-zero only whenever the relative displacements at the corresponding level, along the height of the simulated building, exceed the available clearance, leading to poundings with the adjacent structure. Each of the terms e_i of vector $\underline{e}$, which has a dimension equal to the number of the degrees of freedom, is equal to *0* when no contact is detected in DOF *i*, while it takes the value of *1* when an impact occurs in the corresponding floor. The equations of motion are directly integrated using the Central Difference Method (CDM), computing the displacements at time $(t + \Delta t)$.

2.3 Developed software

A primary aim of this research work was the development of a specialized software tool to efficiently and effectively conduct the necessary numerical simulations and parametric studies of seismically isolated buildings with automatic impact detection and handling capabilities. In particular, a software application has been developed, which is capable of performing efficiently two dimensional (2D) simulations of MDOF systems with shear-beam behaviour under dynamic loadings. In particular, an Object-Oriented Programming (OOP) approach and the Java programming language have been utilized to design and implement a flexible, robust and extendable software application with effective visualization capabilities (Figure 3) that can be used in relevant numerical simulations and parametric analyses.

The developed software application allows the consideration of poundings of a seismically isolated building, either with the surrounding moat wall or with one or more adjacent buildings. Moreover, the software allows both linear and bilinear models to be used for the simulation of the seismic isolation system. The ability to automatically perform large numbers of numerical simulations is also provided, in order to parametrically investigate the effects of certain parameters, such as the structural characteristics, the size of the separation gap and the earthquake characteristics.

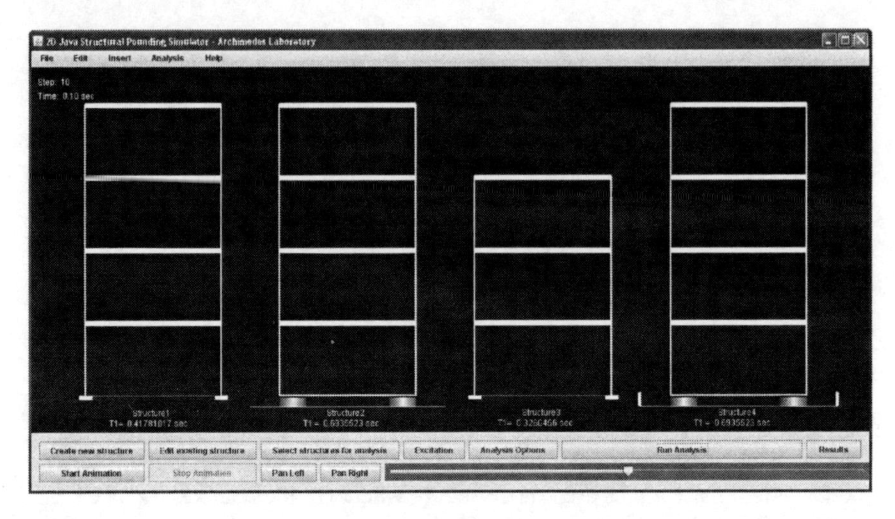

Figure 3: The main window in the graphical user interface (GUI) of the developed software application.

2.4 Example

A typical seismically isolated building is used in the simulations in order to examine the pounding effects on its seismic response. The superstructure is assumed to have 5 floors, each with a lamped mass of 320 *tons*, except of the top floor where a mass of 250 *tons* is considered. Each story has a horizontal stiffness of 600 *MN/m*. An additional mass of 320 tons is assumed to be lumped at the isolation level, while the bilinear properties of the isolation system were taken as follows: $K_1 = 200$ *MN/m*, $K_2 = 25$ *MN/m*, $f_y = 0.1 \times W_{tot}$, where W_{tot} is the total weight of the building. A viscous damping ratio of 2% was assumed for the superstructure, while for the isolation system, in addition to the hysteretic energy dissipation, a 5% viscous damping ratio was considered. The fundamental period of the fixed-supported superstructure is equal to $T_{\text{fixed}} = 0.49$ *sec*.

The structure is considered under two different circumstances. In the first case, the seismically isolated building is considered without the possibility of impacts, assuming a sufficiently wide seismic gap. In the second case a seismic gap width equal to 24 *cm*, which is about 15% smaller than the maximum unobstructed induced relative displacement (27.48 *cm*) of the building under the San Fernando earthquake, is considered at the base, leading to pounding with the surrounding moat wall. The later is assumed to be completely rigid and move with the ground during an earthquake. The San Fernando, California USA earthquake 1971, which is a very strong earthquake (1.17*g*), is used as ground excitation in the performed simulations.

Figure 4 presents the relative displacement time-histories at the base of the seismically isolated building under the San Fernando earthquake for both the case without poundings and the case of a seismic gap equal to 24 *cm*, where the base mat unavoidably hits against the surrounding moat wall, specifically at the time instance of 3.66 *sec*. It is observed that the differences in the two plots are very difficult to be identified, since only a slight reduction of the peak values due to impact can detected.

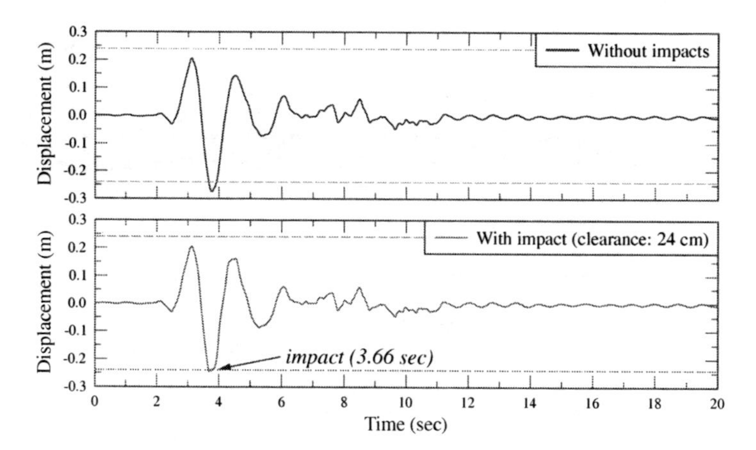

Figure 4: Relative displacement time-histories at the isolation level of the building, for the two cases of without and with poundings.

In contrast to the relative displacements responses, floor accelerations, as expected, are found to be more sensitive to impact occurrences. Although the width of the clearance is only 3.5 *cm* smaller than the maximum unobstructed relative displacement at that level, the effects of pounding are very apparent, especially on the acceleration response at the isolation level where impacts occur. Figure 5 presents the acceleration time-histories of the seismically isolated building during poundings, which are compared to the corresponding time-histories for the case without poundings. Very high accelerations are observed at the time of impact (3.66 *sec*). In particular, the high spikes in the acceleration response reach up to 6 times the corresponding peak floor accelerations without poundings, in which case the response, as shown by the plot, is much smoother. Nevertheless, due to the structure's damping, short time after the impact occurrences, the response tends to become identical to the corresponding response without poundings.

The peak values of the interstory deflections and absolute floor accelerations of the seismically isolated building during impact are plotted in Figure 6 and compared with the corresponding values of the fixed-supported and base-isolated building without impacts. It is observed that, during poundings, interstory deflections at the upper floors are amplified up to 3 times due to poundings with the moat wall, compared to the case without poundings and reach the peak

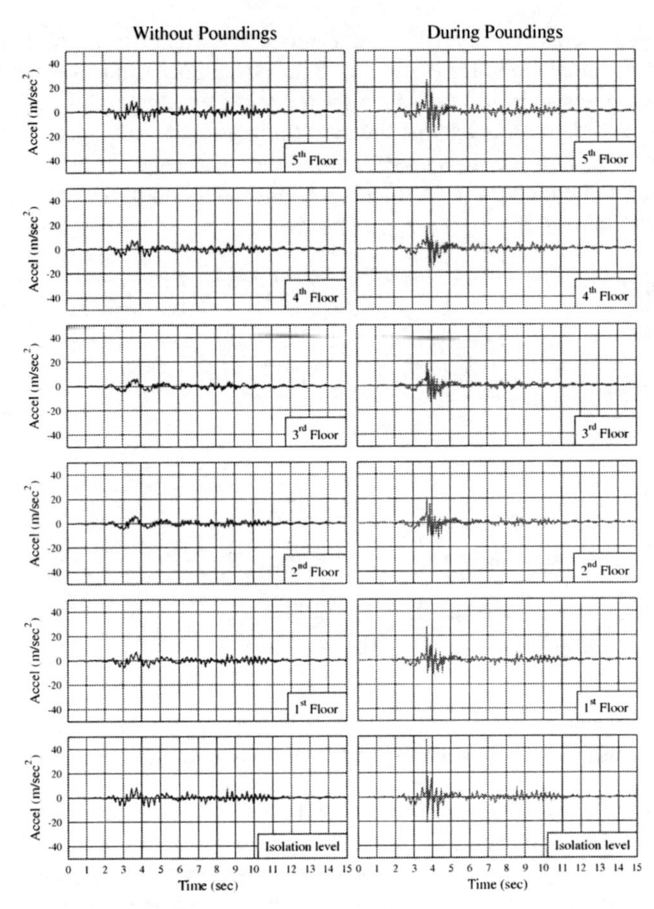

Figure 5: Acceleration time-histories at each floor level of the 5-story seismically isolated building, without and with poundings respectively.

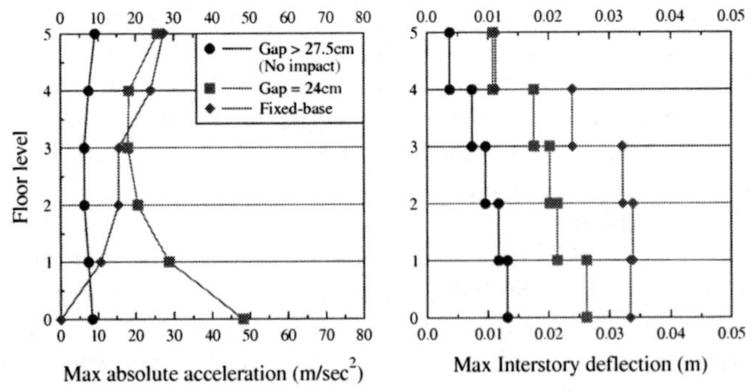

Figure 6: Peak floor accelerations and peak interstory deflections of the 5-story seismically isolated building, for the cases of without and with pounding respectively.

values of the deflections of the corresponding fixed-supported building. Consequently, almost the same shear forces that act on the corresponding fixed-supported building with the same characteristics, act on the particular stories of the superstructure. If not sufficient strength is provided to the structural elements, taking into account these effects of potential poundings during the design of the superstructure of the seismically isolated building, there is a great possibility of causing considerable damage in such cases.

Comparing the computed peak absolute floor accelerations of the building, the influence of poundings in the response is much more pronounced at the lower floors, where the peak floor accelerations become much higher than those for the corresponding fixed-supported building. Due to poundings with the moat wall, the seismically isolated building may experience maximum floor accelerations at the isolation level, where impacts occur, instead of the top-floor, which is the most common case for a MDOF system. It is well known from previous studies [1, 3] that the acceleration response is highly affected by impacts. These high values of floor accelerations that are caused by poundings can damage sensitive equipment that may be accommodated in the building.

In order to examine the effect of the seismic gap size on the response of the seismically isolated building during poundings with the moat wall, a parametric analysis has been conducted. In particular, the width of the seismic gap is varied from 10 to 45 cm with a step of 0.5 cm, considering equal gap sizes on both sides of the building. The plots in Figure 7 present the peak floor accelerations and peak interstory deflections of the 5-story seismically isolated building under the San Fernando earthquake record, in terms of the width of the seismic gap. It is expected that, in general, as the seismic gap increases, both floor accelerations and interstory deflections of the superstructure would decrease. However, the simulation results indicate that for relatively narrow gap sizes the response increases with the width of the available clearance and after a certain value the response of the seismically isolated building begins to decrease, as rationally expected. This observation is more pronounced for the lower floors, which are closer to the impact location. This is due to the fact that for very narrow seismic

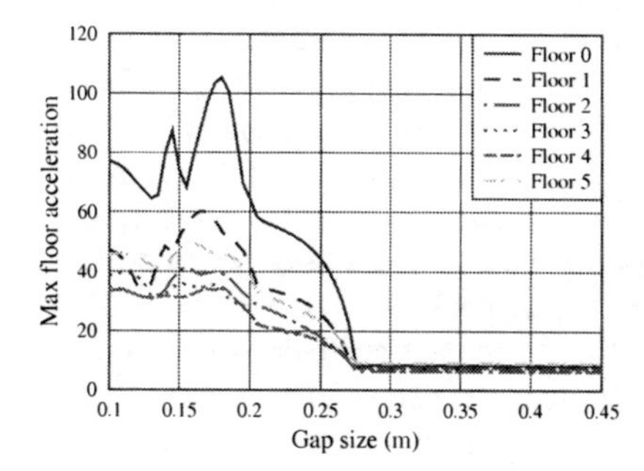

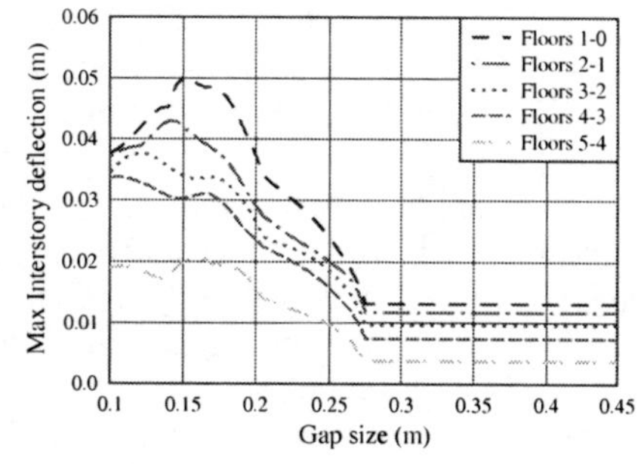

Figure 7: Peak floor responses of the 5-story seismically isolated building in terms of the width of the seismic gap.

gap widths (<15 *cm*), in comparison with the maximum unconstrained induced displacement, the seismically isolated building retains relatively low impact velocities.

Figure 8 presents the total number of impacts and the maximum impact velocity at the isolation level of the 5-storey seismically isolated building, in terms of the seismic gap width. It is observed that the trend of the maximum impact velocity is very similar to the corresponding peak acceleration response of the seismically isolated building at the isolation level, indicating that the amplification of the response due to impact is proportional to the impact velocity. It is also observed that the total number of impacts is not always decreasing with the width of the seismic gap as it was expected, and seems to depend from the earthquake characteristics and the structural properties. Nevertheless, for gap sizes larger than 17 *cm* the peak floor responses, as well as the impact velocity decrease with the width of the seismic gap.

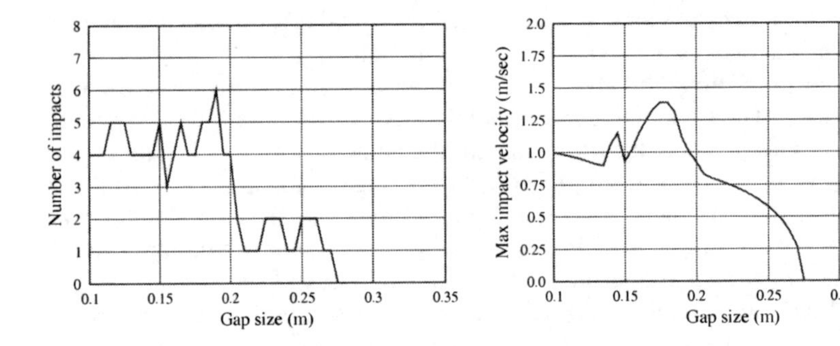

Figure 8: Total number of impacts and maximum impact velocity at the isolation level, in terms of the size of the seismic gap.

Another series of parametric studies is performed, in order to examine the effect of the impact stiffness and the coefficient of restitution on the peak response of a seismically isolated building during poundings, since the value of these impact parameters are usually based on estimations. The same building is considered under the same excitation, while the seismic gap is taken to be 24 *cm*.

The plots in Figure 9 indicate that the effect of impact parameters is localized, since the only response quantity that seems to be substantially affected by the variation of these two parameters is the peak floor acceleration at the seismic isolation level, where impacts occur. In particular, the peak floor acceleration at that level increase very rapidly in contrary to the peak floor accelerations of the upper floors, which are slightly affected by the impact stiffness after a certain value of k_{imp}. Furthermore, the results show that, for values of the coefficient of restitution lower than 0.4, the peak floor acceleration at the isolation level increases and reaches its maximum value when the impact becomes highly overdamped. The rest of the response remains insensitive to the variation of the coefficient of restitution, i.e. the impact damping.

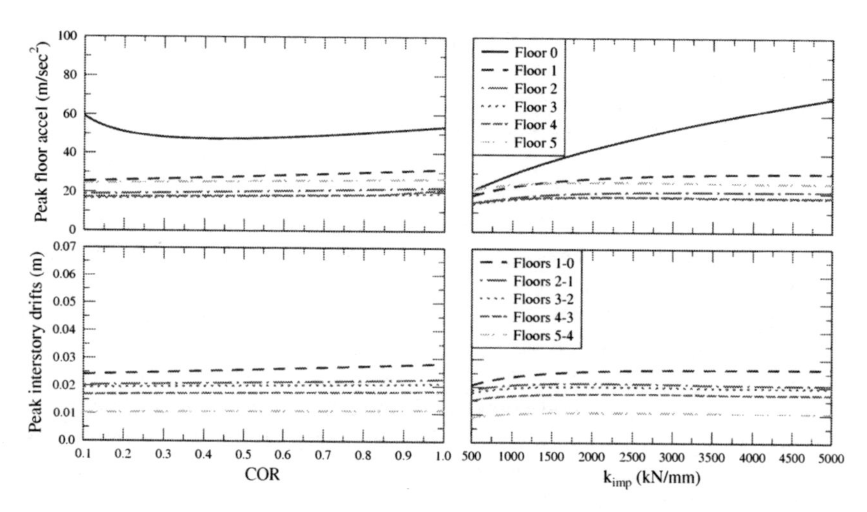

Figure 9: Influence of the impact parameters on the peak floor accelerations and interstory deflections.

3 Conclusions

A methodology has been presented that is used for simulating, in two dimensions, earthquake-induced pounding of seismically isolated buildings. A typical 5-storey building has been used as an example to demonstrate the most important effects of pounding on its structural response during a strong earthquake. The simulations show that poundings are particularly unfavourable for the structure and its contents, since they increase significantly the absolute floor accelerations and interstory deflections of the building. In particular, very high accelerations are observed at the times of impacts due to the sudden changes of the velocity, especially at the isolation level where impacts occur. These high values of induced floor accelerations can damage sensitive equipment that may be accommodated in the building. Furthermore, significant amplification of the storey shear-forces has been observed specially at the upper floors and if not sufficient strength is provided to the corresponding structural elements, there is a great possibility of considerable structural damage, in such cases.

References

[1] Anagnostopoulos, S.A. Pounding of buildings in series during earthquakes. *Earthquake Engineering and Structural Dynamics,* **16** pp. 443-456, 1988.
[2] Papadrakakis, M., Apostolopoulou, C, Zacharopoulos, A, Bitzarakis, S. Three-dimensional simulation of structural pounding during earthquakes. *Journal of Engineering Mechanics,* **122** pp. 423-431, 1996.
[3] Maison, B.F., Kasai, K. Analysis for Type of Structural Pounding. *Journal of Structural Engineering*; **116** pp. 957-977, 1990.

[4] Anagnostopoulos, S.A., Spiliopoulos, K.V. An investigation of earthquake induced pounding between adjacent buildings. *Earthquake Engineering and Structural Dynamics*, **21** pp. 289-302, 1992.
[5] Chau, K.T., Wei, X.X. Poundings of structures modeled as non-linear impacts of two oscillators. *Earthquake Engineering and Structural Dynamics*, **30** pp. 633-651, 2001.
[6] Matsagar, V.A., Jangid, R.S. Seismic response of base-isolated structures during impact with adjacent structures. *Engineering Structures,* **25** pp. 1311-1323, 2003.
[7] Komodromos, P., Polycarpou, P.C., Papaloizou, L., Phocas, M.C. Response of Seismically Isolated Buildings Considering Poundings. *Earthquake Engineering and Structural Dynamics*, **36** pp. 1605-1622, 2007.
[8] Polycarpou, P.C., Komodromos, P. On poundings of a seismically isolated building with adjacent structures during strong earthquakes", *Earthquake Engineering and Structural Dynamics* (Published on-line) DOI: 10.1002/eqe.975, 2009.

Dynamic fluid-structure-soil interaction: applications in earthquake engineering

A. Dimas, N. Bazeos, S. Bousias, T. C. Triantafyllou
& D. L. Karabalis
Department of Civil Engineering, University of Patras, Greece

Abstract

A new discrete model is presented for the evaluation of the dynamic characteristics, i.e. eigenfrequencies and eigenmodes, of tanks of arbitrary shape and fill level. The accuracy and efficiency of the proposed methodology is demonstrated via a number of comparison studies. The above discrete model is combined with structural and soil simulation models for the efficient dynamic analysis of 3-D tanks under earthquake excitation. The obtained results are in excellent agreement to those obtained using detailed analytical and FEM models.
Keywords: discrete sloshing model, arbitrary geometry, arbitrary fill level, seismic excitation, dynamic fluid-structure-soil interaction.

1 Introduction

The safe keeping and uninterrupted flow of liquids or liquid-like materials is of crucial and multifold importance to the industrialized world. Therefore, the interest on the seismic behaviour of modern structures used for storage of such materials has remained strong since the pioneering work of Lamb [1] and is periodically reinforced by the disruption caused by several seismic events, e.g. the 1964 Alaska earthquake. A simple, but accurate and efficient, methodology for the estimation of the hydrodynamic pressures exerted on the walls of a tank, was proposed in the pioneering work of Housner [2]. In these studies on non-deformable vertical prismatic tanks resting on rigid foundations, the solution describing the total hydrodynamic pressure was decomposed into two discrete parts: the "impulsive" and the "convective". The impulsive pressure component is due to a portion of the liquid accelerating with the rigid tank, while the

WIT Transactions on The Built Environment, Vol 113, © 2010 WIT Press
www.witpress.com, ISSN 1743-3509 (on-line)
doi:10.2495/SU100221

convective pressure is exerted by the sloshing motion oscillating at a fundamental frequency. The "impulsive-convective" pressure concept along with the findings of a number of follow up studies, as referenced below, lie at the basis of almost all recent design codes and guidelines, e.g. API Standard 650 [3], Eurocode 8 [4] and ASCE [5].

In addition to the dynamics of a rigid liquid container, a number of other issues, of special interest to structural engineering, have been studied since these early works, most notably: the interaction between the deformable tank walls and the contained liquid, e.g. Veletsos and Shivakumar [6], and the interaction between the supporting medium (soil) and the tank structure, including uplifting and anchoring effects, e.g. Natsiavas and Babcock [7]. In contrast to the vast body of publications concerning vertical prismatic tanks, only few works have appeared on other tank geometries, e.g. Patkas and Karamanos [8]. This extensive analytic and numerical effort is supported by several experimental studies, e.g., Abramson *et al.* [9]. A more complete list of references on this subject can be found in the review articles of Rammerstorfer and Scharf [10], Ibrahim *et al.* [11].

Finally, brief reference is made to numerical solutions obtained by application of the most popular numerical methods, i.e. the finite (FEM) and the boundary (BEM) element methods, since they provide the means for the solution of complicated geometries, boundary conditions and couplings between the liquid and solid domains. A comprehensive discussion and literature review on these matters can be found in Rizos and Karabalis [12].

A search through the literature leaves no doubt that the "impulsive-convective" pressure concept is almost universally used for the analysis and design of vertical prismatic tanks. However, very little or no information exists regarding the application of this concept to tanks of other geometries while the distribution of the hydrodynamic pressure on the walls of tanks of non-prismatic geometry is almost completely missing. This is mainly due to the fact that the development of impulsive-convective spring-mass systems is based, in most cases, on analytical solutions and, thus, the development of such simple solutions for more complicated geometries and boundary conditions is precluded. This work proposes a simple numerical methodology for the computation of the impulsive-convective mass system that can be used in the seismic design of liquid containers of arbitrary shape and fill height. To this end, an eigenvalue analysis is used in conjunction with a simple fluid model which is readily accessible through most of the commercially available general purpose FEM programs, e.g. ANSYS [13]. Even though a FEM analysis can handle a variety of complicated fluid and structural models, as well their various interactions, in an effort to concentrate on the influence of the geometry and to illustrate the generality of the proposed methodology, the fluid flow model used in this work is linear (small free surface amplitude) and inviscid (slip wall condition), while the tank walls are assumed rigid.

2 Numerical formulation

The eigenvalue analysis of a fluid volume can be routinely performed, following an appropriate FEM discretization, by most of the available FEM general purpose programs and results such as the eigenfrequencies ω_i and the corresponding eigenvectors $\{\varphi\}_i$ (i=1,2,…,3N) are obtained, where N is the total number of nodal points. However, it should be noted that only those modes corresponding to a vertical motion of the free surface are of interest and retained for further analysis.

The total hydrodynamic pressure, applied on the tank walls, during a horizontal seismic excitation, can be decomposed into an impulsive part, associated with the fluid motion that has zero relative acceleration with respect to the tank, and a convective part, associated with the fluid sloshing motion. Likewise, the total horizontal force applied on the tank due to the fluid motion can be decomposed into an impulsive and a convective component. The sloshing motion can be represented as a superposition of few eigenmodes of the fluid motion. In general, each eigenmode generates a hydrodynamic wall pressure of unique distribution, but not all of these modes contribute to the development of a nonzero horizontal force on the tank. For example, for a vertically axisymmetric tank, the eigenmodes, which are responsible for the generation of a nonzero horizontal hydrodynamic force and, thus, influence the horizontal motion of the tank structure, are only those that exhibit free-surface antisymmetry with respect to a plane parallel to the tank axis of symmetry. These are the eigenmodes of interest in this work and hereafter are referred as "sloshing modes". The discrete masses M_{Cn}, associated with the sloshing modes n=1,…,∞, are called convective. The convective masses and the impulsive mass M_I, i.e. the remaining portion of the total mass moving in synchronism with the tank, are related via the principle of conservation of mass as

$$M_L = M_I + \sum_{n=1}^{\infty} M_{Cn} \tag{1}$$

where M_L stands for the entire liquid mass. The instantaneous value of the total horizontal hydrodynamic force or base shear, F, applied on the tank due to the liquid motion, can be computed as [6]

$$F(t) = M_I \ddot{x}(t) - \sum_{n=1}^{\infty} M_{Cn} \omega_n^2 u_n(t) \tag{2}$$

where $\ddot{x}(t)$ is the acceleration of the tank structure and $u_n(t)$ is the response of a single degree-of-freedom system, with frequency ω_n, which is computed using as forcing input the acceleration of the tank. In view of the form of eqn (2), the concept of an oscillator with multiple degrees-of-freedom is employed in this work. Therefore, the convective mass M_{Cn} of each sloshing mode is equal to an effective modal mass computed as, e.g. Chopra [14],

$$M_{Cn} = (L_{Cn})^2 / M_n \tag{3}$$

where $M_n = \{\phi\}_n^T [M] \{\phi\}_n$ is the generalized mass of eigenmode n, and $L_{Cn} = \{\phi\}_n^T [M] \{I\}$, with $\{I\}$ being the unit vector in the direction of the seismic excitation. A schematic representation of a convective-impulsive system, for a tank of arbitrary shape under horizontal seismic excitation, is shown in fig. 1 where the convective portion of the liquid is substituted by an equivalent system of masses and springs, while the impulsive portion reduces to a single mass rigidly connected to the tank wall.

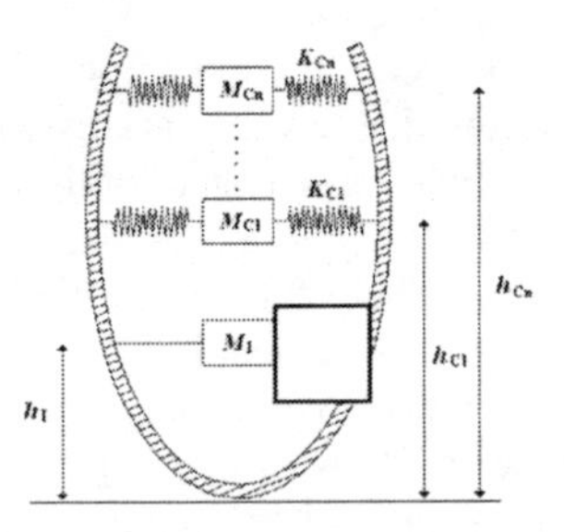

Figure 1: Impulsive-convective model for the representation of fluid motion.

On the basis of the previous discussion, the spring constants for each convective mass can be calculated as

$$K_{Cn} = \omega_n^2 M_{Cn}. \tag{4}$$

The position height, h_{Cn} or h_n, of each mass depends on the convective-impulsive pressure distribution on the tank wall. For vertical prismatic tanks the position heights are available in analytic form, e.g. Refs [2, 11]. For a spherical tank or a horizontal cylinder of diameter D, all discrete masses should be placed at the geometrical center, i.e. $h_{Cn}=h_n=D/2$, since all resultant forces due to fluid pressure are applied at the geometrical center of the container. For arbitrary tank geometries, the corresponding heights can be calculated via a straightforward numerical integration of the fluid pressure distribution and subsequent positioning of the resultant force using standard procedures, e.g. Ref. [6].

Similarly, the total hydrodynamic pressure distribution on the tank wall can be decomposed, into an impulsive part, a convective part, and an axisymmetric part, associated with the axisymmetric eigenmodes of the fluid motion which do not contribute to the total horizontal force, as

$$p(\theta,\varphi,t) = p_I(\theta,\varphi,t) + \sum_{n=1}^{\infty} p_{Cn}(\theta,\varphi,t) + \sum_{m=1}^{\infty} p_{Am}(\theta,t) \tag{5}$$

where θ and φ are the azimuth and meridian angles, respectively, as shown in fig. 2, p_I is the impulsive pressure, p_{Cn} is the convective pressure due to sloshing

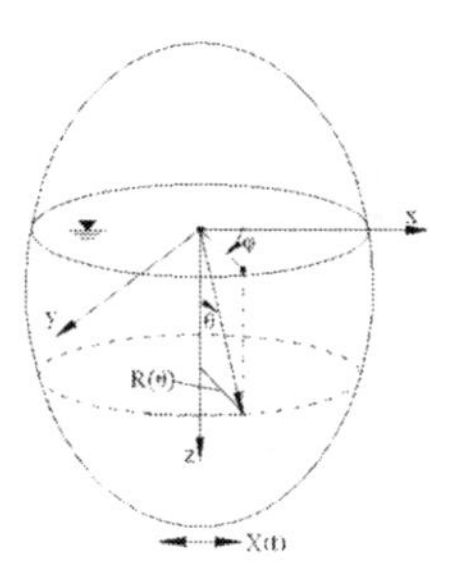

Figure 2: Coordinate system for axisymmetric (with respect to axis z) tanks.

mode n and p_{Am} is the axisymmetric pressure due to axisymmetric mode m. For vertically axisymmetric tanks under horizontal seismic excitation, only the sloshing modes are excited and the axisymmetric pressure is zero. The impulsive pressure is given by the expression

$$p_I(\theta,\varphi,t)=C_I(\theta)\rho R(\theta)\cos(\varphi)\ddot{x}(t) \tag{6}$$

while the convective pressure, for each sloshing mode n, by the expression

$$p_{Cn}(\theta,\varphi,t)=-C_{Cn}(\theta)\omega_n^2\rho R(\theta)\cos(\varphi)u_n(t) \tag{7}$$

with C_I and C_{Cn} being dimensionless pressure profile functions which depend on the tank fill height. The computation of C_I and C_{Cn} is based on a series of dynamic mode superposition analyses. Thus, for any fill height, the response of the fluid-structure FEM model to an arbitrary seismic excitation is computed taking into consideration only the n-th sloshing mode. The resulting hydrodynamic pressure distribution is equal to p_{Cn}, and, therefore, the profile functions C_{Cn} can be computed using eqn (7). Then, a time domain simulation is performed, based on the same seismic excitation as in the dynamic mode superposition analysis, and the total hydrodynamic pressure distribution p is obtained. Finally, the impulsive pressure distribution p_I results from eqn (5) and the corresponding pressure profile function C_I from eqn (6).

3 Numerical example

A series of numerical studies have been performed for the validation of the accuracy and efficiency of the proposed methodology. Due to lack of space, only an example pertaining to a seismic analysis of a spherical tank is presented in this work. However, the interested reader can find a series of related results in Drosos *et al.* [15], Drosos [16] and Drosos [17].

The spherical tank under investigation is shown in fig. 3. Its diameter is 20m, its equator is located 12.5m above the foundation, and contains a liquid with mass density ρ=522Kg/m^3. The supporting system of 11 columns and 11 pairs of diagonal braces is resting on a circular foundation ring of external diameter

20m, approximately, which, in turn, rests on a category C soil medium [4]. Two models are considered for comparison purposes: (a) a detailed FEM where the fluid has been simulated by a mesh of 3D elements (ANSYS-FLUID80) and (b) a simplified model where the mass of the liquid has been decomposed into impulsive and convective parts in full accordance with eqns (1)-(7). The results of this decomposition, as used in the simplified model, are listed in table 1. However, details about the geometry, soil-structure interaction constants, material properties, etc., can be found in [15–17].

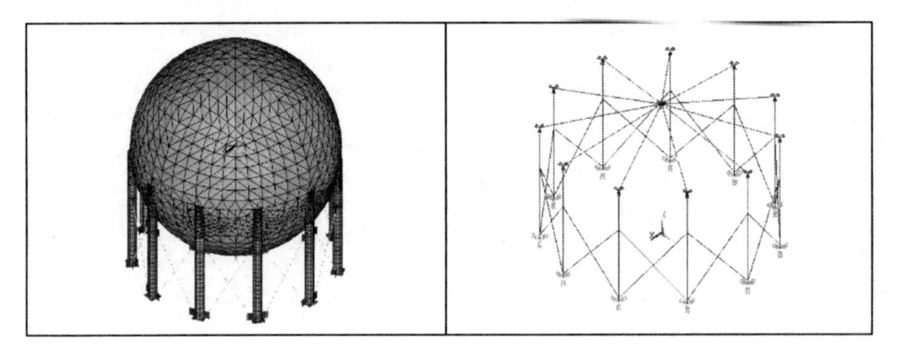

Figure 3: Models of spherical tanks: (a) a detailed FEM model and (b) a simplified model.

Table 1: Impulsive mass and convective mass and spring constants.

	Fill level	
	50%	100%
Impulsive mass + mass of steel shell ($KNsec^2/m$)	445.6+394.5	2186.7+394.5
Convective mass M_{C1} ($KNsec^2/m$)	621.4	0
Spring constant K_{C1} (KN/m)	954	- -

With regard to the constants of table 1 it should be noted: (a) Only the first convective mode has a non-negligible effect on the dynamic behavior of this tank, and is used in the following, (b) the impulsive mass is rigidly attached to center of the sphere along with the mass of the steel shell, (c) the convective mass M_{C1} is attached to the center of the sphere via the corresponding spring K_{C1},

(d) obviously no sloshing occurs in a full tank and thus the corresponding convective mass is zero.

The eigenvalue analysis of the above two models is shown in table 2. Three levels of liquid fill are investigated: (a) empty tank, (b) half full tank, and (c) full tank. For all three cases two types of boundary conditions are considered: (i) the structure is fixed at its foundation, and (ii) soil-structure interaction (SSI) via the usage of spring and dashpot elements appropriate for a ring foundation [17]. The most important eigenvalues of the two models are listed along with the corresponding percentage of participating mass. Apparently there is a very good agreement between the results obtained by the two models. The results for the empty tank indicate that the two models are virtually equivalent with regard to their masses and stiffnesses. The results for the two fill levels, i.e. 50% and 100%, also serve as a proof that the impulsive and convective masses computed by the proposed model can represent the fluid-structure model in an effective and accurate way.

Table 2: Results of eigenvalue analysis of FEM and simplified impulsive-convective models of spherical tank.

Fill	Support conditions	**Simplified model**		**FEM model**	
		Period T (sec)	Participating mass (%)	Period T (sec)	Participating mass (%)
0%	Fixed	0.250	99.10	0.242	96.80
	SSI	0.185	63.70	0.185	60.30
50%	Fixed	0.361	57.70	0.353	57.20
	SSI	0.399	54.00	0.392	56.60
100%	Fixed	0.630	98.70	0.592	83.60
	SSI	0.690	55.00	0.649	87.00

4 Conclusions

A simple and computationally-effective model has been proposed for the simulation of sloshing liquids in tanks of arbitrary shape. The methodology for the computation of the convective and impulsive masses and the associated spring constants is based on standard FEM analyses available in almost all commercially available finite element software. Seismic analyses of elevated spherical tanks, simulating realistic structures, reveal the computational efficiency and accuracy of the proposed models. Obviously, the use of the proposed discrete models results in a dramatic reduction of the size of the dynamic model. Thus, it can become a useful tool for quick, yet accurate, analyses in the design office.

References

[1] Lamb, H., *Hydrodynamics,* Cambridge University Press, Cambridge, 1932.
[2] Housner, G.W., Dynamic pressures on accelerated fluid containers, *Bulletin of the Seismological Society of America*, **47**, pp. 15-35, 1957.
[3] American Petroleum Institute, *Seismic design of storage tanks – Appendix E: Welded Steel Tanks for Oil Storage* (API Standard 650), Washington, D.C., 1995.
[4] Comité Européen de Normalization, *Eurocode 8, Part 4: Silos, tanks and pipelines* (Annex A), CEN ENV-1998-4, Brussels, 1998.
[5] ASCE Committee on Gas and Liquid Fuel Lifelines (Technical Council on Lifeline Earthquake Engineering), *Guidelines for the seismic design of oil and gas pipeline systems,* New York, 1984.
[6] Veletsos, S.A. and Shivakumar, P., Tanks containing liquids and solids (Chapter 15). *Computer Analysis and Design of Earthquake Resistant Structures - A Handbook*, eds D.E. Beskos and S.A. Anagnostopoulos, Computational Mechanics Publications, Southampton, 1984.
[7] Natsiavas, S. & Babcock, C.D., Behavior of unanchored fluid-filled tanks subjected to ground excitation, *Journal of Applied Mechanics*, **55**, pp. 654-659, 1988.
[8] Patkas, L.A. and Karamanos, S.A., Variational Solutions of Liquid Sloshing in Horizontal-Cylindrical and Spherical Containers, *Journal of Engineering Mechanics* (ASCE), **133**, pp. 641-655, 2007.
[9] Abramson, H.N., *The dynamic behavior of liquids in moving containers* (Southwest Research Institute), NASA SP-106, 1966.
[10] Rammerstorfer, F.G. and Scharf, K., Storage tanks under earthquake loading, *Applied Mechanics Reviews* (ASME), **43**, pp. 261-282, 1990.
[11] Ibrahim, R.A., Pilipchuk, V.N. and Ikeda, T., Recent advances in liquid sloshing dynamics, *Applied Mechanics Reviews* (ASME), **54**, pp. 133-177, 2001.
[12] Rizos, D.C. and Karabalis, D.L., Soil-Fluid-Structure Interaction (Chapter 9). *Wave Motion in Earthquake Engineering*, eds E. Kausel and G.D. Manolis, WIT PRESS, Southampton, 2000.
[13] Swanson Analysis Systems, Inc., *ANSYS, User's Manual for Revision 5.0, Procedures*, Houston, PA, 1992.
[14] Chopra, A.K., *Dynamics of Structures*, Prentice Hall, New Jersey, 1995.
[15] Drosos, G.C., Dimas, A.A. and Karabalis, D.L., Discrete models for seismic analysis of liquid storage tanks of arbitrary shape and fill height, *Journal of Pressure Vessel Technology*, **130**, pp. 1-12, 2008.
[16] Drosos, G.C., *The effect of sloshing on spherical tanks* (in Greek), MS Master Thesis, Department of Civil Engineering, University of Patras, 2005.
[17] Drosos, J.C., A *study of seismic protection systems for spherical tanks* (in Greek), MS Master Thesis, Department of Civil Engineering, University of Patras, 2005.

Advanced analysis of a space structure retrofit for an ash-tank

A. Ivan, M. Ivan & I. Both
University Politehnica of Timisoara, Romania

Abstract

In this paper we analyze a structure's response to new loadings in conformity with the new codes. In order to be able to come to a relevant conclusion we analysed three cases: a) strengthening the elements so that it keeps the same configuration, b) strengthening the elements and modifying the moment resisting frames into braced frames and c) reconstruct the entire structure with new elements. By applying the new codes we have analysed the results and we present the conclusions and the best and optimum solution for this particular case.
Keywords: retrofit, comparison to old codes, structure response, moment resisting frames, braced frames, sustainability.

1 Introduction

The Rovinari energetic complex in Romania is one of the biggest producers of electric (1320 MW) and thermal energy using coal as primary source of fuel. That is why in order to comply with the requirements of the Plan for the implementation of the EU Directive 1999/31 on storage of residual waste, it is imperative to introduce certain installations for the evacuation of coarse slag and ash, by using the technology of the self-hardening fluid. There are four energetic blocks each of 330 MW. For each block the ash and the slag are collected in a silo of 500 m^3. This silo leans against a metallic structure, which is to be analysed in this paper. On this structure, supplementary installations for the evacuation of coarse slag and ash have to be mounted. Because of the short time at our disposal and because the silos cannot be dismounted, it is imperative to find a solution to retrofit the existent metal structure without interrupting the production process. The existent metal structure was designed in 1972 and that is

WIT Transactions on The Built Environment, Vol 113, © 2010 WIT Press
www.witpress.com, ISSN 1743-3509 (on-line)
doi:10.2495/SU100231

why a thorough investigation has to be made in order to find out its response to new additional loading: permanent, technological, utile, seismic, wind and snow and to the combinations of these factors.

It is very important to make a full analysis for power plants, since the loadings are changing once the technology changes. Besides this, the codes for the seismic actions were changed and the structural behaviour is different if we consider the magnitude (i.e. the intensity of the base force) and also the displacement (concerning the development of the plastic hinges).

2 Structural configuration

The structure studied in this case (Figure 1) was designed in 1972. At that time the codes for earthquake design were not very accurately studied as they are nowadays. Still the structure was also designed for seismic actions, thus resulting a reserve for the design at permanent and variable loads. The increase of the combination factors for permanent and variable loads, according to new codes, does not lead to a strengthening of existing structure. But the response of the structure to lateral seismic forces is significantly different compared to the one assumed in the initial design of the structure.

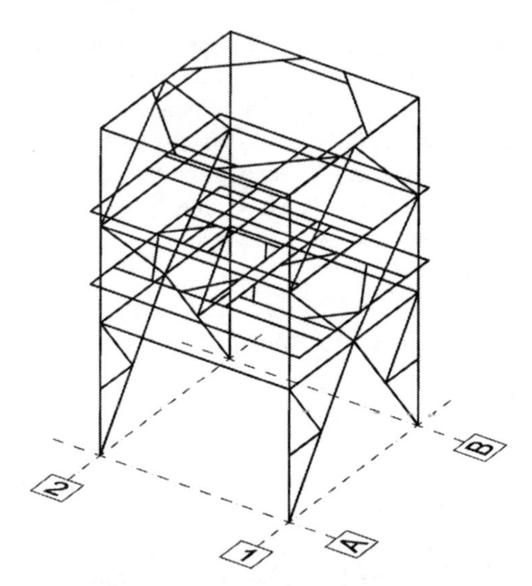

Figure 1: Spatial structure.

The structure has four frames: two are moment resisting frames and two are braced frames. Horizontally, the structure has 3 platforms at +7.6 m, +11.3 m, + 15.5 m and also the elements from a previous platform at the height of +6.3 m. Considering the entire structure 80% of the elements are superficially corroded, consequently rust has to be removed. It is obvious that the cleaning of the rust involves a lot of labour. The evaluation of this work will be made after the presentation of the solutions made.

The objectives of this paper are: to evaluate the seismic performance of the existing individual building structure, to describe the approach in selecting the necessary corrective measures in order to meet the requirements of the new standards and codes [1].

3 Conceptual design

3.1 The first analysis

The first analysis was performed in order to evaluate the response of the existing structure with no structural modification, but with new loadings and considering the previous code for the seismic action. The analysis was a linear elastic one with lateral forces.

The result for this analysis was: none of the elements exceeded the admissible resistance and furthermore the strength reserve for the columns is considerably greater than for the dissipative elements of the structure.

For the seismic actions, the ratio, between the effective stress and the admissible stress for beams in the frames, is 0.5 maximum, while for the columns is of 0.26. Basically, the elements of the structure are more stressed because of the combination of loads that does not include the seismic action considered to be the one from previous codes.

3.2 The second analysis

The second analysis considers the structure response to the actual seismic actions according to the new codes. The ground acceleration a_g was taken as 0.12g, the elastic response spectrum for horizontal components of ground acceleration is presented in Figure 2.

Taking into consideration [1] and the checking performed [2, 3], it resulted that all the columns need to be strengthened as they were checked for a combination that included the over strength for non-dissipative elements. Because of the new seismic provisions other elements have to be strengthened as

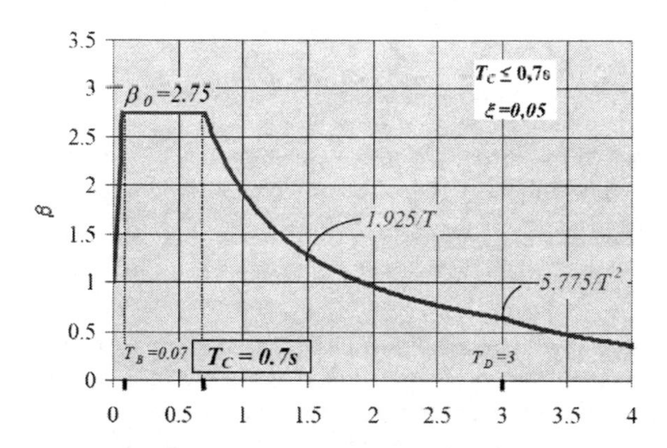

Figure 2: Elastic response spectrum.

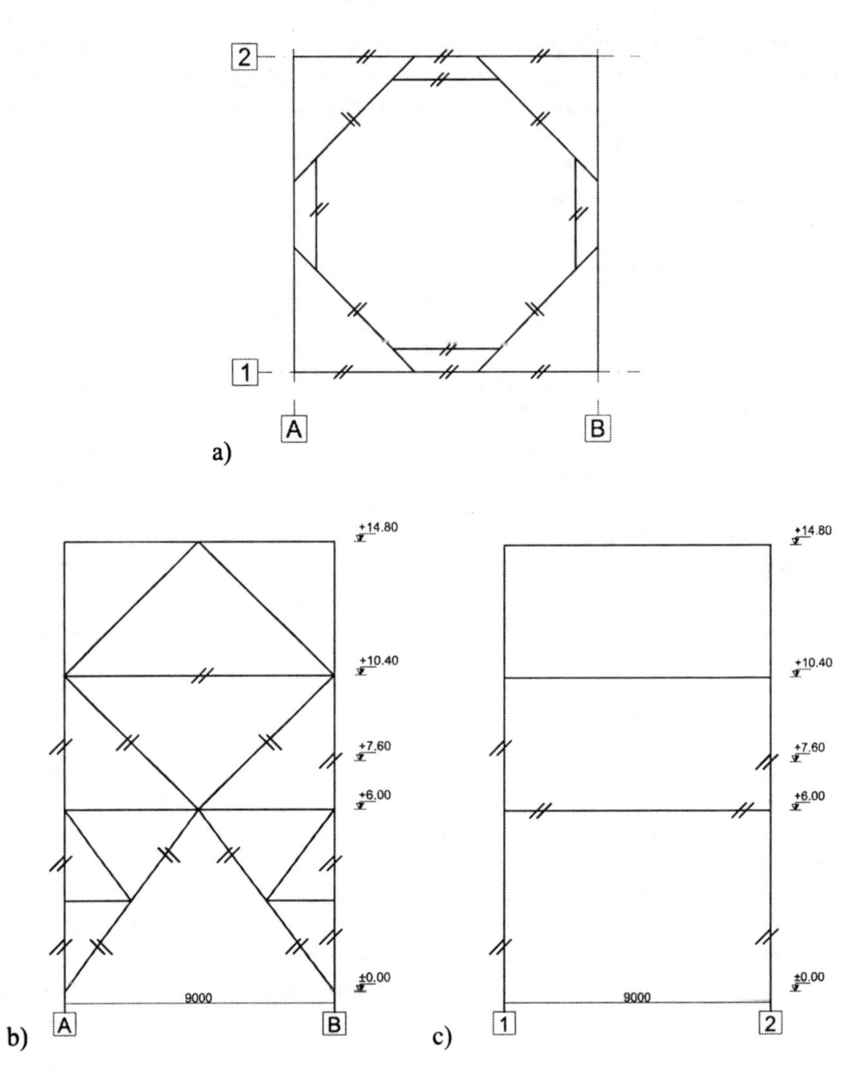

Figure 3: a) Floor +15.5 m, b) braced frame, c) MRF.

well: the bracings at the lower and intermediate part of the building (figure 3b), the beams that support the ash tank at the top of the building (figure 3a) and the beam in the MRF (figure 3c).

3.3 The third analysis

The third analysis includes the modification of the structural system by the addition of some new structural elements namely: bracings in the Moment Resisting Frame (Figure 4). This helps to reduce the moments in the columns for the major axis but it also changes the fundamental period of structure. If for the case of initial

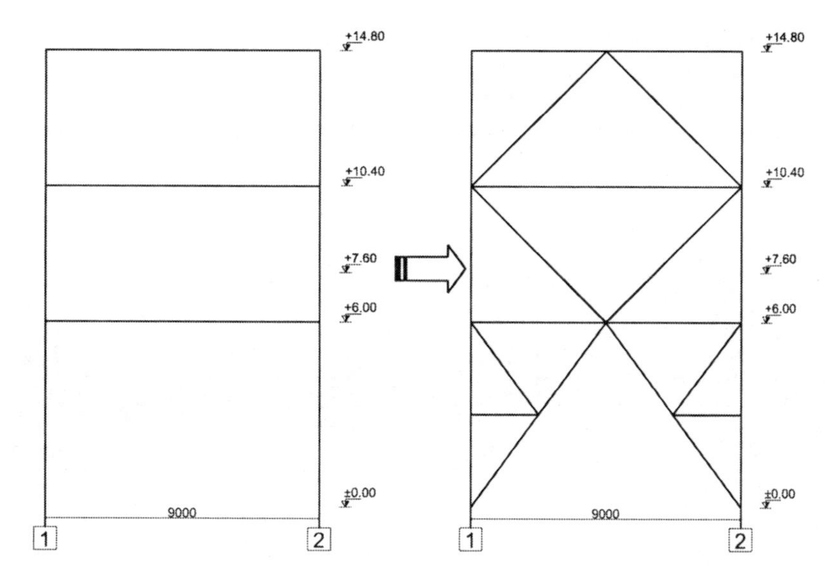

Figure 4: Structural modification.

structure, the period was around 1.6 seconds and the movement was in the plane of the moment resisting frame, for the complete braced structure, the first mode of vibration has a period of 0.68 s which is near the eigen period of the ground.

It also has to be reminded that this operation is technologically allowed. If the railway that goes under the structure would still have to remain functional or the tanks that have been recently installed were too large, this option would not have been taken into account.

The results from this analysis indicated a smaller value for the ratio between the effective stress and the capacity for all the elements. With these considerations the structure still needs to be strengthened, but the steel quantity needed for strengthening is smaller. These elements are presented in Figure 5.

In this solution the beams of the moment resisting frames become over dimensioned and the bracings are designed from the slenderness conditions. That why this is not the optimum solution.

3.4 The fourth analysis

The last choice of analysis was to consider the demolition of the old structure and the erection a new one [4, 5]. For this new structure the elements with flaws in the initial design (over-strengthened bracings, weak joints) are removed and the elements and joints are redesigned according to the new codes. The results are presented in the following sections.

4 Strengthening solutions

The elements that need to be strengthened are presented in this section.

The bracings at the lower and the intermediate part need more material, but also their slenderness has to be improved because of their length. Another pair of angle profile was used as shown in Figure 6.

The strut at +10.4 m is a U300 and although it was subjected to tension only, it has stresses greater than the admissible limit and a too high slenderness.

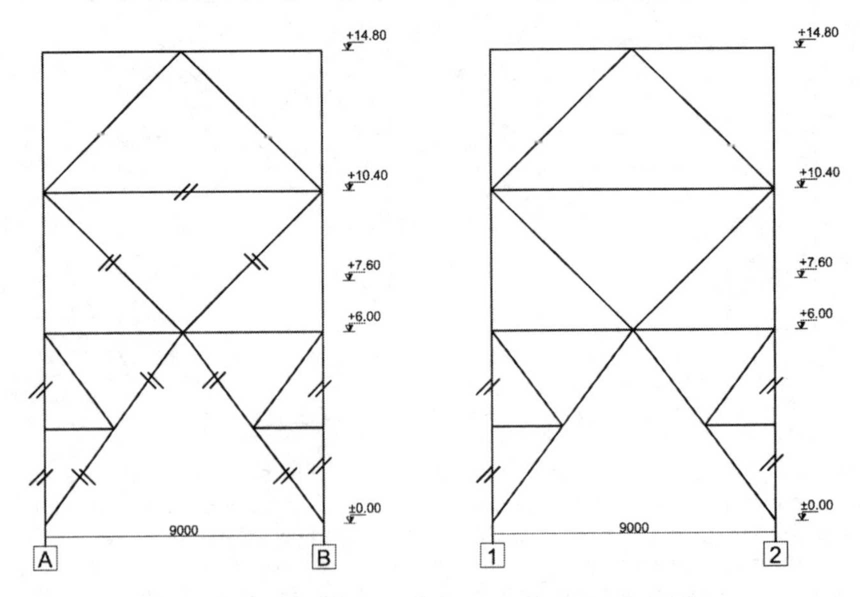

Figure 5: Overstressed elements for braced structure.

	Initial section	Strengthened section
Bracings	2xL 120X11/15	4xL 120X11/15

Figure 6: Bracing strengthening.

Because of that, the bracings at the top level (HEB 300) transmitted great forces in the columns. By strengthening this strut (Figure 7) the deformations in the columns and the stresses at +10.4 m were reduced significantly.

The beams at the top level which support the tank are greatly influenced by the seismic actions. The permanent and live loading the ratio between the stress and resistance is around 0.35 while in combination with the seismic action this value goes over 1.0 thus requiring improvements (Figure 8).

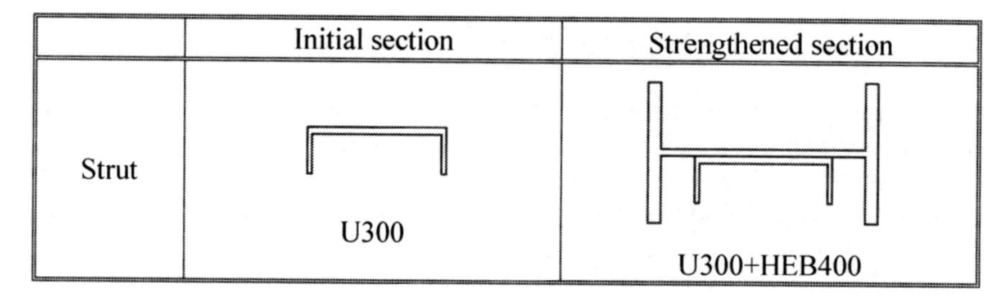

Figure 7: Strut strengthening.

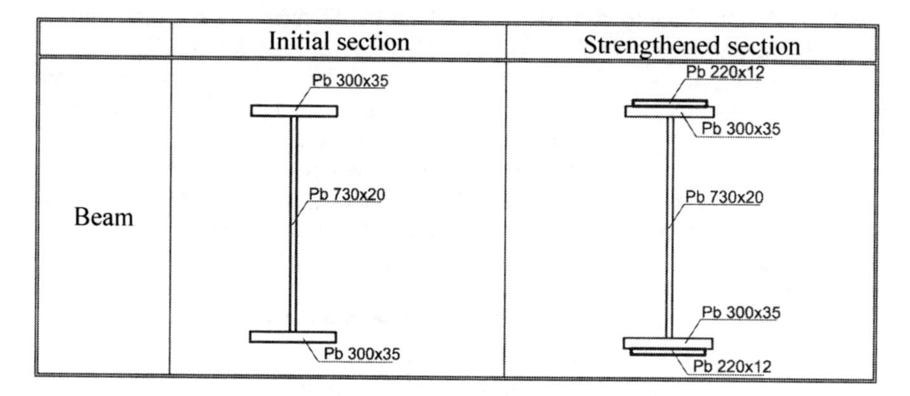

Figure 8: Beam strengthening.

The columns were made of welded plates and the rust is very deep within the material in the supports. The stresses would not be over limits if the initial section was intact, as it stands now for the sections at the base of the columns resulting the strengthening is a necessity (Figure 9).

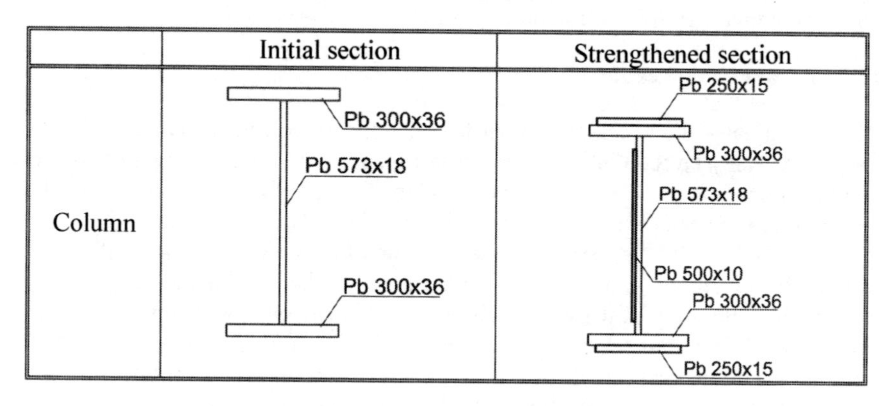

Figure 9: Column strengthening.

5 Sustainability considerations

Sustainability means an increase in the economical development (maintain and improve profitability), social policy (improving safety and health, improving quality of the built environment) and environment protection (energy usage, operational usage, embodied energy, emissions from manufacturing, processes, waste to surroundings, and pollution) [6].

5.1 Economic prosperity (profit)

The economic prosperity refers to maintaining or improving profitability. This section will not take into account the profitability of using certain technology for desulphurization.

For what the existing structure is concerned it can no longer be considered sustainable. Rust has affected the ash tubes and any intervention upon the structure, would require an evaluation which would lead to the modifications above shown.

The Steel Recycling Institute gathered information that electric arc furnaces can obtain the same steel by using 80% of scrap [6]. It results that for a new structure, 53 tones can be obtained by recycling the old structure.

The material consumptions for each case are presented in Table 1.

Table 1: Material consumptions.

Case	Total weight of structure	Strengthening parts	Initial structure	Usage of initial structure
S	99.67 tones	33.46 tones	66.21 tones	66.4%
SM	81.13 tones	14.92 tones	66.21 tones	81.6%
NS	67.68 tones	-	-	78.3% (recycled)

S – Strengthening; SM – Structural modification; NS – New Structure.

5.2 Social capital (people)

It was evaluated the necessary labour for one tone of steel in each analysed case. As it can easily be seen in Table 2 a greater quantity of labour is required for the new components that strengthen the structure than for the refurbishment of the existing elements and for their preparation in order to be strengthened.

In Table 2, “new” represents the labour done for elements that strengthen the structure and “initial” represents the labour done for the existing elements.

In what the safety of the workers is concerned there cannot be any differences between the situations taken into account. The health of the workers is negatively influenced in the case in which the existing structure has to be refurbished and prepared to be strengthened because of the rust and the dust that have to be eliminated.

Table 2: Amount of labour.

Case	Man·hour (new)	Man·hour (initial)	New parts	Initial structure	Man·hour (1x3)	Man·hour (2x4)	Total (5+6)
0	1	2	3	4	5	6	7
S	11.41	2.28	33.46 tones	66.21 tones	381.8 man·hour	150.9 man·hour	532.7 man·hour
SM	11.41	2.28	14.92 tones	66.21 tones	170.2 man·hour	150.9 man·hour	321.1 man·hour
NS	11.41	-	67.68 tones	-	772.2 man·hour	-	772.2 man·hour

Usually the quality of the building environment is superior for a new structure due to technological improvements applied to that structure. In our case the analysed structure is just the support for the supplementary installations required by the new non-pollution technology. The existing structure itself is not improved by the implemented technology.

5.3 Environment protection

Among the most important points to be considered for the environment protections are the following:

- •Energy –building energy use;
- •Embodied energy –materials;
- •Operational energy –thermal efficiency;
- •Pollution;
- •Waste to landfill.

In our case, the first three items do not make a big difference, since they refer to a building for which the heat transfer towards the environment is almost nil. Inside the structure there will not be any heating devices.

The pollution data taken into account are from Table 1. According to the statistics for one tone of steel it is produced 0.9 tones of carbon emissions. It results that between the S and SM options the reduction of carbon emissions is 16.69 tones. If a new structure is manufactured it will give a production of 60.91 tones of carbon emissions. The other pollution factors (transportation, erecting with cranes) can be considered equal for all three cases, since the excess of material that needs to be transported might be compensated with the pollution for devices needed for the cleaning of the existing elements which are corroded.

Waste to landfill is predominant for the cleaning of the corroded elements since this rust cannot be recycled. The quantity of this waste is function to the corrosion on the entire structure.

6 Results synthesis

From the analysis performed in section 2, we can observe that there are only three cases that need to be taken into account since the first case does not meet the requirements for the actual code regarding strength and safety. The case where the structure remains the same will need more material for strengthening the elements, especially elements in the MRF, but also the bracings at the lower part of the structure. Bracings at the lower and intermediate part need more material but furthermore because of their length, their slenderness has to be reduced. The strut at height +10.4 has to be strengthened and thus a more rigid assemble of elements is obtained.

For the new configuration, the forces are transmitted at lower height of the columns distributing the moment with a smoother slope between the platforms.

The beams at the top level which support the tank are greatly influenced by the seismic actions, since for the permanent and variable loads, the ratio between the stress and resistance is around 0.35 while in combination with the seismic action this value goes over 1.0 requiring improvements.

The initial configuration needs more than twice the material used for the structural modification.

For strengthening and the refurbishment of the elements there is necessary more hours of labour than for the case of structural modification and less than for the case of erecting a new structure. Considering the health of the workers, the erection a new structure is the most advisable one.

Pollution is greater for a new structure than the other two cases.

7 Conclusions

Since this structure was designed mainly to the new seismic loading codes it has to be taken into consideration different technological process situations and restrictions also. Three structural configurations were analysed with three different results and responses.

The analysis has shown that the factors for efficiency, besides strength and serviceability, can alternate in the classification of these three cases. By summarizing the positive aspects of each case, the best solution is the change of the structural system. This can be achieved only if the railway under the structure is no longer necessary. If this is not possible the second solution could be the reconstruction of a new structure as the manufacture is almost the same, but the energy and the use of material is greater than for the strengthening of the structure without having to change the structural configuration.

Taking into consideration the data presented in the paper, one can choose the best solution for the seismic retrofit of such a structure only based on an accurate quantification of economic, social and environment aspects involved.

References

[1] Eurocode 8 *Design of structures for earthquake resistance* Part 3: *Assessment and retrofitting of buildings*

[2] Sarno, L.Di & Elnashai A.S., *Seismic retrofitting of steel and composite building structures*, Report 02-01, September 2002

[3] Bruneau, M., Uang, C.M. & Whittaker, C.M. (1997). *Ductile design of steel structures*. McGraw Hill, New York, New York, USA.
[4] Ivan, M., Ivan, A., Stanciu, M. & Popa, V., *The influence of geometric imperfections generated by erection and production errors on the behaviour of steel tower,* Proceedings of the 10th International Conference on Metal Structures, Orizonturi Universitare, Timisoara, pp. 50-56, 2003
[5] Ivan, M & Ivan, A., *The rehabilitation of the runway of the 160kN overhead travelling crane from the machine room of Iron Gates I,* Proceedings of the 10th International Conference on Metal Structures, Orizonturi Universitare, Timisoara, pp. 44-50, 2003
[6] Barrett, R.B., *Developing a Sustainable Business*, NASCC 2009, Phoenix - 1st April 2009

Nonlinear seismic behaviour of experimentally identified stiffness and damping characteristics of cold formed steel infill frames

N. Bourahla[1], B. El-Djouzi[2] & N. Allal[2]
[1]*Civil Engineering Department, University Saad Dahlab Blida, Algeria*
[2]*R&D Unit, ALRIM, Algiers, Algeria*

Abstract

This paper evaluates the effect of wall panels, made of cold formed steel with sheathing on both sides used as infill, on the seismic behaviour of steel frames in terms of the overall additional stiffness and damping capacity. For this purpose, a series of ambient vibration testing was first carried out on a bare steel skeleton and then on a fully finished five-storey building. Based on the global characteristics of the panels, which were identified by experimental matching, a finite element modeling was made to reproduce the behaviour of the entire structure. The numerical model of the structure was subjected to a ground acceleration time histories recorded in the region during the Boumerdes 2003 earthquake. By using this procedure, inadequacies in the FE modeling are highlighted and suggestions are made for better modelling practice for similar structures. The results obtained elucidate some aspects of the stiffening and energy dissipating capacity role of the infill panels.
Keywords: cold formed steel, infill frame, earthquake response, ambient vibration, modeling technique, energy dissipation capacity.

1 Introduction

Cold-formed steel members and components have been produced and widely used as structural elements for building in many places around the world for at least three decades of the century. Modern design specifications have taken substantial steps in providing design analysis methodology, but these are

WIT Transactions on The Built Environment, Vol 113, © 2010 WIT Press
www.witpress.com, ISSN 1743-3509 (on-line)
doi:10.2495/SU100241

becoming more complex, and can nowadays involve greater labour than rigorous analysis using numerical methods implemented in dedicated computer packages.

The design analysis of such structures is often complex, as their behaviour can be influenced by effects, which arise from the slenderness of members, walls and cross-sections. Prime among these effects are the various types of buckling which can occur, and which may interact with each other to promote failure at loads substantially less than those, which would be obtained in the absence of these effects. The complications induced by such effects occurred more often under severe earthquake loadings. Therefore, extensive experimental and analytical research work have been carried out to gain a better insight of the dynamic behaviour of shear walls and to establish design tables containing the ultimate capacities for static and seismic stress for different wall assemblies [1, 2]. Based on the results of large series of tests, design procedures and analytical methods were also developed to allow for the design of walls carrying horizontal and vertical loads [3, 4].

The overall seismic behaviour of cold formed structures has been also investigated in post-elastic domain. The performance of the building, as a whole, depends on the wall panels, which is governed by the performance of the connectors e.g.: sheeting-to-sheeting connectors, and sheeting-to-framing connectors. On the other hand the global behavior of the 3D structure of the building is significantly influenced by nonstructural elements, traditionally not considered explicitly in the design procedures (Dubina [5]). Finite element (FE) models were developed and proposed in recent years to predict the failure mechanisms and possibility of progressive collapse (Bae *et al.* [6]). Along with the lack of the implementation data of the guidelines, the FE models also needed to be investigated in terms of their accuracy and efficiency.

The intent of this paper is to evaluate the performance of a five storey hybrid building in terms of lateral displacement, energy dissipation capacity, and structural damage with a particular emphasis on the cold formed steel panels used as infill.

2 Description of the structure

A structure of a five-storey housing building was considered in this study. The layout of the ground floor is 30.0 m long and 23.0 m large. The skeleton of the building is made of hot rolled structural steel elements.

The frame infill is made of cold formed profiles with sections 400S200-43, 400S162-43 and 400S162-27, placed at 650 mm intervals. The studs and tracks are stiffened using 10 mm thick magnesium board (MagBoard) fixed to both sides (figs. 1 and 2). Material properties of the cold-formed steel used in this structure are for members of 1.14 mm thickness and lighter having minimum yield strength of 228MPa. All members 1.4 mm thickness and heavier were formed from steel with minimum yield strength of 345MPa.

This structure is designed to resist the dead load, live load, wind load and seismic load for Seismic Zone III (RPA99v2003, [7]).

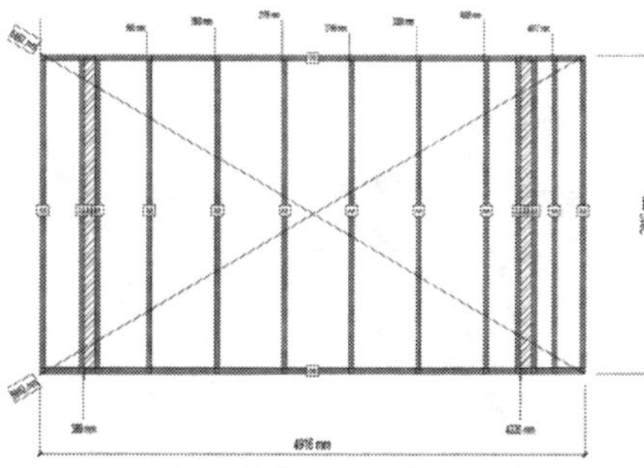

a- Panel frame w/o opening

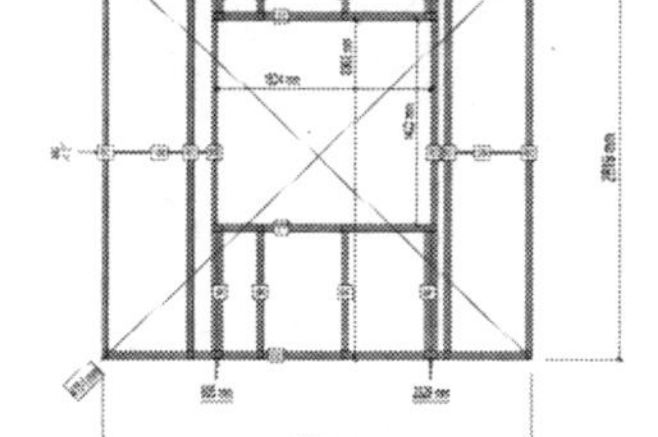

b- Panel frame with opening

Figure 1: Typical cold formed steel frame of wall panel.

a- skeleton made of hot rolled steel

b- Infill panels made of cold formed steel

Figure 2: Global views of the building under construction.

3 Numerical modelling

The bare skeleton of the building is modelled using a 3-D finite element model. The beams and the columns are idealised as flexural elements and the diagonal as axial elements. The model was validated by an experimental modal analysis.

The floor decks and especially the non-bearing walls are modelled in order to evaluate the contribution of these elements in the overall nonlinear seismic behaviour of the building in terms of additional initial stiffness, modal damping and hysteresis energy dissipation capacity. Therefore, an equivalent simple nonlinear shear link connected to a rigid bracing system is introduced to account for the overall lateral stiffness and strength of a panel (fig. 3).

The multilinear plastic-pivot hysteresis model of the FEA software package, SAP2000 [8] was used to account for the nonlinear behavior of the cold formed steel panel. The hysteretic model incorporates stiffness degradation, strength deterioration and non-symmetric response (fig. 4).

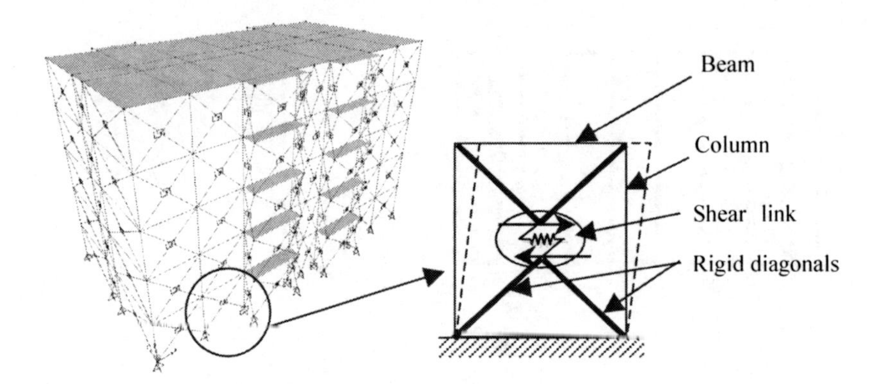

Figure 3: The 3-D model with an illustration of the shear link model of a panel.

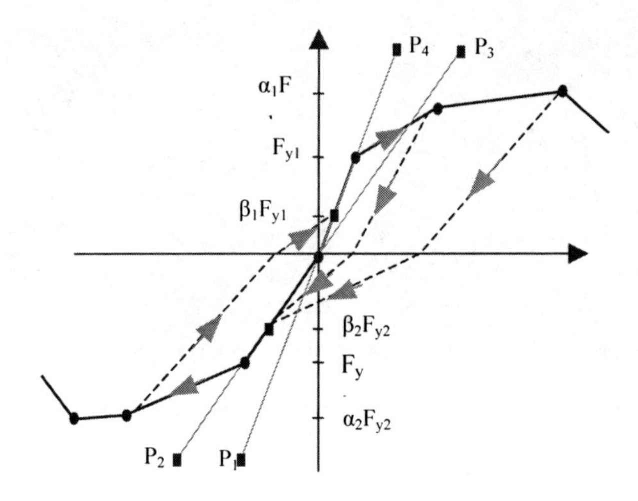

Figure 4: Multilinear plastic pivot hysteresis model [9].

The parameters of the hysteresis models are estimated through a number of load-deformation curves obtained experimentally under cyclic displacement histories, collected from a number of recent studies available in the literature [3, 9, 10].

The structure has six different types of panels with different dimensions and openings. The relative elastic stiffness's were normalized using equivalent FE shell elements models, then the actual values of the initial stiffness was estimated by matching the building frequencies obtained experimentally using the ambient vibration testing.

4 Ambient vibration testing

The elastic, mass and damping characteristics of the structure to be analysed must be known to a sufficient degree of accuracy in order to evaluate the actual structural capacity of the construction. The elastic dynamic properties, particularly the natural frequencies and the corresponding mode shapes are a combined measure of the structural characteristics of the construction. These model characteristics can be successfully estimated, especially in elastic range, using the well known ambient vibration testing. In this paper we present briefly the main issues pertaining to this particular modal testing; frequency response function (FRF) measurement techniques, testing procedure, and modal parameter estimation method.

For this particular case preliminary modal analyses were first carried out and the fundamental modes were predetermined. On the base of these results optimum sensor locations were chosen nevertheless additional measurement points were also included to take account of any other modes that were not predicted by the analytical model. Measurement near the nodal point of any of the modes will omit that particular mode and aliasing effect is to be prevented by avoiding intersection regions of the fundamental modes to be identified.

In total 4 measurement points were performed. Fig. 5 shows the locations and the orientations of the sensors on the 4th floor.

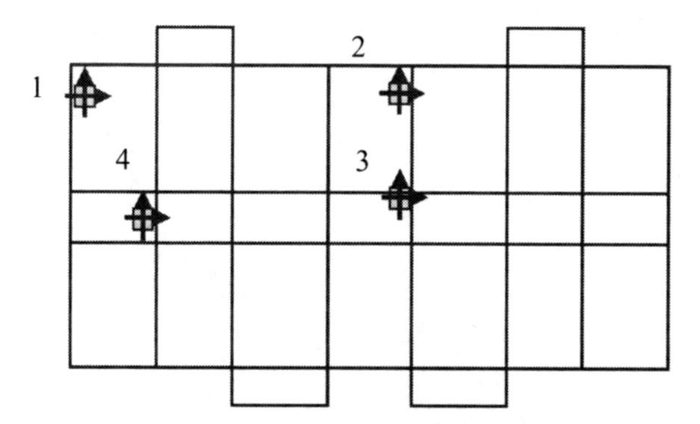

Figure 5: Sensor locations on the structure.

The tests were performed using three degrees of freedom seismometer type Lennartz electronic (Le3Dlite) and a data acquisition system type City Shark II. The measured signals were processed using the GEOPSY program (Wathelet [11]) capable to perform most of the signal processing operations for the analysis of ambient vibration data. The sensors were located and oriented according to the previously defined test programme. The recording time for each sequence was set to 5 mn and found to be largely sufficient to obtain smooth FRF curves.

4.1 Ambient vibration measurement on the bare skeleton of the building

The natural frequencies of the bare skeleton were first identified using a "peak cursor" on the frequency response functions. Due to the flexibility of the floors (without decks), the individual vibration modes do not exhibit translational motions of rigid diaphragms, but generally the mode shapes have deformed floors especially along the transversal direction.

The first curve in fig. 6 shows the FRF of the transverse vibrations measured on the centre of the top floor. The clearly distinct two first peaks at 1.56 Hz and 2.32 Hz correspond to the first and second translational mode determined by the model in the transverse direction. Because of the very stiff braced bays at the edges of the building, the flexible floor deformed excessively at the centre. The second FRF curve was measured by a sensor located at point 4 to detect the vibration modes of the braced bays. The latter shows a peak at a frequency equal to 10.25 Hz which corresponds to a higher vibration mode of the model as shown on the same figure. The dominant longitudinal vibration mode was successfully identified using the sensor located on point 4 at a frequency equal to 3.2 Hz together with a second peak at 4.51 Hz corresponding to a torsional mode.

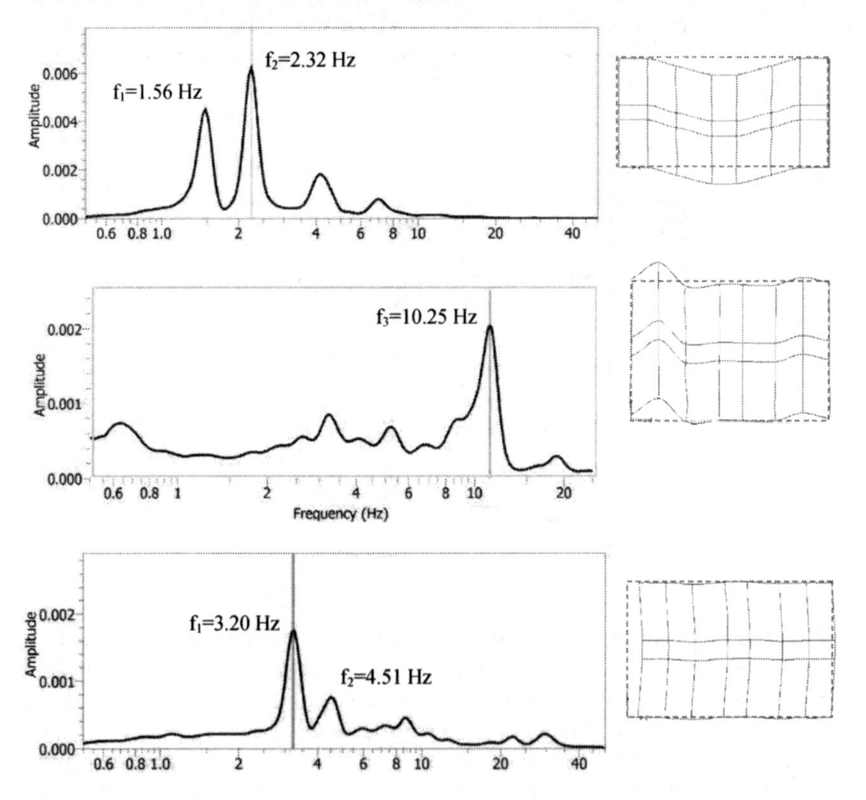

Figure 6: The FRF curves and the corresponding mode shapes (skeleton).

In general, excellent correlation between experimental and analytical natural frequencies is achieved in most cases as can be seen from Table 1.

Table 1: Natural frequencies and corresponding damping ratios of the bare skeleton.

Mode	Analytical Frequency (Hz)	Experimental Frequency (Hz)	Direction	Damping (%)
1	1.59	1.56	Transversal	0.5
2	2.45	2.32	Transversal	0.4
3	3.57	3.20	Longitudinal	0.4
4	4.43	4.51	Longitudinal	0.3
5	11.8	11.3	Torsional	0.3

4.2 Ambient vibration measurement on the completed building

In order to account for the overall rigidity contribution of the cold formed decks (for floor) and the cladding, series of ambient vibration testing similar to the previous one have been carried out on a fully finished building.

We noticed that the first frequency equal to 3.48 Hz corresponds to a dominant longitudinal mode. The building becomes stiffer in the transverse direction with a frequency equal to 4.65 Hz and the mode shapes have non-deformable floors.

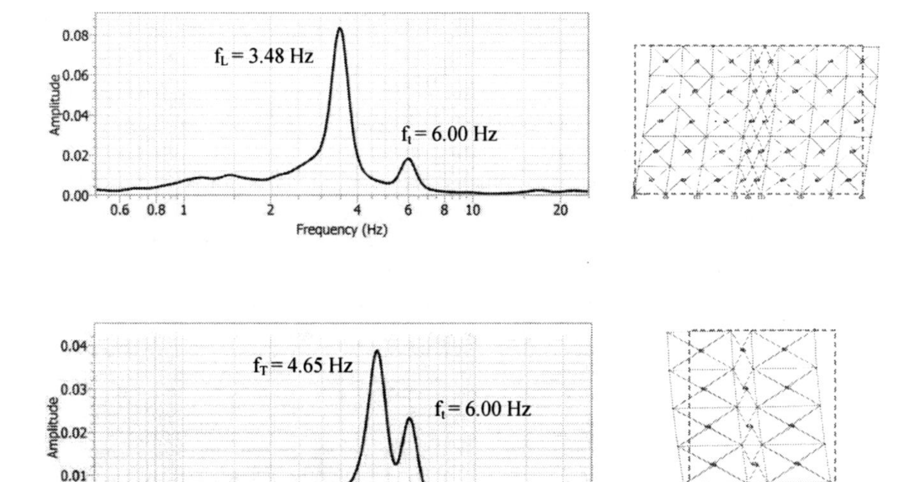

Figure 7: FRF curves and corresponding mode shapes (full building).

5 Evaluation of the stiffness and damping contribution of the cold formed steel infill

In first instance, an evident increase of the overall rigidity is achieved by the non-bearing cold-formed steel panels in the horizontal directions. The substantial increase of the frequency along the transverse direction (more than 200%) is attributed to the fact that the lateral panels have no openings beside that the decks stiffen the floors and eliminate the flexible modes of the floors which makes the skeleton bracing system more efficient. As can be seen in table 2 the modal damping has increased for all modes.

Table 2: Natural frequencies and corresponding damping ratios of the completed building.

Mode	Analytical Frequency (Hz)	Experimental Frequency (Hz)	Direction	Damping (%)
1	3.69	3.48	Transversal	0.6
2	4.63	4.65	Transversal	1.4
3	5.26	6.00	Longitudinal	1.0

6 Nonlinear dynamic analysis

Nonlinear dynamic analyses are carried out to investigate the overall behavior of the building under earthquake ground motion. For the purpose of the present analysis the model described previously is subjected to a ground acceleration recorded during Boumerdes earthquake in 2003, at a station located in Dar-El-Beida (about 40 km from the building site). The duration of the strong motion used for the analysis is 20 seconds. The response analysis is carried out using the E-W component with a PGA equal to 0.52g. Modal damping has been included in the model and this was set at 5% for all modes as is usual practice for steel structures. The step-by-step numerical integration is carried out at time interval 1/1000 sec.

For a comparison reason, a linear dynamic analysis is performed assuming that all elements remain in the elastic range. The global response measured in terms of the top floor displacement indicates that the overall behaviour is dominated by the fundamental frequency (fig. 8).

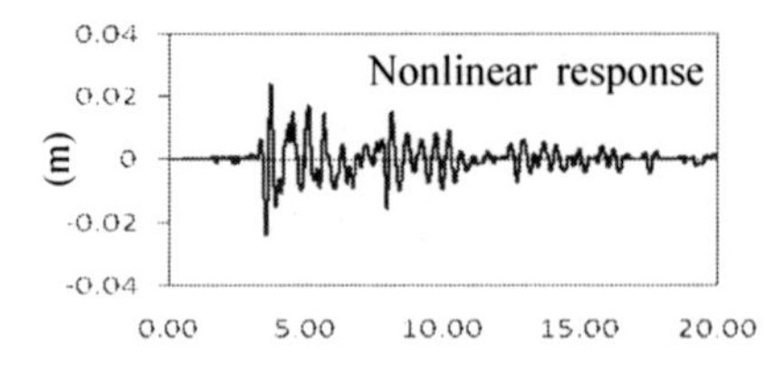

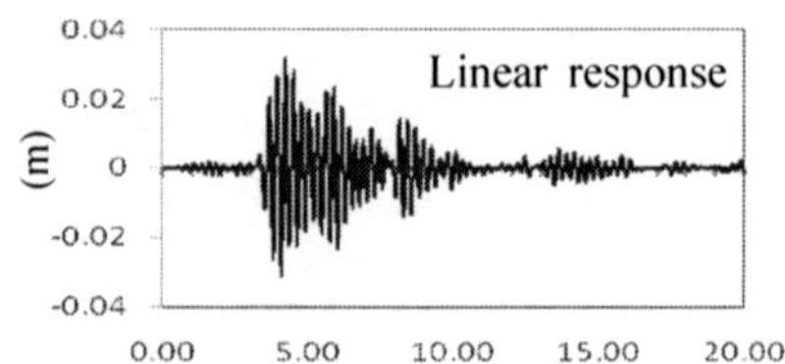

Figure 8: Time history of the top floor lateral displacement.

Under the same earthquake ground motion conditions to which the linear model has been subjected, the structure with plastic pivot multi-linear characteristics exhibit different behaviour. At an early stage several panels undergo plastic deformations at the lower storey and propagate to the upper storeys during the severe phase of the ground acceleration. The building showed evidence of a considerable variation of the overall stiffness, which become visible on the response time history characterized by an elongated waveform (Fig. 9). The maximum displacement of the top floor is about 30% lower than the linear model and more importantly, the vibration are damped out after the main peak acceleration with almost no permanent deformation at the end of the response.

The low yield strength of the panels provide the building with an additional energy dissipation capacity at an early stage of the response which help damp out intense vibration and detune the structure from possible resonance. The energy dissipated by hysteresis effect of the infill panel is more than 30% of the modal damping energy of the structure. The curve of the input energy is characterized by small fluctuations, which reveal that the energy imparted by the ground motion to the structure is dissipated almost instantaneously which indicate an efficient damping capacity (Fig. 9). It should be noted also that the yielding is not concentrated only in the lower storey. The panel shear–deformation curves show significant hysteresis loops even on the panels of the upper storey.

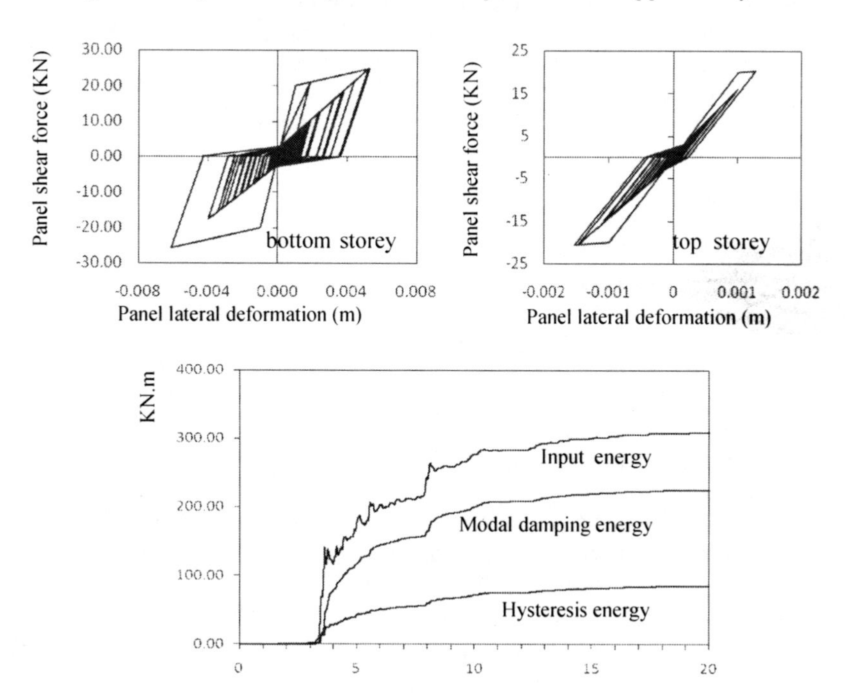

Figure 9: Energy time histories and shear-deformation curves of the panel walls.

7 Conclusion

The non-bearing walls are commonly discarded in the design practice of buildings. This research showed that in the case of flexible steel frames with wall panels, made of cold-formed steel with sheathing on both sides used as infill, the latter can contribute significantly to the overall rigidity and energy dissipation capacity. The ambient vibration testing used at different construction stages is practical in evaluating the global characteristics of the wall panels.

An efficient simplified modeling technique for the analysis of structures incorporating wall panels is developed. The model presented is based on the multi-linear plastic pivot hysteresis curve, but any suitable relationship could be used. This makes it particularly attractive for nonlinear response history analysis, but could be of great help for engineers to perform nonlinear static analysis (push-over) both in the design stage or the seismic evaluation of existing structures.

References

[1] Serrette R., Hall G. and Ngyen H., *Shear wall values for light weight steel framing*, final report, AISI, Washington, 1996.

[2] Serrette R., Hall G. and Encalada J., *Additional shear wall values for light weight steel framing*., Final report, AISI, Washington, 1997.

[3] Langea J. and Naujoksb B., Behaviour of cold-formed steel shear walls under horizontal and vertical loads, *Thin-Walled Structures,* **44 (2006)**, pp. 1214-1222, 2006.

[4] Al-Kharat M. and Rogers C. A., Inelastic performance of cold-formed steel strap braced walls, *Journal of Constructional Steel Research*, **63(4)**, pp. 460-474, 2007.

[5] Dubina D., Behavior and performance of cold-formed steel-framed houses under seismic action, *Journal of Constructional Steel Research,* **64 (2008)**, pp. 896–913, 2008.

[6] Bae S., LaBoube R. A., Belarbi A. and Ayoub A., Progressive collapse of cold-formed steel framed structures, *Thin Wall Structures,* **46 (2008)**, pp. 706-719, 2008.

[7] RPA99v2003, *Regles parasismiques algeriennes*, National Earthquake Engineering Centre CGS, Algiers, 2003.

[8] Computer and Structures, Inc. CSI analysis reference manual for Sap2000. Berkeley, 2004.

[9] Gad E.F., Duffield C.F., Hutchinson G.L., Mansella D.S. and Starkb G., Lateral performance of cold formed steel framed domestic structures, *Engineering Structures*, **21 (1999)**, pp. 83–95, 1999.

[10] Fulop L.A., and Dubina D., Performance of wall-stud cold-formed shear panels under monotonic and cyclic loading Part II: Numerical modelling and performance analysis, *Thin-Walled Structures* **42 (2004)**, pp. 339–349, 2004.

[11] Wathelet M., GEOPSY geophysical Signal Database for Noise Array Processing, Software, LGIT, Grenoble, France, 2005.

Comparison of seismic performance of strengthened historical masonry buildings under different structural designs

O. Isler & V. Oztas
Department of Theory of Structure, Istanbul Technical University, Turkey

Abstract

The subject of this study is the investigation of historical masonry buildings under earthquake loads. During any earthquake the behaviour of the structural system of masonry buildings is very complicated, and the shear resistances are very low. The masonry building's wall materials, thickness, height and workmanship have a significant affect on the stability and shear resistance of the buildings. The dimensions of wall openings for windows and doors and their placing in the wall, and continuous walls or non continuous walls are also important for stability. In order to prepare a weak masonry building for a possible and heavily damaging earthquake, the structural system needs strengthening to increase its seismic performance. In this study, in order to strengthen masonry buildings, two different strengthening structural models are considered. Firstly, the masonry building is strengthened with the additional RC shear walls, secondly; the walls of the masonry building are strengthened with the FRP/GFRP grid bonded. For example, an existing and historical masonry building chosen from Istanbul is considered. Their structural systems, with strengthened and non strengthened examples, are analyzed under earthquake loads. The results of the analyses are investigated and the obtained seismic performances of the different strengthened structural systems are compared between each other. Consequently, seismic performance and lateral displacements are improved by strengthening with additional RC shear walls and/or FRP/GFRP grid bonded systems. The analyses also show the displacements of the strengthened systems are reduced and these are improved 9 and 4 times according to the present building, respectively. Finally, to maintain

WIT Transactions on The Built Environment, Vol 113, © 2010 WIT Press
www.witpress.com, ISSN 1743-3509 (on-line)
doi:10.2495/SU100251

outdoor views and for easy application, the FRP/GFRP grid bonded strengthened systems are recommended instead of the additional RC shear wall system.
Keywords: masonry building, earthquake, strengthen.

1 Introduction

Unreinforced masonry (URM) walls in masonry structures have architectural and statical properties. URM walls constitute volume, protect structures against external effect and divide structure into compartments [1]. The versatility of URM walls is an advantage according to use and construction of masonry structures. In masonry structures, damage may occur because of earthquake forces, but various strengthening methods are used in order to renovate these damaged structures.

In this study, in order to strengthen masonry buildings, two different strengthening structural models and existing models are considered. Firstly, the existing model is analyzed. Secondly, the masonry building is strengthened with the additional RC shear walls. Finally, the walls of the masonry building are strengthened with the GFRP.

GFRP is composed of carbon, aramid or glass fibers with epoxy resin. GFRP has high durability, high tensile strength, resists against fatigue, and is applicable in various forms.

2 Analyzing of a historical masonry building

In this study, a three-story historical masonry building is analyzed with Sta4-CAD software. The dimensions of the structure are 18.11 m × 17.10 m and URM wall thicknesses 33cm at the basement, 22cm at the other stories. This structure is in the first earthquake zone and other parameters are shown in the following table.

Table 1: Parameters of structure.

Building Properties	Value
Soil group	Z3
Number of flat	3
earthquake acceleration coefficient (Ao)	0,4
Structure type coefficient (R)	2
Soil periods.(Ta/Tb)	0,15/0,6
Live load coefficient	0.6
ground safety stress (t/m²)	15.0
ground bearing coefficient (t/m³)	1500.0
Gravity of concrete (t/m³)	2,5
Earthquake code of Turkey [2]	TDY2007

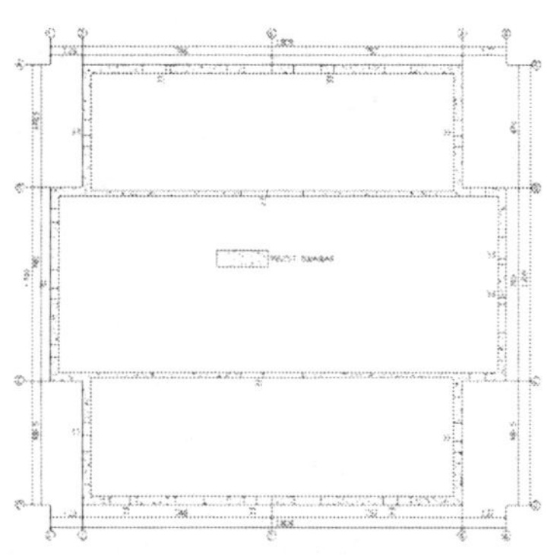

Figure 1: Examined historical building and plan.

2.1 Analysis of existing structure

Firstly, the existing structural system is analyzed with software and the results are examined.

The spectrum coefficient for masonry structures S(T)=2.5 and earthquake load reduction coefficient for masonry system is Ra(t)=2.

Earthquake loads for x and y direction are found.

$$V_{tx} = W.A(t)/Ra(t) > 0{,}10.\ A_o.I.W \qquad 2259.01 > 180.72\ t$$
$$V_{ty} = W.A(t)/Ra(t) > 0{,}10.\ A_o.I.W \qquad 2259.01 > 180.72\ t$$

Modal analysis minimum load ratio multiplies with equivalent earthquake loads.

Maximum earthquake load is selected from multiplicand value and modal analysis value.

X direction earthquake load selection:

$$0.90 \times 2259.012 = 2033.111 > 1487.672 \rightarrow 2033.111\ t \quad (4.5a)$$

Y direction earthquake load selection:

$$0.90 \times 2259.012 = 2033.111 > 1533.656 \rightarrow 2033.111\ t \quad (4.5b)$$

Table 2: Maximum displacements (cm).

Kat	9. yükleme		10. yükleme		11. yükleme		12. yükleme	
(dyf)	δx (m)	θz (rad)	δx (m)	θz (rad)	δy (m)	θz (rad)	δy (m)	θz (rad)
4	0.1487000	0.0000026	0.1487000	0.0000026	-0.258344	-0.000106	-0.258344	-0.000106
3	0.0988657	0.0000021	0.0988657	0.0000021	-0.183651	-0.000057	-0.183651	-0.000057
2	0.0456890	0.0000012	0.0456890	0.0000012	-0.088422	-0.000020	-0.088422	-0.000020
1	0.0033021	0.0000001	0.0033021	0.0000001	-0.006277	-0.000001	-0.006277	-0.000001

Earthquake displacements are found.

Maximum displacements for existing structure are:

$$\delta x = 0.148700,\ \delta y = 0.258344m.$$

2.2 Analysis of masonry structure strengthened with reinforced concrete shear wall

30cm RC shear walls are entered to the system at the x and y directions. St420 steel bars and C30 RC are used at the strengthened system.

For C30, E =318.000 kg/cm^2 fu =300 kg/cm^2
St420 tensile stress = 4200 kg/cm^2

Equivalent RC cross-section and modulus of elasticity of RC and masonry shear wall which are worked together are found from breaking load of experiment result which are made by Franklin, S., Lynch J., Abrams D, (Performance of Rehabilitated URM Shear Walls: Flexural Behaviour of Piers) [3]

Breaking load of existing masonry wall= FT = 29 kN
Breaking load of RC strengthened masonry wall = FB= 68 kN

$$k1 = FB / FT = 68 / 29 = 2.34$$
$$Eeş1 = ET + EB/2.34 = 3000 + 318000/2.34 = 138897 \text{ kg/cm}^2$$
$$beş1 = bT + bB / 2.34 = 33 + 30 / 2.34 = 45 \text{ cm}$$

RC strengthened structural system is analyzed using equivalent RC cross-section and elasticity coefficient and examined results

Spectrum coefficient for strengthened masonry structures S(T)=2.5 and earthquake load reduction coefficient for strengthened masonry system Ra(t)=2

Earthquake loads for x and y direction are found

$$Vtx=W.A(t)/Ra(t) > 0,10.Ao.I.W \quad 2694.02 > 215.52$$
$$Vty=W.A(t)/Ra(t) > 0,10.Ao.I.W \quad 2694.02 > 215.52$$

Modal analysis minimum load ratio multiplies with equivalent earthquake loads.

Earthquake load select from big one between multiplicand value and modal analysis value.

X direction earthquake load selection:

$$0.90 \times 2694.018 = 2424.617 > 1913.870 \rightarrow 2424.617 \text{ t}$$

Y direction earthquake load selection:

$$0.90 \times 2694.018 = 2424.617 > 2110.676 \rightarrow 2424.617 \text{ t}$$

Table 3: Maximum displacements (cm).

Kat	9. yükleme		10. yükleme		11. yükleme		12. yükleme	
(dyf)	δx (m)	θz (rad)	δx (m)	θz (rad)	δy (m)	θz (rad)	δy (m)	θz (rad)
4	0.0383761	0.0002163	0.0383761	0.0002163	-0.028532	0.0004148	-0.028532	0.0004148
3	0.0262679	0.0001638	0.0262679	0.0001638	-0.021903	0.0003142	-0.021903	0.0003142
2	0.0133390	0.0000967	0.0133390	0.0000967	-0.013202	0.0002061	-0.013202	0.0002061
1	0.0032673	0.0000352	0.0032673	0.0000352	-0.004790	0.0000842	-0.004790	0.0000842

Earthquake displacements are found.
Maximum displacements for existing structure are:

$$\delta x = 0.0383761, \quad \delta y = 0.0285320 \text{ t.}$$

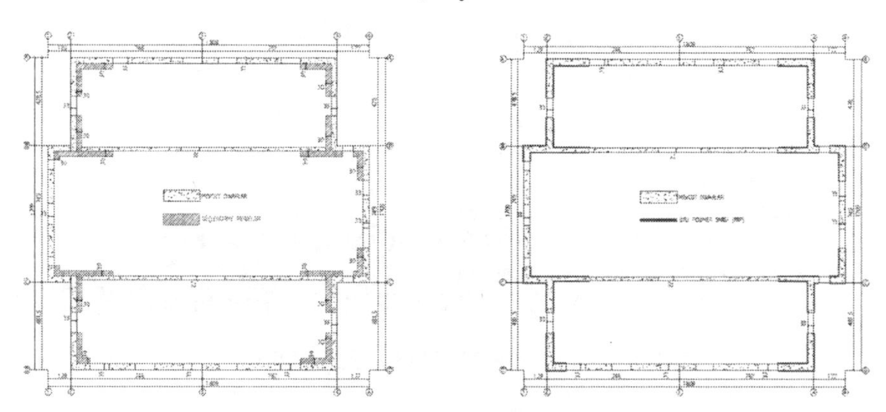

Figure 2: RC and GFRP strengthened structure plan.

2.3 Analysis of masonry structure strengthened with Glass Fiber Reinforced Polymer (GFRP)

GFRP is applied to two faces of masonry walls instead of RC walls. GFRP is applied full face on the walls, not diagonally. Properties of GFRP are explained in the following table.

Table 4: GFRP properties.

GFRP TYPE	Tensile strength (kg/cm2)	Modulus of elasticity (kg/cm2)	Ultimate stress (%)
MbraceFibre C1-30	34300	2300000	1.5

Equivalent RC cross-section and modulus of elasticity of GFRC and masonry shear wall which are work together are found from breaking load of experiment result which are made by Özsaraç and Torubalcı [4].
Breaking load of existing masonry wall = FT = 11.33 kN.
Breaking load of GFRP strengthened masonry wall = FF = 29.33 kN.

$$k2 = FF / FT = 29.33 / 11.33 = 2.59$$
$$Eeş2 = ET + EF / 2.59 = 3000 + 2300000 / 2.59 = 891000 \text{ kg/cm}^2$$
$$beş2 = bT + bF / 2.59 = 33 + (0{,}2x2) / 2.59 = 34 \text{ cm.}$$

GFRP strengthened structural system is analyzed with using equivalent GFRP cross-section and elasticity coefficient and examined results.

Spectrum coefficient for strengthened masonry structures S(T)=2.5 and earthquake load reduction coefficient for strengthened masonry system Ra(t)=2

Earthquake loads for x and y direction are found

$$Vtx=W.A(t)/Ra(t)> 0{,}10.\ Ao.I.W \quad 2260.11 > 180.81\ t$$
$$Vty=W.A(t)/Ra(t)> 0{,}10.\ Ao.I.W \quad 2260.11 > 180.81\ t$$

Modal analysis minimum load ratio multiplies with equivalent earthquake loads. Earthquake load select from big one between multiplicand value and modal analysis value.

X direction earthquake load selection:

$$0.90 \times 2260.106 = 2034.096 > 1627.951 \rightarrow 2034.096\ t$$

Y direction earthquake load selection:

$$0.90 \times 2260.106 = 2034.096 > 1702.915 \rightarrow 2034.096\ t$$

Table 5: Maximum displacements (cm).

Kat	9. yükleme		10. yükleme		11. yükleme		12. yükleme	
(dyf)	δx (m)	θz (rad)	δx (m)	θz (rad)	δy (m)	θz (rad)	δy (m)	θz (rad)
4	0.0946083	0.0006067	0.0946083	0.0006067	-0.062306	0.0004859	-0.062306	0.0004859
3	0.0645475	0.0004544	0.0645475	0.0004544	-0.046252	0.0003993	-0.046252	0.0003993
2	0.0339478	0.0002673	0.0339478	0.0002673	-0.026059	0.0002765	-0.026059	0.0002765
1	0.0099263	0.0001049	0.0099263	0.0001049	-0.008675	0.0001222	-0.008675	0.0001222

Maximum displacements are found.

Maximum displacements for existing structure:

$$\delta x = 0.094608,\ \delta y = 0.062306\ m.$$

3 Conclusion

Consequently, analysis of an existing structure, an RC strengthened structure and a GFRP strengthened structure are compared from the point of earthquake loads and displacements.

Earthquake loads of existing structure

$$F_x = F_y = 2033\ t$$

Earthquake load of RC strengthened structure

$$F_x = F_y = 2424\ t$$

Earthquake load of GFRP strengthened structure

$$F_x = F_y = 2034\ t$$

Maximum displacements compared to three conditions

Maximum displacements of existing structure

$$\delta x= 0.1487,\ \delta y= 0.2583\ m$$

Maximum displacements of RC strengthened structure

$$\delta x= 0.0383,\ \delta y= 0.0285\ m$$

Maximum displacements of GFRP strengthened structure

$$\delta x= 0.0946,\ \delta y= 0.0623\ m$$

Because of the weak wooden structural system of the slabs, large value displacements are found in the analysis.

Maximum displacements are compared in table 6.

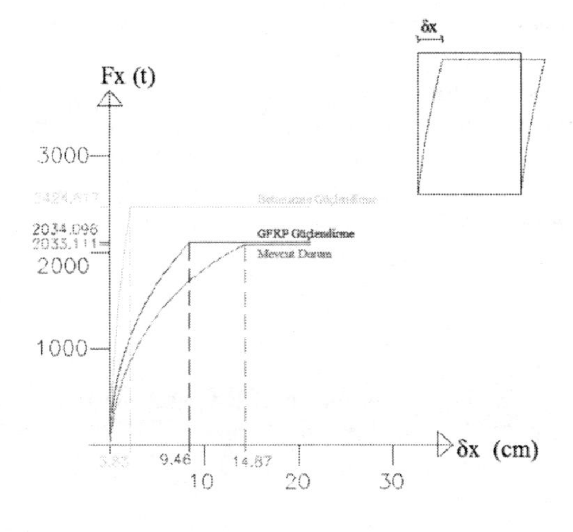

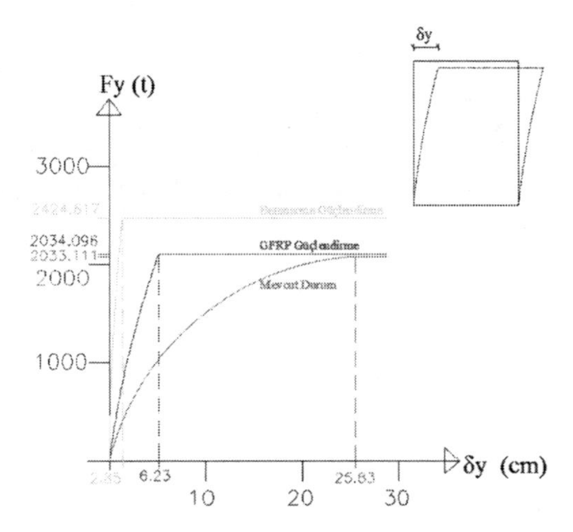

Figure 3: Earthquake loads-x and y direction displacement curves.

Table 6: Maximum displacements.

Strengthened System	Existing structure (cm)	RC strengthened structure (cm)	GFRP strengthened structure (cm)
δx	14.87	3.83	9.46
δy	25.83	2.85	6.23

Earthquake loads and maximum displacements for x and y directions are displayed on graphical representations

The analyses show the displacements of the strengthened systems are reduced, and these are improved 9 and 4 times according to the present building respectively. Finally, RC shear walls strengthened system provides new code criteria. However, if it is necessary to keep outdoor views and for easy application, the FRP/GFRP grid bonded strengthened systems are recommended instead of the additional RC shear wall system.

References

[1] Torunbalcı N, İsler O. Performance of R/C and Masonry Structures During the 2003 Bingöl Earthquake in Turkey', 13th World Conference on Earthquake Engineering,, Vancouver, B.C., Canada, August 1-6, 2004, Paper No. 3145.

[2] DBYBHY. 2007 Earthquake Code of Turkey, Bayındırlık ve İskan Bakanlığı, Ankara.

[3] Franklin, S., Lynch J., Abrams D, 2001. Performance of Rehabilitated URM Shear Walls: Flexural Behaviour of Piers, Department of Civil Engineering, University of Illinois at Urbana-Champaign, Illinois.

[4] Özsaraç, S., Torubalcı, N. 2008. Experimental study of GFRP strengthened of load-carrying wall at masonry structures, Istanbul Technical University.

Section 8
Behaviour of structures

Scaling of the modal response of a simply-supported rectangular plate

R. D. Hampton[1], T. H. Li[2] & L. K. Byers[1]
[1]*Civil and Mechanical Engineering Department, US Military Academy, USA*
[2]*Weapons & Materials Research Directorate, US Army Research Laboratory, USA*

Abstract

Actual military transports (such as ground vehicles or ships) are often not acceptable for shock-response testing, for such reasons as cost, vehicle availability, personnel availability, or time. If testing can be conducted on scaled-down models, whether of the actual transports or of critical subsections, the results might be usable to approximate the shock response of the full-scale hardware, provided that appropriate scaling relationships can be developed. Since many military vehicles and vessels have large structural portions consisting of homogeneous flat plates, these elements serve as a logical starting point for shock-response scaling efforts. This paper develops scaling relationships for thin rectangular plates in simple support, subject to transverse point-shock loads, under the assumptions of linearity, homogeneity, and geometric similarity of the plate faces. The plates are also assumed to have the same respective modal damping ratios, for corresponding modes. It is found that simple scaling factors exist, between corresponding modal frequencies, and between corresponding damped modeshapes. Further, via suitable discretization, these factors can be used to provide simple scaling relationships between the shock response of points on a scaled-down plate and that of corresponding (scaled) points on the scaled-up plate.
Keywords: shock response, rectangular plates, modal analysis, model scaling.

WIT Transactions on The Built Environment, Vol 113, © 2010 WIT Press
www.witpress.com, ISSN 1743-3509 (on-line)
doi:10.2495/SU100261

1 Introduction

Shock-response testing of actual, full-sized hardware is often impractical or impossible, for reasons of cost or availability. The hardware to be tested may be prohibitively expensive, limited in number, unavailable (e.g., in use), inaccessible, or even nonexistent (e.g., under development). And even should the hardware—or a suitable full-scale model—be available and released for testing, an appropriate test facility may itself be unavailable, nonexistent, or excessively time-intensive or expensive to use. In such cases it may be possible to conduct shock tests of scaled-down test models, and to scale up the results to approximate the results expected from full-sized models or the actual hardware.

The present paper describes how shock-response tests of a scaled-down model could be used to approximate anticipated test results of larger hardware, in the case of a simply-supported, homogeneous, damped rectangular plate. The larger and smaller plates must be proportional in length and width, but the thicknesses need be neither identical nor related by the same proportion. Both plates are assumed to have the same respective modal damping ratios, for corresponding modes; and the shock is assumed to be transverse, point-loaded, and ideal (Dirac-delta). Plastic deformation must be local only—restricted to the immediate region of the impact, and of negligible effect on remote-point vibratory response.

2 Problem statement

Referring to fig.'s 1 and 2, consider two flat, simply-supported, homogeneous rectangular plates (plate #1: the smaller, or "test" plate; and plate #2, the larger, or "full-scale" plate), with identical modal damping ratios for each respective pair of modes. The respective Poisson's ratios (ν), Young's moduli of elasticity (E), and area densities (ρ) need not be identical. Assume that numerical shock-response data is available for the test plate—whether determined experimentally or otherwise. Identify its length, width, and thickness by a_1, b_1, and h_1, respectively. Let the full-scale plate have respective length, width, and thickness of a_2, b_2, and h_2, with dimensions related to those of the test plate as follows:

$$\begin{bmatrix} a_2 & b_2 & h_2 \end{bmatrix} = \begin{bmatrix} \alpha a_1 & \alpha b_1 & \beta h_1 \end{bmatrix} \tag{1}$$

It is desired to relate mathematically the vibratory shock responses of the two plates. In particular, let the qth plate $(q = 1, 2)$ have k^{th} undamped modal frequency, k^{th} damped modal frequency, and k^{th} modeshape designated, respectively, by $\omega_{n,k,q}$, $\omega_{d,k,q}$, and $W_{k,q}$. (The subscripts n and d indicate "undamped" and "damped," respectively.) Then the first two objectives of this paper are as posed below:

- Determine the relationship between undamped modal frequencies $\omega_{n,k,1}$ and $\omega_{n,k,2}$. (The same relationship will obtain between damped modal frequencies $\omega_{d,k,1}$ and $\omega_{d,k,2}$.)
- Determine the relationship between modeshapes $W_{k,1}$ and $W_{k,2}$.

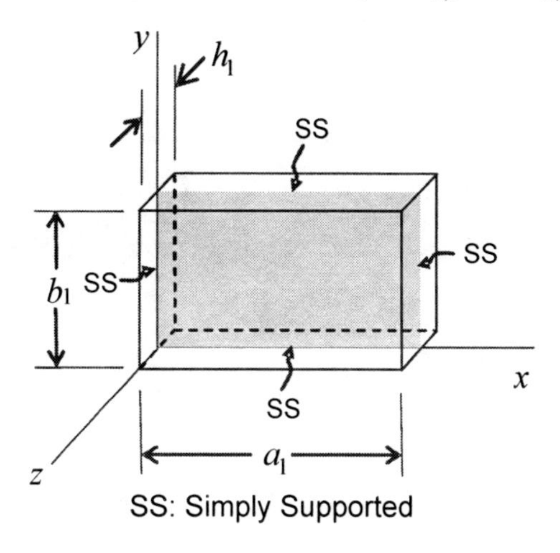

Figure 1: Plate #1: Thin test plate, with indicated edges in simple support (SS).

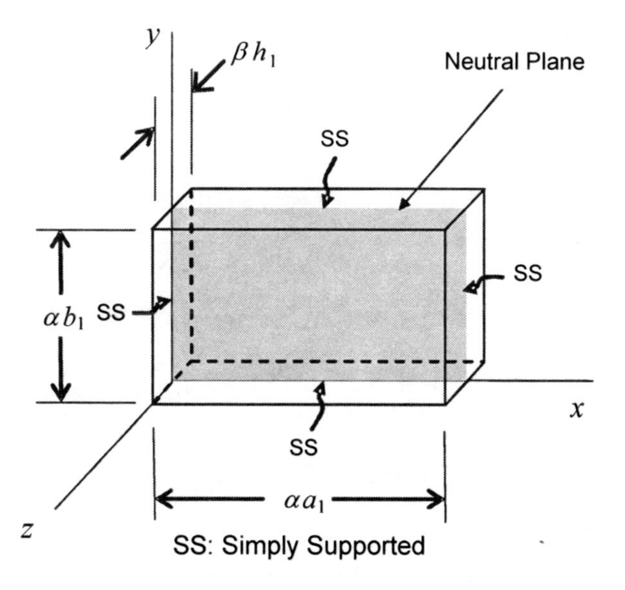

Figure 2: Plate #2: Full-scale plate, with integral scaling as indicated.

Refer now to fig. 3. Consider the case of the same two plates in free response to known initial conditions. The third objective is as follows:

- Under the assumption of modal damping, determine the relationship between the two plates' respective free responses $\left(w_{j,1}\right)_{\text{free}}$ and $\left(w_{j,2}\right)_{\text{free}}$.

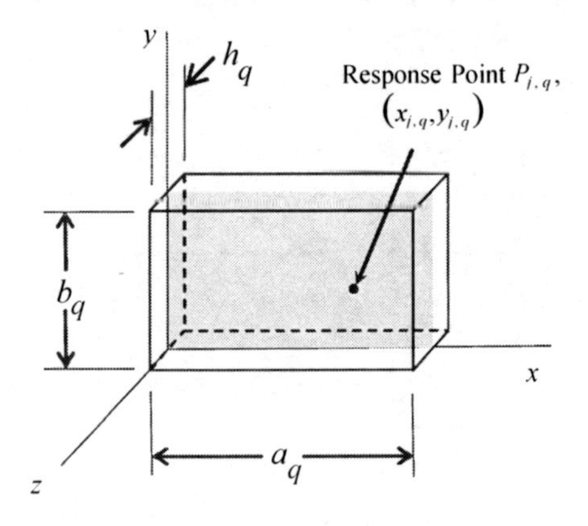

Figure 3: Response point for idealized plate.

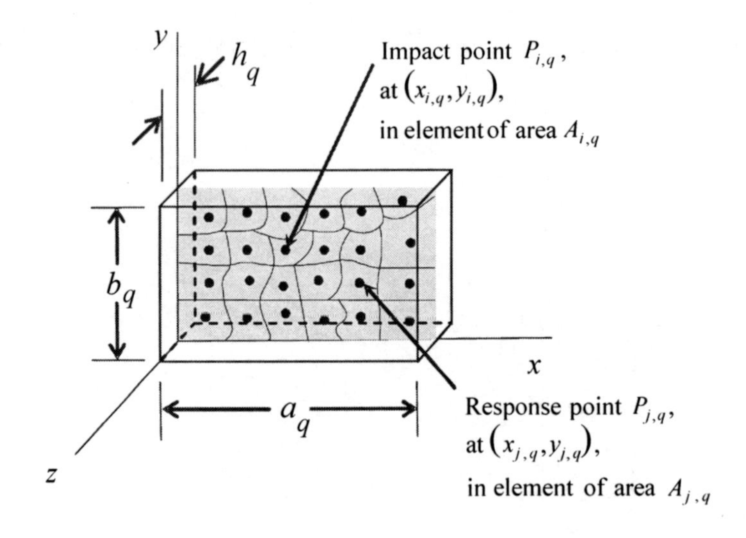

Figure 4: Discretized plate, with elements shown.

Refer now to fig. 4. For the same two plates consider the case of an ideal (Dirac-delta) impulse (having strength γ_q for plate q) applied transversely at corresponding points $P_{i,q}$ on the respective plates, where the respective impact points are "proportionally" located. That is, for $P_{i,1}$ located on the test-plate midplane at $(x_{i,1}, y_{i,1})$ let corresponding point $P_{i,2}$, with coordinates $(x_{i,2}, y_{i,2})$ on the midplane of the full-scale plate, be located at $(\alpha x_{i,1}, \alpha y_{i,1})$. Designate the respective transverse responses at corresponding points $P_{j,q}$ (also on the respective midplanes) by $(w_{j,q})_{\text{forced}}$. If the two plates are now discretized as indicated (fig. 4), a mathematical relationship can be determined between the forced responses for the two plates. This is the final objective of the paper:

- Determine the relationship between the forced responses $(w_{j,1})_{\text{forced}}$ and $(w_{j,2})_{\text{forced}}$, for a discretized, lumped-parameter model, with the two plates subjected transversely to ideal (Dirac-delta) impulsive point-shock loads of known strengths.

The following section summarizes pertinent governing equations of the rectangular plates described above. The subsequent four sections address respectively the four objectives given above. A concluding section summarizes the results, indicates some possible applications, and suggests directions for future work.

3 Modal response of a simply-supported rectangular plate

Consider a flat, linear, homogenous rectangular plate (fig. 1) of length a (x-direction), width b (y-direction), uniform thickness h, Young's modulus E, Poisson's ratio ν, and mass density ρ per unit area. The plate is assumed to be simply supported around its perimeter. Assume the plate to be subject to a time-varying transverse external force intensity (i.e., per unit area of plate surface):

$$q = q(x, y, t), \tag{2}$$

with the x and y axes lying in the undeformed neutral plane.

The partial differential equation for the undamped plate is well-known (e.g., [1–3]):

$$\rho w_{tt} + D\nabla^4 w = q(x, y, t); \tag{3}$$

where

$$w = w(x, y, t) \tag{4}$$

is the transverse (z-direction) displacement of the neutral surface, the flexural rigidity D is defined by

$$D = Eh^3 / \left(12 \left(1 - \nu^2\right)\right), \tag{5}$$

the subscript notation indicates partial differentiation:

$$w_{tt} := \partial^2 w / \partial t^2, \tag{6}$$

and

$$\nabla^4 := \nabla^2 \nabla^2 \tag{7}$$

is the biharmonic differential operator (the dual application of the Laplacian operator ∇^2). For rectangular coordinates the Laplacian is expressed by

$$\nabla^2 = \partial^2 / \partial x^2 + \partial^2 / \partial y^2 . \tag{8}$$

The simply supported boundary conditions are represented by

$$w = w_{xx} + \nu w_{yy} = 0 . \tag{9}$$

An analytical, modal solution to the homogeneous form of the undamped differential equation is well known, for this plate geometry and set of boundary conditions [1–3]. In particular,

$$w_{\text{free}}(x, y, t) = \sum_{m=1}^{\infty}\sum_{l=1}^{\infty} W_{m,l}(x, y)\eta_{m,l}(t), \tag{10}$$

where the undamped natural frequency for a particular set of values for m and l is

$$\omega_{n,m,l} = \pi^2 \sqrt{D/\rho}\left[(m/a)^2 + (l/b)^2\right], \tag{11}$$

the associated (normalized) modeshape is

$$W_{m,l} = \sin(m\pi\, x / a)\sin(l\pi\, y / b), \tag{12}$$

and the corresponding undamped modal coordinate is

$$\eta_{m,l} = \sin(\omega_{m,l} t + \phi_{m,l}), \tag{13}$$

where the phase angles $\phi_{m,l}$ depend on the initial conditions.

If the undamped natural frequencies are arranged in increasing order, then the double indices m, l can be replaced with a single index k, and eqns (10) through (13) can be rewritten as follows [4]:

$$w_{\text{free}}(x, y, t) = \sum_{k=1}^{\infty} W_k(x, y)\eta_k(t) \tag{14}$$

where the kth undamped natural frequency is

$$\omega_{n,k} = \pi^2 \sqrt{D/\rho}\left[(m/a)^2 + (l/b)^2\right], \tag{15}$$

the associated (normalized) modeshape is

$$W_k(x, y) = \sin\left(\frac{m\pi x}{a}\right)\sin\left(\frac{l\pi y}{b}\right) = \sin\left(\frac{m\pi x}{a}\right)\sin\left(y\sqrt{\omega_{n,k}\sqrt{\frac{\rho}{D}} - \left(\frac{m\pi}{a}\right)^2}\right), \tag{16}$$

and the corresponding undamped modal coordinate is

$$\eta_k = \sin(\omega_{n,k} t + \phi_k). \tag{17}$$

With modal damping ς_k, the modal coordinate becomes

$$\eta_{d,k} = e^{-\varsigma_k \omega_{n,k} t} \sin\left(\omega_{n,k} t + \phi_k\right). \tag{18}$$

The modal free response, then, is

$$w_{\text{free}}(x, y, t) = \sum_{k=1}^{\infty} C_k W_k(x, y) \eta_{d,k}(t), \tag{19}$$

where the constants C_k, along with the phase angles ϕ_k, depend on the initial conditions.

4 Scaled modal frequencies

From (15), the undamped natural frequencies for the qth plate are given by

$$\omega_{n,k,q} = \pi^2 \sqrt{D_q / \rho_q} \left[\left(m / a_q\right)^2 + \left(l / b_q\right)^2 \right] ; \tag{20}$$

where, from eqn (5), $$D_q = E_q h_q^3 / \left(12 \left(1 - \nu_q^2\right)\right). \tag{21}$$

In terms of their respective volumetric densities $\bar{\gamma}_q$, the undamped modal frequencies for the test- and full-scale-plates are, respectively,

$$\omega_{n,k,1} = h_1 \pi^2 \sqrt{E_1 / \left(12 \bar{\gamma}_1 \left(1 - \nu_1^2\right)\right)} \left[\left(m / a_1\right)^2 + \left(l / b_1\right)^2 \right] \tag{22}$$

and $$\omega_{n,k,2} = \beta h_1 \, \pi^2 \sqrt{E_2 / \left(12 \bar{\gamma}_2 \left(1 - \nu_2^2\right)\right)} \left[\left(m / \left(\alpha a_1\right)\right)^2 + \left(l / \left(\alpha b_1\right)\right)^2 \right]. \tag{23}$$

It follows readily that the undamped natural frequencies are related as follows:

$$\omega_{n,k,2} = \left(\left(\beta / \alpha^2\right) \sqrt{E_2 \bar{\gamma}_1 \left(1 - \nu_1^2\right) / \left(E_1 \bar{\gamma}_2 \left(1 - \nu_2^2\right)\right)} \right) \omega_{n,k,1} \, . \tag{24}$$

For identical modal damping ratios ς_k, the damped natural frequencies are similarly related:

$$\omega_{d,k,2} = \left(\left(\beta / \alpha^2\right) \sqrt{E_2 \bar{\gamma}_1 \left(1 - \nu_1^2\right) / \left(E_1 \bar{\gamma}_2 \left(1 - \nu_2^2\right)\right)} \right) \omega_{d,k,1} \, . \tag{25}$$

Defining the frequency gain, Γ_ω, by

$$\Gamma_\omega = \left(\beta / \alpha^2\right) \sqrt{E_2 \bar{\gamma}_1 \left(1 - \nu_1^2\right) / \left(E_1 \bar{\gamma}_2 \left(1 - \nu_2^2\right)\right)}, \tag{26}$$

the scaling relationships for the modal frequencies become:

$$\omega_{n,k,2} = \Gamma_\omega \, \omega_{n,k,1} \cdot \tag{27}$$

and

$$\omega_{d,k,2} = \Gamma_\omega \, \omega_{d,k,1} \tag{28}$$

Observe that Γ_ω is constant with modal frequency (i.e., it is invariant with k).

5 Scaled modeshapes

From eqn (16), the undamped modeshapes for the test- and full-scale plates are, respectively,

$$W_{k,1}(x,y) = \sin(m\pi x / a_1)\sin(l\pi y / b_1), \tag{29}$$

and

$$W_{k,2}(x,y) = \sin(m\pi x / a_2)\sin(l\pi y / b_2) \tag{30}$$

Note that eqns (29) and (30) have meaning as physical modeshapes only for coordinates describing actual points on the test and full-scale plates, respectively. That is, for eqn (29) to be physically meaningful the coordinates must be limited to values of x from 0 to a_1; and of y, from 0 to b_1. For eqn (30), the coordinates must be limited to values of x from 0 to αa_1; and of y, from 0 to αb_1. (Mathematically, of course, there are no such restrictions.) The designation $W_{k,1}^{ext}(x,y)$ will be used in this paper when indicating a domain for $W_{k,1}(x,y)$ that extends beyond the test-plate boundaries.

$W_{k,2}$ can be expressed in terms of $W_{k,1}$ as follows. Let α first be expressed as the ratio of two positive integers, for reasons to be seen shortly:

$$\alpha = \lambda / \mu \, . \tag{31}$$

Then, substitution from eqn (31) into eqn (1) yields

$$a_2 / \lambda = a_1 / \mu \tag{32}$$

Accordingly,

$$\sin\left(m\pi x / (a_2 / \lambda)\right) = \sin\left(m\pi x / (a_1 / \mu)\right); \tag{33}$$

equivalently,

$$\sin(\lambda m\pi x / a_2) = \sin(\mu m\pi x / a_1). \tag{34}$$

Since λ and μ are positive integers ([5], p. 811),

$$\sin\left(\lambda \frac{m\pi x}{a_2}\right) = \binom{\lambda}{1}\cos^{\lambda-1}\left(\frac{m\pi x}{a_2}\right)\sin^1\left(\frac{m\pi x}{a_2}\right) - \binom{\lambda}{3}\cos^{\lambda-3}\left(\frac{m\pi x}{a_2}\right)\sin^3\left(\frac{m\pi x}{a_2}\right)$$
$$+ \binom{\lambda}{5}\cos^{\lambda-5}\left(\frac{m\pi x}{a_2}\right)\sin^5\left(\frac{m\pi x}{a_2}\right) \mp \cdots \tag{35}$$

$$= \sum_{r=0}^{\text{int}((\lambda-1)/2)} (-1)^r \binom{\lambda}{2r+1} \cos^{\lambda-(2r+1)}\left(\frac{m\pi x}{a_2}\right) \sin^{2r+1}\left(\frac{m\pi x}{a_2}\right) \tag{36}$$

$$= \sin\left(\frac{m\pi x}{a_2}\right)\left[\sum_{r=0}^{\text{int}((\lambda-1)/2)} (-1)^r \binom{\lambda}{2r+1} \cos^{\lambda-(2r+1)}\left(\frac{m\pi x}{a_2}\right) \sin^{2r}\left(\frac{m\pi x}{a_2}\right)\right], \tag{37}$$

where
$$y = \text{int}(x) \tag{38}$$

represents the largest integer such that
$$y \le x \tag{39}$$

Similarly,

$$\sin\left(\mu\frac{m\pi x}{a_1}\right) = \binom{\mu}{1}\cos^{\mu-1}\left(\frac{m\pi x}{a_1}\right)\sin^1\left(\frac{m\pi x}{a_1}\right) - \binom{\mu}{3}\cos^{\mu-3}\left(\frac{m\pi x}{a_1}\right)\sin^3\left(\frac{m\pi x}{a_1}\right)$$

$$+\binom{\mu}{5}\cos^{\mu-5}\left(\frac{m\pi x}{a_1}\right)\sin^5\left(\frac{m\pi x}{a_1}\right) \mp \cdots \tag{40}$$

$$= \sum_{r=0}^{\text{int}((\mu-1)/2)} (-1)^r \binom{\mu}{2r+1} \cos^{\mu-(2r+1)}\left(\frac{m\pi x}{a_1}\right) \sin^{2r+1}\left(\frac{m\pi x}{a_1}\right) \tag{41}$$

$$= \sin\left(\frac{m\pi x}{a_1}\right)\left[\sum_{r=0}^{\text{int}((\mu-1)/2)} (-1)^r \binom{\mu}{2r+1} \cos^{\mu-(2r+1)}\left(\frac{m\pi x}{a_1}\right) \sin^{2r}\left(\frac{m\pi x}{a_1}\right)\right]. \tag{42}$$

By an analogous development, one can also readily derive the following:

$$b_2 / \lambda = b_1 / \mu; \tag{43}$$

$$\sin(\lambda m\pi x / b_2) = \sin(\mu m\pi x / b_1); \tag{44}$$

$$\sin\left(\lambda\frac{l\pi y}{b_2}\right) = \sin\left(\frac{l\pi y}{b_2}\right)\left[\sum_{r=0}^{\text{int}((\lambda-1)/2)} (-1)^r \binom{\lambda}{2r+1} \cos^{\mu-(2r+1)}\left(\frac{l\pi y}{b_2}\right) \sin^{2r}\left(\frac{l\pi y}{b_2}\right)\right]; \tag{45}$$

$$\sin\left(\mu\frac{l\pi y}{b_1}\right) = \sin\left(\frac{l\pi y}{b_1}\right)\left[\sum_{r=0}^{\text{int}((\mu-1)/2)} (-1)^r \binom{\mu}{2r+1} \cos^{\mu-(2r+1)}\left(\frac{l\pi y}{b_1}\right) \sin^{2r}\left(\frac{l\pi y}{b_1}\right)\right]. \tag{46}$$

Define now the following:

$$\Gamma_{\lambda x,k}(x) = \left[\sum_{r=0}^{\text{int}((\lambda-1)/2)} (-1)^r \binom{\lambda}{2r+1} \cos^{\lambda-(2r+1)}\left(\frac{m\pi x}{a_2}\right) \sin^{2r}\left(\frac{m\pi x}{a_2}\right)\right], \tag{47}$$

$$\Gamma_{\mu x,k}(x) = \left[\sum_{r=0}^{\text{int}((\mu-1)/2)} (-1)^r \binom{\mu}{2r+1} \cos^{\mu-(2r+1)}\left(\frac{m\pi x}{a_1}\right) \sin^{2r}\left(\frac{m\pi x}{a_1}\right)\right], \tag{48}$$

$$\Gamma_{\lambda y,k}(y)=\left[\sum_{r=0}^{\text{int}((\lambda-1)/2)}(-1)^r\binom{\lambda}{2r+1}\cos^{\mu-(2r+1)}\left(\frac{l\pi y}{b_2}\right)\sin^{2r}\left(\frac{l\pi y}{b_2}\right)\right],\tag{49}$$

and $$\Gamma_{\mu y,k}(y)=\left[\sum_{r=0}^{\text{int}((\mu-1)/2)}(-1)^r\binom{\mu}{2r+1}\cos^{\mu-(2r+1)}\left(\frac{l\pi y}{b_1}\right)\sin^{2r}\left(\frac{l\pi y}{b_1}\right)\right].\tag{50}$$

Then eqns (37), (42), (45), and (46) can be written, respectively, as:

$$\sin(\lambda m\pi x/a_2)=\Gamma_{\lambda x,k}(x)\sin(m\pi x/a_2),\tag{51}$$

$$\sin(\mu m\pi x/a_1)=\Gamma_{\mu x,k}(x)\sin(m\pi x/a_1),\tag{52}$$

$$\sin(\lambda l\pi y/b_2)=\Gamma_{\lambda y,k}(y)\sin(l\pi y/b_2),\tag{53}$$

and $$\sin(\mu l\pi y/b_1)=\Gamma_{\mu y,k}(y)\sin(l\pi y/b_1)\tag{54}$$

Since, by eqns (34) and (44), the four expressions in eqns (51) and (52) are mutually equivalent,

$$\sin\left(\frac{m\pi x}{a_2}\right)=\frac{\Gamma_{\mu x,k}(x)}{\Gamma_{\lambda x,k}(x)}\sin\left(\frac{m\pi x}{a_1}\right)\tag{55}$$

Similarly, the expressions in eqns (53) and (54) are mutually equivalent, so that

$$\sin(l\pi y/b_2)=\left(\Gamma_{\mu y,k}(y)/\Gamma_{\lambda y,k}(y)\right)\sin(l\pi y/b_1)\tag{56}$$

Substitution from eqns (55) and (56) into eqn (30) leads directly to the desired scaling relationship for the modeshapes:

$$W_{k,2}(x,y)=\Gamma_{W,k}(x,y)W_{k,1}^{ext}(x,y),\tag{57}$$

where the kth modeshape gain, $\Gamma_{W,k}$, is

$$\Gamma_{W,k}(x,y)=\left(\Gamma_{\mu x,k}(x)\Gamma_{\mu y,k}(y)\right)\div\left(\Gamma_{\lambda x,k}(x)\Gamma_{\lambda y,k}(y)\right).\tag{58}$$

6 Scaled free response

It is now possible to write a full expression of a scaled free response. For a test-plate free response described by

$$w_{1,\text{free}}(x,y,t)=\sum_{k=1}^{\infty}C_{k,1}W_{k,1}(x,y)e^{-\varsigma_k\omega_{n,k,1}t}\sin\left(\omega_{d,k,1}t+\phi_{k,1}\right),\tag{59}$$

the corresponding free response of the full-scale plate is

$$w_{2,\text{free}}(x,y,t)=\sum_{k=1}^{\infty} C_{k,2}\Gamma_{W,k}(x,y)W_{k,1}^{ext}(x,y)e^{-\varsigma_k \Gamma_\omega \omega_{n,k,1} t}\sin\left(\Gamma_\omega \omega_{d,k,1} t+\phi_{k,2}\right), \quad (60)$$

where the modal gains (scaling factors) Γ_ω and $\Gamma_{W,k}(x,y)$ are as defined by eqns (26) and (58), supported by eqns (47) through (50). The coefficients $C_{k,q}$ and the phase angles $\phi_{k,q}$ depend on the initial conditions, as noted previously.

7 Scaled point response, for discretized plate with point loading

For the test plate, the modal response at point $P_{j,1}$ to an ideal impulse of strength γ_1 at point $P_{i,1}$ is

$$\left(w_{j,1}\right)_{\text{forced}}=\frac{\gamma_1}{m_{i,1}}\sum_{k=1}^{p}\frac{u_{j,k,1}v_{k,i,1}}{\omega_{d,k,1}}e^{-\varsigma_k \omega_{n,k,1} t}\sin\left(\omega_{d,k,1} t+\phi_{k,1}\right), \quad (61)$$

where $u_{j,k,1}$ is the (j,k) element of the normalized modeshape matrix U_1 for the discretized plate [4], and $v_{k,i,1}$ is the (k,i) element of that matrix's pseudoinverse, U_1^+. (As described in [4], the normalization is accomplished by making each discretized modeshape of unit length.) Similarly, for the full-scale plate, the modal response at point $P_{j,2}$ to an ideal impulse of strength γ_2 at point $P_{i,2}$ is

$$\left(w_{j,2}\right)_{\text{forced}}=\frac{\gamma_2}{m_{i,2}}\sum_{k=1}^{p}\frac{u_{j,k,2}v_{k,i,2}}{\omega_{d,k,2}}e^{-\varsigma_k \omega_{n,k,2} t}\sin\left(\omega_{d,k,2} t+\phi_{k,2}\right), \quad (62)$$

where $u_{j,k,2}$ and $v_{k,i,2}$ are analogously defined.

From eqn (30),

$$W_{k,2}(\alpha x,\alpha y)=\sin\left(m\pi\alpha x/a_2\right)\sin\left(l\pi\alpha y/b_2\right). \quad (63)$$

Substitution from (1) into (63) leads directly to

$$W_{k,2}(\alpha x,\alpha y)=\sin\left(m\pi x/a_1\right)\sin\left(l\pi y/b_1\right), \quad (64)$$

so that, applying eqn (29) to eqn (64),

$$W_{k,2}(\alpha x,\alpha y)=W_{k,1}(x,y). \quad (65)$$

(Notice the difference between eqn (57) and the above equation. The former relates the scaled-plate modeshape at any point to the extended modeshape of the test-plate for the same point; the latter equates the scaled-plate modeshape at a scaled point to the test-plate modeshape at the corresponding unscaled point.)

By using the scaling relationships of eqns (28) and (66), and with corresponding points on the two plates proportionally located (as defined in section 2), one can now modify eqn (62) to express the modal forced response of the full-scale plate in terms of the modal parameters of the test plate. Key substitutions are made as follows:

$$u_{j,k,2} = u_{j,k,1}, \tag{66}$$

$$v_{j,k,2} = v_{j,k,1}, \tag{67}$$

and

$$m_{i,2} = m_{i,1}\left(\alpha^2 \beta \rho_2 / \rho_1\right) \tag{68}$$

Accordingly, $$\left(w_{j,2}\right)_{\text{forced}} = \frac{\gamma_2 \Gamma_R}{m_{i,1}} \sum_{k=1}^{p} \left(\frac{u_{j,k,1} v_{k,i,1}}{\omega_{d,k,1}} \right) e^{-\varsigma_k \Gamma_\omega \omega_{n,k,1} t} \sin\left(\Gamma_\omega \omega_{d,k,1} t + \phi_{k,2}\right), \tag{69}$$

where the parameter Γ_R is a shock-response gain with definition

$$\Gamma_R = \rho_1 / \left(\alpha^2 \beta \rho_2 \Gamma_\omega\right) \tag{70}$$

Equivalently, $$\Gamma_R = \rho_1 / \left(\rho_2 \beta^2\right) \sqrt{E_1 \bar{\gamma}_2 \left(1 - v_2^2\right) / \left(E_2 \bar{\gamma}_1 \left(1 - v_1^2\right)\right)} \tag{71}$$

For proportionally located impact and response points, eqn (60) becomes

$$w_{2,\text{free}}(\alpha x, \alpha y, t) = \sum_{k=1}^{\infty} C_{k,2} W_{k,1}(x, y) e^{-\varsigma_k \Gamma_\omega \omega_{n,k,1} t} \sin\left(\Gamma_\omega \omega_{d,k,1} t + \phi_{k,2}\right), \tag{72}$$

By limiting the free response to p modes, and summing the respective free and forced responses (for the test plate, from eqns (59) and (61); for the full-scale plate, from eqns (72) and (69)), the total responses of the discretized plates can now be expressed. For the test plate, the total response at point $P_{j,1}$ to an ideal impulse of strength γ_1 at point $P_{i,1}$ is:

$$w_1\left(x_{j,1}, y_{j,1}, t\right) = w_{1,\text{free}}\left(x_{j,1}, y_{j,1}, t\right) + w_{1,\text{forced}}\left(x_{j,1}, y_{j,1}, t\right) \tag{73}$$

$$= \sum_{k=1}^{p} C_{k,1} W_{k,1}\left(x_{j,1}, y_{j,1}\right) e^{-\varsigma_k \omega_{n,k,1} t} \sin\left(\omega_{d,k,1} t + \phi_{k,1}\right)$$

$$+ \left(\gamma_1 / m_{i,1}\right) \sum_{k=1}^{p} \left(u_{j,k,1} v_{k,i,1} / \omega_{d,k,1}\right) e^{-\varsigma_k \omega_{n,k,1} t} \sin\left(\omega_{d,k,1} t + \phi_{k,1}\right) \tag{74}$$

$$= \sum_{k=1}^{p} \left(C_{k,1} W_{k,1}\left(x_{j,1}, y_{j,1}\right) + \left(\gamma_1 / m_{i,1}\right)\left(u_{j,k,1} v_{k,i,1} / \omega_{d,k,1}\right)\right) e^{-\varsigma_k \omega_{n,k,1} t} \sin\left(\omega_{d,k,1} t + \phi_{k,1}\right). \tag{75}$$

Similarly, for the full-scale plate, the total response at point $P_{j,2}$, located at $\left(x_{j,2}=\alpha x_{j,1},\ y_{j,2}=\alpha y_{j,1}\right)$ to an ideal impulse of strength γ_2 at point $P_{i,2}$, located at $\left(x_{i,2}=\alpha x_{i,1},\ y_{i,2}=\alpha y_{i,1}\right)$ is:

$$w_2\left(x_{j,2},y_{j,2},t\right)=w_{2,\text{free}}\left(x_{j,2},y_{j,2},t\right)+w_{2,\text{forced}}\left(x_{j,2},y_{j,2},t\right) \tag{76}$$

$$=\sum_{k=1}^{p}C_{k,2}W_{k,2}\left(x_{j,2},y_{j,2}\right)e^{-\varsigma_k\omega_{n,k,2}t}\sin\left(\omega_{d,k,2}t+\phi_{k,2}\right)$$

$$+\left(\gamma_2/m_{i,2}\right)\sum_{k=1}^{p}\left(u_{j,k,2}v_{k,i,2}/\omega_{d,k,2}\right)e^{-\varsigma_k\omega_{n,k,2}t}\sin\left(\omega_{d,k,2}t+\phi_{k,2}\right) \tag{77}$$

$$=\sum_{k=1}^{p}\left(C_{k,2}W_{k,2}\left(x_{j,2},y_{j,2}\right)+\left(\frac{\gamma_2}{m_{i,2}}\right)\left(\frac{u_{j,k,2}v_{k,i,2}}{\omega_{d,k,2}}\right)\right)e^{-\varsigma_k\omega_{n,k,2}t}\sin\left(\omega_{d,k,2}t+\phi_{k,2}\right). \tag{78}$$

In terms of test-plate parameters, the total response of the full-scale plate is:

$$w_2\left(x_{j,2},y_{j,2},t\right)=w_{2,\text{free}}\left(x_{j,2},y_{j,2},t\right)+w_{2,\text{forced}}\left(x_{j,2},y_{j,2},t\right) \tag{79}$$

$$=\sum_{k=1}^{p}C_{k,2}W_{k,1}\left(x_{j,1},y_{j,1}\right)e^{-\varsigma_k\Gamma_\omega\omega_{n,k,1}t}\sin\left(\Gamma_\omega\omega_{d,k,1}t+\phi_{k,2}\right)$$

$$+\left(\gamma_2\Gamma_R/m_{i,1}\right)\sum_{k=1}^{p}\left(u_{j,k,1}v_{k,i,1}/\omega_{d,k,1}\right)e^{-\varsigma_k\Gamma_\omega\omega_{n,k,1}t}\sin\left(\Gamma_\omega\omega_{d,k,1}t+\phi_{k,2}\right). \tag{80}$$

Upon collecting terms, one obtains

$$w_2\left(x_{j,2},y_{j,2},t\right)=\sum_{k=1}^{p}\left(C_{k,2}W_{k,1}\left(x_{j,1},y_{j,1}\right)+\left(\frac{\gamma_2\Gamma_R}{m_{i,1}}\right)\left(\frac{u_{j,k,1}v_{k,i,1}}{\omega_{d,k,1}}\right)\right)e^{-\varsigma_k\Gamma_\omega\omega_{n,k,1}t}\sin\left(\Gamma_\omega\omega_{d,k,1}t+\phi_{k,2}\right) \tag{81}$$

For the common case of zero initial conditions, eqns (75) and (81) reduce, respectively, to:

$$w_1\left(x_{j,1},y_{j,1},t\right)=\left(\gamma_1/m_{i,1}\right)\sum_{k=1}^{p}\left(u_{j,k,1}v_{k,i,1}/\omega_{d,k,1}\right)\ e^{-\varsigma_k\omega_{n,k,1}t}\sin\left(\omega_{d,k,1}t\right) \tag{82}$$

and $$w_2\left(x_{j,2},y_{j,2},t\right)=\left(\gamma_2\Gamma_R/m_{i,1}\right)\sum_{k=1}^{p}\left(u_{j,k,1}v_{k,i,1}/\omega_{d,k,1}\right)\ e^{-\varsigma_k\Gamma_\omega\omega_{n,k,1}t}\sin\left(\Gamma_\omega\omega_{d,k,1}t\right). \tag{83}$$

The frequency gain Γ_ω (eqn (26)) and the shock-response gain Γ_R (eqns (70) and (71)) are known functions of plate geometric and material parameters.

8 Initial verification of scaling equations

In order to verify frequency-gain equation (26) two linear finite-element models were developed, one scaled geometrically relative to the other; otherwise the two plates are identical. Each model describes an isotropic, homogeneous, aluminum test-plate. For each the plate material is 6061-T6 aluminum, with Young's modulus 69 GPa, mass density 2700 kg/m^3, Poisson's ratio 0.33, and modal damping ratio 0.02 (all modes). The unscaled plate is simply supported along each of its four edges, with dimensions 1.0 m x 0.75 m x 25 mm; the scaled plate is twice as long, twice as wide, and one-and a half times as thick. (I.e., $(\alpha,\beta)=(2,1.5)$—refer to eqn (1)). Each FEA model comprises uniform rectangular (parallelepiped) elements: 50 element divisions along the length (x-direction, measured in the neutral plane from the lower left corner), and 38 along the width (y-direction, measured correspondingly). The modal frequencies determined by FEA were found to scale according to the analytical frequency gain Γ_ω determined above, eqn (26)—for detailed results, see table 1. The shock-response gain, Γ_R (eqns (70) and (71)), remains to be verified.

Table 1: Comparison of frequency scaling between FEA and analytical models.

Modal Frequencies (Hz) Comparison: First 20 Modes scaled by alpha = 2, beta = 1.5			
Original	Scaled by ANSYS	Scaled analytically	Difference (%)
168.61	63.235	63.228	0.01150
350.65	131.52	131.50	0.02254
492.28	184.64	184.60	0.02071
654.03	245.34	245.26	0.03094
674.12	252.91	252.80	0.04574
977.17	366.69	366.44	0.06959
1031.6	386.98	386.85	0.03180
1078.7	404.67	404.51	0.03916
1213.1	455.21	454.92	0.06377
1401.4	525.98	525.51	0.08917
1515.6	568.94	568.36	0.10202
1624.6	609.51	609.21	0.04821
1786.4	670.22	669.91	0.04610
1939.1	728.15	727.15	0.13825
1946.6	730.76	729.99	0.10576
1967.5	738.41	737.81	0.08125
2269.3	852.07	850.97	0.12876
2291.6	859.85	859.35	0.05833
2483.4	932.84	931.26	0.16991
2612.9	981.04	979.85	0.12081

9 Application

In order to use these results one could first determine analytically the damped and undamped modal frequencies, the modeshapes, and the lumped masses for the discretized test plate, via the procedure of [4]. Next, one could determine the frequency- and shock-response gains, via the definitions found above in eqn

(26) and either eqn (70) or eqn (71). One could then use eqn (83) to evaluate the shock response at the desired response point for a scaled plate (whether scaled up or down); no finite element analysis would be required—nor would any experimental testing, except as needed for verification.

10 Conclusion

This paper has developed equations by which the shock response of one simply-supported rectangular plate can be related to that of a scaled plate also under simple support, when the scaling is such that the faces of the two plates are geometrically similar. The plates are assumed to be linear; and the shock, to be ideal (Dirac-delta) transverse point-shock loading. The shock impulses may be of different strengths; and the plates may have different Young's moduli, Poisson's ratios, and volumetric densities. Modal damping is assumed, with identical damping ratios for corresponding modes. The analysis led to analytical expressions for the modal-frequency-, modeshape-, and shock-response gains.

Future work includes verifying further the above scaling relationships, finding corresponding relationships for plates with different boundary conditions, and determining situations in which these simple assumptions and geometries could be useful for real equipment and more realistic shock scenarios. Verifying these relationships could begin with finite-element analysis, but determining the degree of applicability to real equipment will likely require shock tests to physical hardware.

Acknowledgements

The authors are grateful to the Army Research Laboratory (via the U.S. Military Academy's Mathematical Sciences Center of Excellence), and to the U.S. Military Academy's Department of Civil and Mechanical Engineering, for their funding of this work.

References

[1] Leissa, A.W., *Vibration of Plates*, NASA SP-160, 1969.
[2] Leissa, A.W., "The Free Vibration of Rectangular Plates," *Journal of Sound and Vibration*, **31(3)**, pp. 257—293, 1973.
[3] Harris, C.M. & Piersol, A.G., editors, *Harris' Shock and Vibration Handbook, Fifth Edition*, McGraw-Hill, New York, 2002, chap. 23.
[4] Hampton, R.D., Li, T.H., & Nygren, K.P., "Analytical Shock Response of a Transversely Point-Loaded Linear Rectangular Plate," *Proceedings of the 14th ARL/USMA Technical Symposium*, November 2006.
[5] Korn, G.A. & Korn, T.M., *Mathematical Handbook for Scientists and Engineers, Second Edition*, McGraw-Hill, New York, 1968.

Dynamic behaviour of a steel plate girder railroad bridge with rail joints

H. M. Kim[1], S. I. Kim[2] & W. S. Hwang[2]
[1]*Department of Railroad Structure Research, Korea Railroad Research Institute, Korea*
[2]*Department of Civil and Environmental Engineering, Inha University, Korea*

Abstract

This study investigates the effects of rail joints on the dynamic behaviour of railroad bridges through a dynamic experiment. The subject bridge is a typical steel plate girder railroad bridge. The bridge is 53.4m long, and consists of four simply supported spans (4@12.10m). Rail joints are positioned in the third span. It is 1.8m away from the starting point of the third span. The regular service train applied the dynamic loading. The test train is composed of one diesel locomotive and seven passenger carriages, which is commonly running on the existing railroad lines in Korea. The range of measured speed is from 19.1 to 124.0km/hr. Accelerometers are installed at the bottom flange of the centre of the third span to acquire a bridge response caused by the rail joint. For comparison, another accelerometer is installed at the same position of the first span, which has no rail joint. From the measured results, Peak vale (PV), root mean square (RMS) and dynamic amplification factor (DAF) value analysis are performed in order to compare the instantaneous maximum and mean amplitude of acceleration at each span. The result shows a significant increase in the acceleration of the span with joints. In addition, DAF is from 0.22 to 1.92 for the measured speed zone. The result also shows that rail joints on a bridge produce an impact effect. Thus, it increases the dynamic response of the bridge and deteriorates ride comfort.

Keywords: rail joint, acceleration, impact load, dynamic amplified factor, steel plate girder railroad bridge.

WIT Transactions on The Built Environment, Vol 113, © 2010 WIT Press
www.witpress.com, ISSN 1743-3509 (on-line)
doi:10.2495/SU100271

1 Introduction

The dynamic behaviour of a railroad bridge is affected by not only the interaction between the bridge and track, but also the interaction between the train and the track. This is because train running along the fixed path of the track changes according to the condition of the contact surface. The irregularity of the track surface produces dynamic load. In this condition, the variation of dynamic load is large. In particular, defects or rail joints on the track with discontinuous contact surface results in considerable high impact load. The calculation methods of dynamic load caused by defective tracks have been studied by Jenkins [1] and Alias [2]. However, there are insufficient empirical data on what direct effects rail joints have on the behaviour of the bridge. In order to prevent discontinuous operating conditions caused by rail joints, most high-speed railroads adopt continuous welded rail (CWR). However, many existing bridges using standard-length rails (SLR) have rail joints on them. The steel plate girder bridge is a typical structure of short-span, high-strength and high-frequency bridges. It is commonly connected to sleepers without ballast. So, it is most vulnerable to acceleration.

The present study measured the vertical acceleration of bridges with joints and those without in steel plate girder bridges, and examined the effect of rail joints on the dynamic response of bridges through analyzing the PV, RMS and DAF values from measured acceleration.

2 General theory of impact load caused by a rail joint

Rail joints that have a geometrically discontinuous section with the adjoining rail are largely divided into upward bump, downward bump, joint gap and bump with a break. Depending on the form of discontinuity, the pattern of impact increase is different. It is known that impact from speed is high in order of upward bump and bump with a break, and in these forms of discontinuity impact is proportional to speed. In downward bump and joint gap, however, impact does not increase along with the rise of speed.

Impact load consist of a short-time peak (P1 load) and a delayed peak (P2 load). The short time peak is associated with battering of the rail-end corner by the unsprung mass of the wheel set. The delayed peak is associated with rail bending, which is a more resilient deformation mode than corner batter. Each force is described by the following equations, as given by Jenkins [1].

$$P_1 = P_0 + 2\alpha v \frac{\sqrt{k_H m_{T1}}}{\sqrt{1 + m_{T1} / m_u}} \tag{1}$$

$$P_2 = P_0 + 2\alpha v \frac{\sqrt{m_u}}{\sqrt{m_u + m_{T2}}} \left[1 - \frac{c_T \pi}{\sqrt{k_{T2}(m_u + m_{T2})}} \right] \sqrt{k_{T2} m_u} \tag{2}$$

where

= static wheel-rail contact force [kN]
= total dip angle at joint [rad]
= train speed [m/s]
= linearize Hertzian contact stiffness [N/m]
= effective track mass for P1 calculation [kg]
= unsprung mass [kg]
= equivalent track mass for P2 calculation [kg]
= equivalent track stiffness for P2 calculation [N/m]
= equivalent track damping for P2 calculation[Ns/m]

In addition, an approximate formula is given by Alias (2) as following equation.

$$P = P_0 + \lambda 2\alpha v\sqrt{km_u} \tag{3}$$

where

= dimensionless factor depending on track damping (≤ 1)
= track stiffness [N/m]

The above equations indicate that speed and dip angle have most significant influence on dynamic loads at a rail joint. Other properties of track have relatively weak effect on dynamic loads.

3 Dynamic testing of a steel plate girder railroad bridge

3.1 Bridge description

The subject bridge is a typical steel plate girder railroad bridge located at 290.927 km from Seoul to Busan. The bridge is 53.4m long, and consists of 4

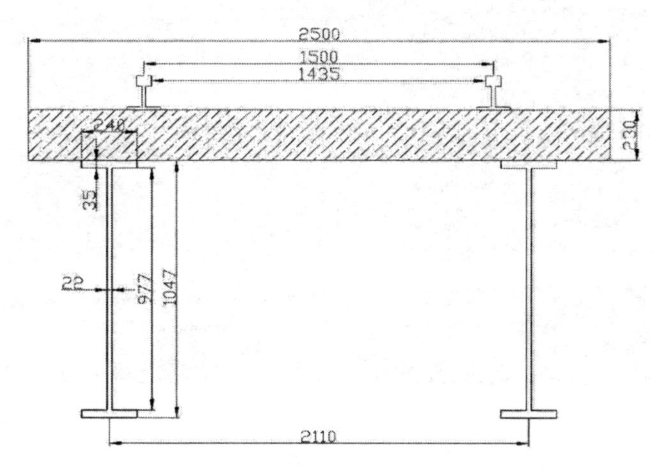

Figure 1: Central section of the mid-span.

simply supported spans (4@12.10m). The length of each span is 12.10m. It is composed of two welded plate girders built compositely with wooden sleepers without ballast, as shown in figure 1. Horizontal bracings are installed only on the upper part of the girder and vertical bracings are installed on the left and right point parts and 3.66m from each point part to the centre of the girder, so a total of 4 vertical bracings are installed. This is the same for all of the 4 spans. The height of each girder is 1.047m, and the space between two girders is 2.11m. As can be seen in figure 2, a rail joint is positioned 1.8m apart from the starting point of the 3rd span. It is situated on a sleeper and at the same position on the left and right rails, as shown in figure 3. The joint has a gap 10mm axially, 0.5mm vertically and 0.3mm horizontally.

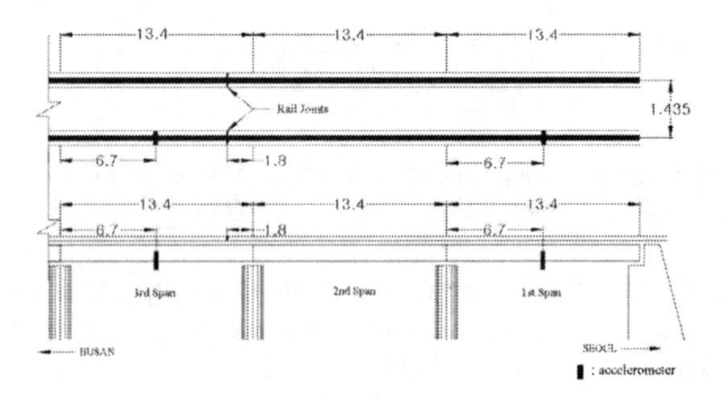

Figure 2: Positions of rail joints and accelerometers.

Figure 3: Rail joints.

Figure 4: Test train.

3.2 Train description

A test train consists of one diesel locomotive and 7 passenger carriages, as shown in figure 4. It is commonly running on the existing railroad lines in Korea. The diesel locomotive is 20.287m long, 3.128m wide, and 4.254m high. It is composed of two bogies and each bogie has three axles. The load per axle is 215kN. The distance between two axles is 1.854m, also the distance between the centres of the bogies is 12.497m. Each of the passenger carriages connected to the diesel locomotive is 23.5m long, 3.2m wide and 3.7m high. It is composed of two bogies and each bogie has two axles. Load per axle is 98kN. The distance between two wheel axles is 2.3m, and the distance between the centres of the bogie is 15.9m.

3.3 Experimental program

Accelerometers are installed at the bottom flange of centre of 3rd span to acquire bridge response caused by rail joint. For comparison, another accelerometer is installed at the same position of 1st span which has not rail joint. The accelerometers used in the experiment can measure within the maximum range of 5G. Because this was an experiment on acceleration caused by the operation of an ordinary train, we attached wheel load gauges at 10m before entering the 1st span. Signal data were collected during the train passing the subject bridge. A total number of measured data for 45 hours is 30. All data are filtered through the 100Hz low pass technique. This study compare the acceleration response in the 1st span without a rail joint and the 3rd span with a rail joint to examine the effects of rail joints for each speed band. In measuring, the data sampling rate was set as large as 2000 to prevent the omission of acceleration peaks in the section where the rail joint exists. The regular service train applied the dynamic loading.

4 Experimental observations

4.1 Test results

As an example, a vertical acceleration time history in the 1st and 3rd span for train running at speed of 122km/hr are as in figure 5 and figure 6. It shows a significant increase in the acceleration of the span with joints. It also shows that not only instantaneous maximum acceleration but also general acceleration is high.

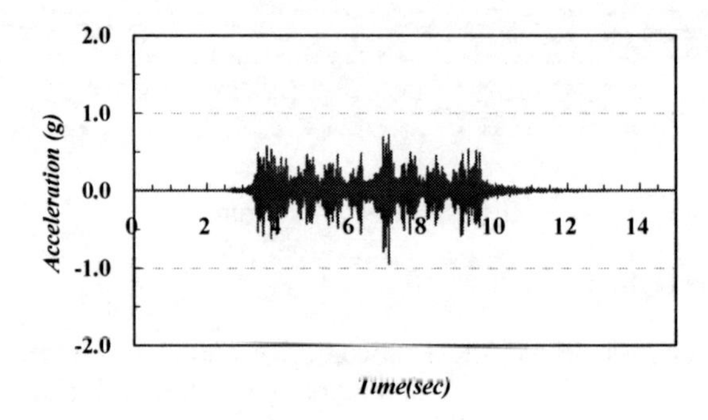

Figure 5: Acceleration time history: first span.

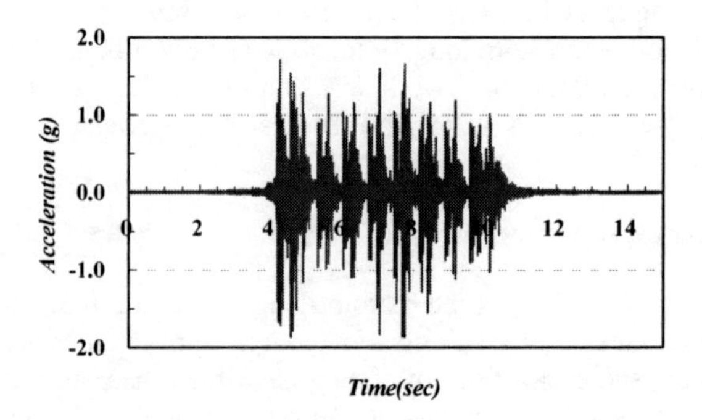

Figure 6: Acceleration time history: third span.

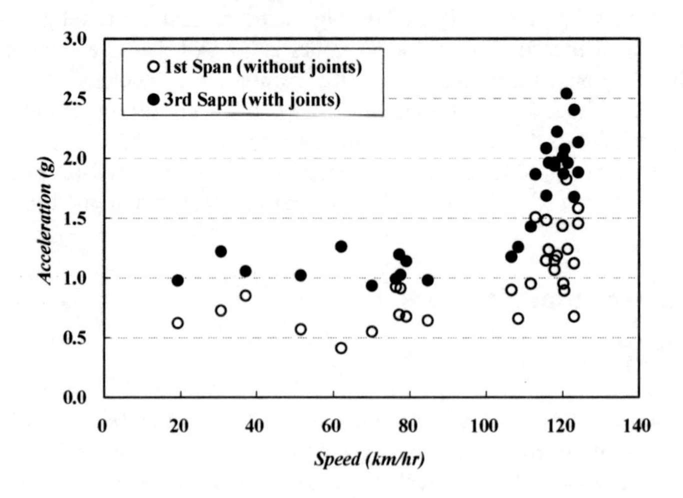

Figure 7: Comparison of PV.

Table 1: PV, RMS, and DAF value of measured acceleration.

V (km/hr)	1st Span (without		3rd Span (with joints)		DAF
	Peak(g)	RMS(g)	Peak(g)	RMS(g)	
19.14	0.62	0.08	0.98	0.11	0.44
30.66	0.73	0.08	1.22	0.10	0.22
37.17	0.85	0.10	1.05	0.13	0.37
51.51	0.57	0.09	1.02	0.14	0.52
62.06	0.41	0.11	1.26	0.27	1.48
70.08	0.55	0.09	0.93	0.16	0.88
76.31	0.92	0.13	0.99	0.28	1.12
77.12	0.69	0.11	1.19	0.19	0.78
77.47	0.91	0.13	1.02	0.29	1.19
78.94	0.68	0.14	1.14	0.32	1.20
84.61	0.64	0.12	0.98	0.21	0.78
106.55	0.90	0.14	1.17	0.28	1.04
108.33	0.66	0.15	1.26	0.30	1.09
111.65	0.95	0.18	1.43	0.32	0.74
112.84	1.51	0.19	1.87	0.35	0.83
115.63	1.14	0.17	2.08	0.35	1.06
115.64	1.48	0.20	1.69	0.39	0.98
116.33	1.23	0.18	1.96	0.42	1.35
117.85	1.14	0.17	1.94	0.43	1.57
117.85	1.07	0.17	1.96	0.40	1.36
118.45	1.18	0.17	2.22	0.42	1.44
119.91	1.43	0.22	2.01	0.47	1.17
120.16	0.95	0.16	1.87	0.39	1.49
120.41	0.89	0.17	2.08	0.42	1.40
120.91	1.82	0.22	2.54	0.42	0.88
121.28	1.24	0.19	1.96	0.40	1.15
122.95	0.68	0.14	2.40	0.40	1.92
122.95	1.12	0.19	1.67	0.37	0.94
123.99	1.58	0.22	1.88	0.40	0.85
123.99	1.45	0.23	2.13	0.44	0.92

The peak value (PV) of vertical acceleration for each speed band was calculated based on the result of filtering measured acceleration at 100Hz. The result is presented in table 1 and figure 7. It is from 1.07 to 3.55 times higher in the span with a rail joint than in that without. The result shows that the instantaneous maximum acceleration of the subject bridge increases significantly by impact load generated by the rail joint.

Root Mean Square (RMS) is an equation to calculate the mean amplitude of signals alternating between plus and minus like vibration. If the representative value is calculated using the maximum amplitude of acceleration signals measured, it may be distort analysis from abnormal signals. RMS can be used as a supplementary indicator for preventing this apprehension. RMS is described by the following equations.

$$RMS = \sqrt{\frac{1}{N}\sum_{i=1}^{N} a_i^{\,2}} \tag{4}$$

where
N = Total number of data
= Each value of acceleration

The RMS of acceleration for each speed band is analyzed based on the result of filtering acceleration measured in each span at 100Hz. The result is presented in table 1 and figure 8. It is from 1.22 to 2.92 times higher in the span with a rail joint than in that without. This result shows that the mean amplitude of acceleration in the steel plate girder bridge increases significantly by impact load generated by the rail joint

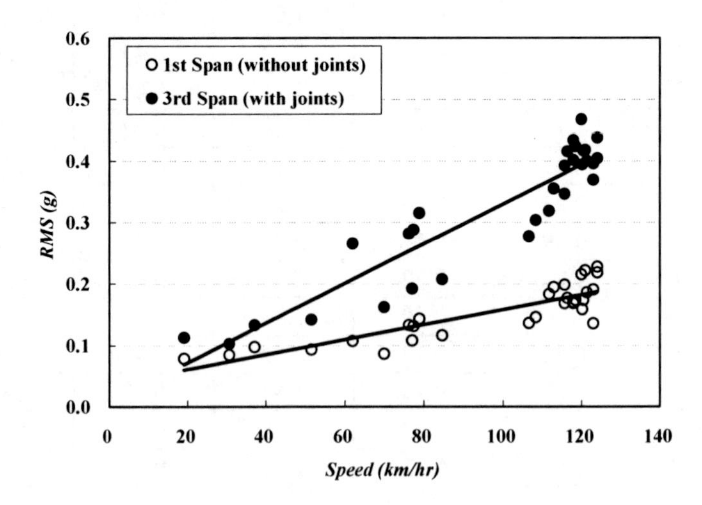

Figure 8: Comparison of RMS.

4.2 Discussion of results

The basis for determining the dynamic factors were provided by Office for Research and Experiments (ORE) specialists committee D23 [6]. When service trains pass over a bridge, the resulting oscillations increase the load by a quantity φ made up of the sum of two components. The first component is ϕ', which is the proportion applicable for a track in perfect geometrical condition. Another components is ϕ", which is the proportion representing the effect of track irregularities. ϕ" can be adopted the parameter of comparison with measured result. It is described by the following equation 4. φ'' of the subject bridge is calculated 0.99.

$$\varphi''=\frac{1}{100}\left[56e^{-(L_\phi^2/100)}+50\left(\frac{L_\phi n_0}{80}-1\right)e^{-(L_\phi^2/400)}\right] \tag{5}$$

where

= the effective length (span L for a simply supported beam)
= the first natural frequency of the beam

Also, base on the RMS of the beam at mid-span values of the parameter dynamic amplified factor (DAF) representing the increase in acceleration due to a rail joint can be calculated at different speed by following equation 5.

$$DAF=\frac{a_{RMS,joint}}{a_{RMS,jointless}}-1 \tag{6}$$

where

= RMS acceleration of the span with a rail joint
= RMS acceleration of the span without a rail joint

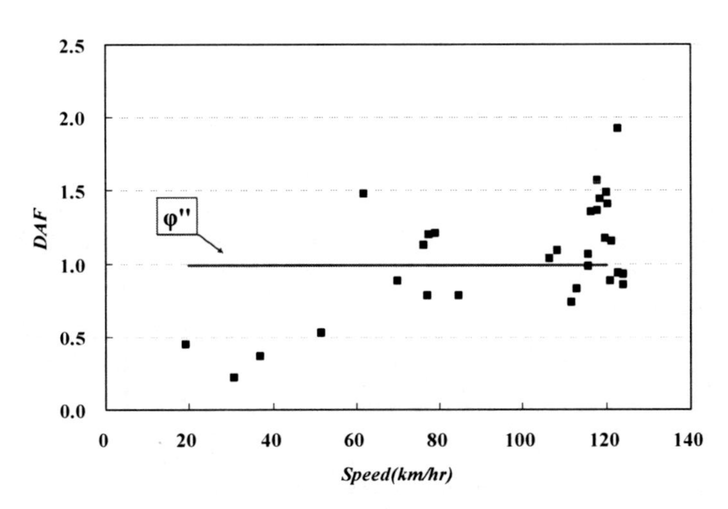

Figure 9: DAF according to speed.

The calculated DAF from measured data are presented in table 1 and figure 9 with the φ'' value. Increasing speed of the test train results in a gradual increase in the DAF. Moreover, the DAF is apt to be larger than φ'' value in the high speed zone.

5 Summary and conclusions

The dynamic behaviour of a railroad bridge is affected by not only the interaction between the bridge and track, but also interaction between the train and the track. This is because train running along the fixed path of the track changes according to the condition of the contact surface. The irregularity of the track surface produces dynamic load. In this condition, the variation of dynamic load is large. In particular, defects or rail joints on the track with discontinuous contact surface results in considerable high impact load. The calculation methods of dynamic load caused by defect of tracks have been studied by Jenkins [1] and Alias [2]. However, there are not sufficient empirical data on what direct effects rail joints have on the behaviour of the bridge. In order to prevent discontinuous operating condition caused by rail joints, most of high-speed railroads adopt continuous welded rail (CWR). However, many of existing bridges using standard-length rails (SLR) have rail joints on them. The steel plate girder bridge is a typical structure of short-span, high-strength and high-frequency bridges. It commonly connected to sleepers without ballast. So, it is most vulnerable to acceleration.

The present study measured the vertical acceleration of bridge with joints and those without in steel plate girder bridges, and examined the effect of rail joints on the dynamic response of bridges through analyzing the PV, RMS and DAF value from measured acceleration.

References

[1] Jenkins H.H, Stephnson J.E., Clayton G.A., Morland G.W. and Lyon D. The effect for track and vehicle parameters on wheel/rail vertical dynamic forces, *Railroad Engineering Journal 3*, 1974.
[2] Alias J. *Lavoice ferree*, Eyrolles, Paris, 1977.
[3] European Rail Research Institute (ERRI) Specialists' Committee D 214. *Rail bridges for speeds>200km/h (Final report – RP 9)*, ERRI, 1999.
[4] *UIC leaflet 776-1R: Loads to be considered in railroad bridge design*, 4th edition, France, 1994.
[5] Esveld C. *Moder Railway Track*, MRT-Productions, Netherlands, 2001.
[6] Office for Research and Experiments (ORE) Question D 23. *Determination of Dynamic Force in Bridges (RP 16): Theoretical Study of Dynamic Forces in Bridges*, ORE, 1970.
[7] Office for Research and Experiments (ORE) Question D 161.1. *Dynamic Effects of 22.5 t Axle Loads on the Track (RP 4)*, 1987.
[8] L. FrýBA, *Dynamics of Railway Bridges*, Thomas Telford, 1996.

Numerical simulations on adiabatic shear behaviour of 921A steel pure shear hat-shaped specimens

J. C. Li, X. W. Chen & G. Chen
Institute of Structural Mechanics,
China Academy of Engineering Physics, China

Abstract

The adiabatic shear behaviour of a specimen is observed, based on a series of numerical simulations of SHPB test using 921A steel pure shear hat-shaped specimens in the present paper. The initiation and the propagation of an adiabatic shear band (ASB) and the relative distribution of temperature field in the specimen are analyzed. It is found that ASB is formed as the spread of an unstable zone with high temperature and high strain in the two ends of the Shear Zone. The spread speed of an ASB is related to the loading rate V_0, whilst the width of ASB is nearly independent of the loading rate. All the ASBs are deformed bands corresponding to various values of V_0.
Keywords: SHPB, pure shear hat-shaped specimen, adiabatic shear deformation, numerical simulation.

1 Introduction

Adiabatic shear deformation usually occurs in the high-speed deformations such as high-speed impact and penetration, etc. Adiabatic rising-temperature will lead the thermal softening effect and thus adiabatic shear band (ASB) occurs in the material. An ASB usually has a width of 10–100μm, in which the shear strain, shear strain rate and temperature can achieve a value of $1–10^2$, $10^5–10^7 s^{-1}$, and $10^2–10^3$K, respectively.

Meyers et al. [1] designed a pure shear hat-shaped specimen as shown in Figure 1, in which the Shear Zone deforms as a pure shear deformation when the specimen undergoes compressive loading. This hat-shaped structure has been used widely to study the adiabatic shear behaviour of material. Accompanied

WIT Transactions on The Built Environment, Vol 113, © 2010 WIT Press
www.witpress.com, ISSN 1743-3509 (on-line)
doi:10.2495/SU100281

with the development of computer techniques, numerical simulation method becomes more and more important in the researches on adiabatic shear deformation and numerous relative simulations have been conducted, of which, the outstanding work is Batra and his colleagues'. Moreover, previous researches on adiabatic shear deformation are mainly based on stress / strain field.

The adiabatic shear behaviour of material affects directly its performance under high strain rates. Chen et al. [2] conducted SHPB tests and corresponding simulations using 921A steel pure shear hat-shaped specimens and found that, while under a comparatively high loading rate, the large-deformation zone in the specimen centralizes gradually to the designed Shear Zone, and fracture occurs in the Shear Zone with increasing the loading rate to a certain value.

In general, the deformation process of specimen can't be observed in detail in SHPB test; furthermore, it is hard to conduct SHPB test under an ultrahigh loading rate. However, it becomes possible to conduct SHPB test under an ultrahigh loading rate in numerical simulation; also, the history of deformation and failure of the specimen can be observed conveniently. In the present paper, integrated with Chen et al. [2]'s tests, a series of numerical simulations of SHPB test using 921A steel pure shear hat-shaped specimens are conducted with ANSYS/LS-DYNA, and the adiabatic shear behaviour of specimen is observed.

2 FEM model

All the bars in Chen et al. [2]'s tests are made of alloy steel, and the geometry of 921A steel pure shear hat-shaped specimen is shown in Figure 1.

Firstly, a numerical test is conducted with ANSYS/LS-DYNA to replay the SHPB test done by Chen et al. [2]. Regarding its axial symmetry, an axial symmetric FEM model is employed in the numerical simulation and all the elements are defined as shell-162 axial symmetric elements. Especially, the meshes of the Shear Zone are fined to adapt the simulation, and the minimum mesh size of the Shear Zone is modified as about 30μm. As the width of ASB usually doesn't exceed 100μm, and the characteristic size of a crystal is 10–60μm, thus the meshes could satisfy the relative requirement of simulation.

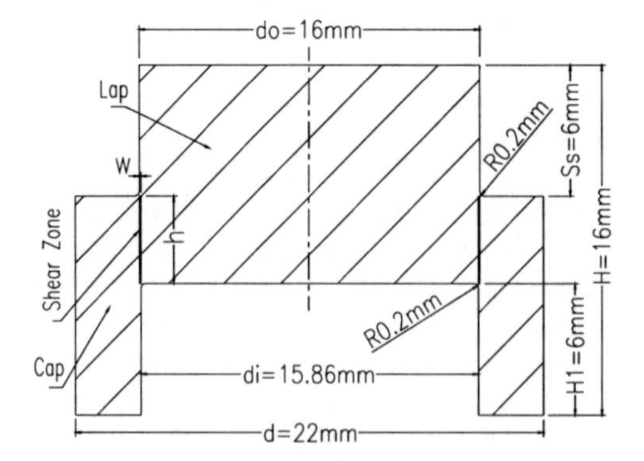

Figure 1: Geometry of a 921A steel pure shear hat-shaped specimen.

The deformation is adiabatic and thus heat exchange is neglected in the simulations. The material of all bars is defined as isotropic elastic, and the specimen is described by Johnson-Cook model (Johnson and Cook [3]) and Gunnison EOS for simulating its adiabatic shear deformation well. Johnson-Cook model includes both the definitions of flow stress and fracture strain of material, and in its fracture model accumulative damage is used to analyze the failure of material. Based on Chen et al. [2]'s tests, the largest fracture strain is used as the criterion of element erosion and its value is defined as 1.13. Element is deleted immediately when it erodes. In general, it is assumed that 90 per cent of the deformation energy of material transforms into heat energy, and the rest transforms into strain energy of material, thus the temperature rising in material could be calculated from the increase of stress and strain using the dependence of $dT=0.9\sigma d\varepsilon/(\rho C_p)$, where ρ is the density of material and C_p the specific thermal capacity. Corresponding specimen material parameters are listed in Table 1.

Table 1: Specimen material parameters of Johnson-Cook model.

ρ (kg/m^3)	G (GPa)	E (GPa)	υ	C_p (J/(kg•k))	T_r (k)
7800	0.7776	205	0.28	400.90	300
T_m (k)	A (MPa)	B (MPa)	C	n	m
1765	760	500	0.014	0.53	1.13
D_1	D_2	D_3	D_4	D_5	
1.13	0	0	0	0	

Regarding the same loading conditions, same strain waves and deformations of specimen as that of Chen et al. [2] are obtained in numerical simulations, thus the validation of the numerical model is confirmed. Then simulations under different loading rates are conducted using the same model. Detailed data of simulation model, strain waves and deformations of specimen can be seen in previous paper of the authors in DYMAT2009 (Li et al. [4]).

3 Adiabatic shear deformations of the specimens

Numerical simulations of SHPB test under different loading rates are conducted by using the method mentioned above. Different strain waves in the bars and dynamic deformations of the specimens under different loading rates V_0 are obtained (seen in [4]). Here only the adiabatic shear behaviours of the specimens will mainly be discussed.

Wright and Batra [5] conducted one-dimensional simulations of adiabatic shear deformation and found the phenomenon of stress collapse, i.e. the stress in an ASB keeps almost invariable for a certain period while it approaches to the largest value, and then descends rapidly. Wright and Ockendon [6] further analyzed stress collapse basing on asymptotic analysis method and suggested three criteria of the initiation of ASB, i.e. the effective plastic strain reaches 0.5, effective stress reaches the largest value or stress collapse occurs, respectively. Analysis on adiabatic shear deformation will be conducted integrated with these three criteria in this paper.

As adiabatic shear deformation occurs mainly in the Shear Zone, only the deformation of Shear Zone and its surrounding regions is emphatically observed. Corresponding analytical zone is shown in Figure 2, where X-coordinate represents the distance from the central symmetrical axis, and Y-coordinate the distance from the bottom of the Cap. Five elements in different locations of the Shear Zone are selected as the characteristic analytical elements, and they are marked as 22832, 22306, 22300, 22293 and 23238, respectively, from the top of the Shear Zone to its bottom. In Figure 2, the Y-coordinate value of the transverse zone denoted by broken line is 8, and the X-coordinate value of the longitudinal zone denoted by dash-dotted line is also 8. Moreover, the transverse zone intersects with the longitudinal zone in element 22300 which locates in the centre of Shear Zone.

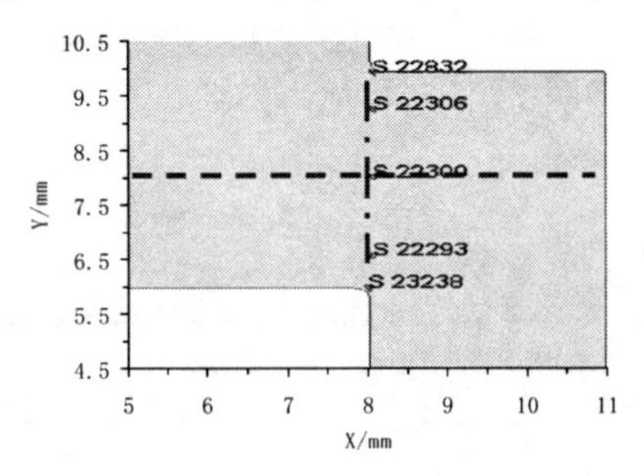

Figure 2: The analytical zone.

3.1 Histories of stress and strain in the Shear Zone

According to the criteria of the initiation of ASB in [6], whether adiabatic shear deformation has occurred could be identified based on the histories of the stress and strain in the Shear Zone. Histories of effective stress in element 22300 under different loading rates V_0 are shown in Figure 3, where only the period in which the effective stress varies obviously is shown for the clarity of figure.

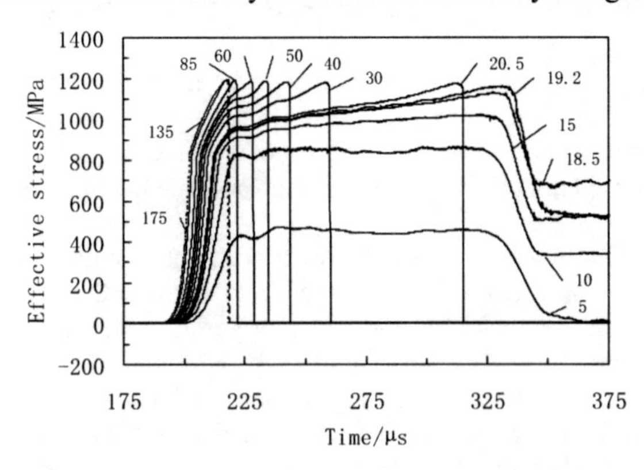

Figure 3: Histories of effective stress in element 22300 under different V_0s (m/s).

www.witpress.com, ISSN 1743-3509 (on-line)

From Figure 3 it can be seen that within the loading rate range of V_0 <19.2m/s, effective stress in element 22300 doesn't reach its largest value, and lower loading rate corresponds to smaller effective stress. Stress collapse doesn't occur in the element, and the decline of stress in the end of pulse is due to the unloading of extra load; when V_0 =19.2m/s, effective stress approaches to the largest value and stress collapse occurs; while V_0 ≥20.5m/s, the Shear Zone finally fractures thoroughly, and effective stress declines immediately to zero when it reaches the largest value because of the element erosion. Moreover, as V_0 increases, both the time that effective stress reaches its largest value and the initiation time of element erosion move up gradually.

Correspondingly, it can be found from the histories of effective plastic strain in element 22300 that, when V_0 =5m/s, there is no plastic strain because the specimen is still elastic; while V_0 ≥10m/s, plastic deformation initiates in the element. However, within the range of V_0 <19.2m/s, the value of effective plastic strain is less than 0.5, and lower loading rate corresponds to smaller effective plastic strain. When V_0=19.2m/s, effective plastic strain exceeds 0.5. Regarding higher loading rates, effective plastic strain declines immediately to zero when it reaches the critical value of fracture strain because of the erosion of element. Moreover, histories of effective plastic strain in the element are consistent with that of effective stress, i.e. same conclusions for the analysis of adiabatic shear deformation could be obtained from either effective stress or effective plastic strain, and latter analysis in this paper is mainly based on the effective plastic strain.

It can also be found from Figure 3 that while the loading rate is comparatively low (e.g. V_0=18.5m/s), ASB may occurs in its two ends although adiabatic shear deformation does not initiate in the centre of Shear Zone. Therein it needs to analyze the histories of effective plastic strain in the ends of Shear Zone. Not losing the universality, elements 22306, 22293 are selected as the example, and the histories of effective plastic strain in these two elements under comparatively low loading rates are shown in Figure 4.

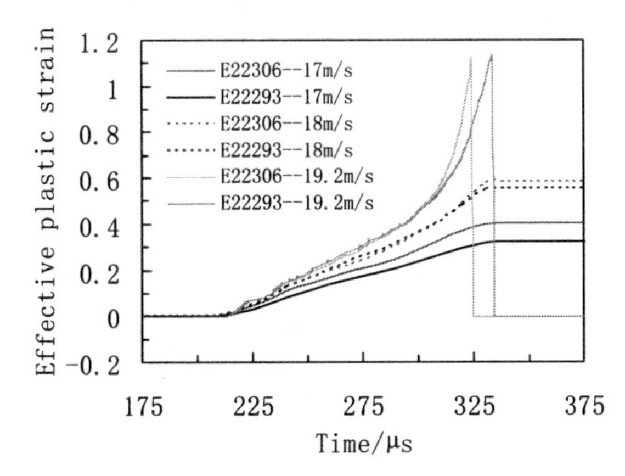

Figure 4: Histories of effective plastic strain in the elements in the ends of Shear Zone under different V_0s.

It can be seen from Figure 4 that when V_0 =17m/s, all the effective plastic strain doesn't reach the value of 0.5; however, when V_0=18m/s, effective plastic strain exceeds 0.5; and when V_0 =19.2m/s, both elements erode and effective plastic strain declines to zero correspondingly. Thus, it can be known that, within the range of V_0 <18m/s, adiabatic shear deformation has not occurred in the specimen; while V_0 ≥18m/s, ASB initiates in the two ends of Shear Zone, however, fracture doesn't occur in the specimen; when V_0 =19.2m/s, ASB runs through all the Shear Zone, and adiabatic shear fracture begins to initiate in its two ends; while V_0 ≥20.5m/s, all the ASB fractures thoroughly.

Moreover, it can also be found from the histories of effective plastic strain ε_p that, the initiation of ASB in the Shear Zone becomes earlier and earlier with increasing the loading rate. Besides, the duration from ε_p=0.5 to element erosion becomes shorter and shorter, i.e. the carrying capacity of specimen trails off gradually with increasing the loading rate. Here element 22832 which locates in the top of Shear Zone, is selected as an example to analyze the variety of ASB initiation and element erosion with increasing the loading rate, and the elements in the Shear Zone are all similar to it. Initiation time of ε_p =0.5 and element erosion of element 22832 versus loading rates are shown in Figure 5, the interval between these two time is defined as the duration from the initiation of ASB to element erosion.

It can be seen from Figure 5 that, both the initiation time of ASB and element erosion move up with increasing the loading rate. While the loading rate is comparatively low (V_0 <40m/s), both the initiation time of ASB and element erosion move up distinctly; however, while V_0≥40m/s, the trend of moving up of the initiation time slows down rapidly, and then the time approaches to a saturated value.

Moreover, from Figure 5 it can also be known that the duration from the initiation of ASB to element erosion shortens obviously with increasing the loading rate, and gradually approaches to a saturated value, i.e. 2μs, after V_0 ≥40m/s. It indicates that the carrying capacity of specimen trails off with increasing the loading rate and approaches to its lower limit.

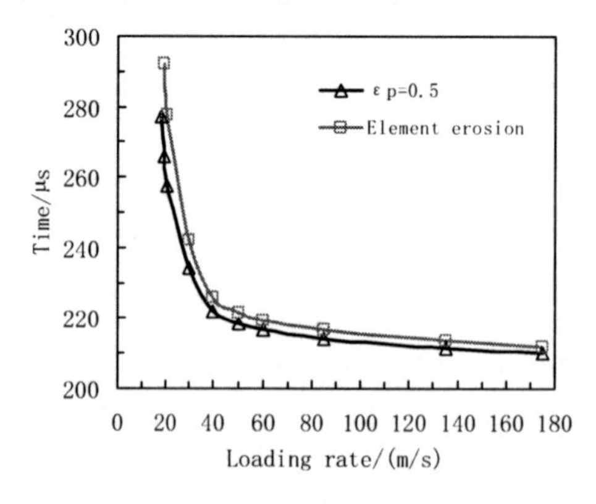

Figure 5: Initiation time of ε_p =0.5 and element erosion of element 22832 versus loading rates.

3.2 History of rising-temperature in the specimen

In this section the rising-temperature in specimen will be discussed. According to the deformations of specimen in the simulations [4], the case of V_0=20.5m/s is selected as an example, and the transverse zone and longitudinal zone shown in Figure 2 are mainly observed. Temperature distribution of the analytical zone is shown in Figure 6, where (a) is temperature curve of the transverse zone and (b) that of the longitudinal zone, in which the eroded elements are no longer shown.

From Figure 6(a) it can be seen that when the specimen deforms under V_0 =20.5m/s, temperature rising in the transverse zone mainly centralizes in a narrow band which consists of the Shear Zone and its surrounding regions (7.5mm $<X<$ 8.5mm). Before the time t=210.5μs, no obvious temperature rising occurs in the specimen; then temperature in the Shear Zone and its surrounding regions begins to rise; before t=280μs, temperature in the zone within a distance of 0.5mm from the centre of Shear Zone rises continuously, and temperature rising in the Shear Zone is faster than that in its surrounding regions. A gauss distribution of temperature is formed. After t=280μs, temperature in the Shear Zone rises rapidly and reaches about 700K when the element erodes, whereas temperature in the surrounding regions almost no longer change. Furthermore, in the whole process of deformation, there is no obvious temperature rising in the regions out the distance of 0.5mm from the Shear Zone. Thus, the deformation of specimen confirms the qualification of adiabatic shear deformation well.

Correspondingly, it can be found from Figures 6 (b, c) that, when t=210.5μs, temperature rising initiates in the two ends of Shear Zone, then the zone with temperature rising extends towards the inner regions of Shear Zone. When t=218.5μs obvious temperature rising occurs in the whole Shear Zone, and subsequently the zone with temperature rising extends transversely and becomes wider. When t=280μs shear fracture occurs in the two ends of Shear Zone, and the transverse extension of the zone with temperature rising stops. When the two ends of Shear Zone fractures, its temperature rises rapidly to about 700K. The shear fracture follows to evolve towards the inner region of Shear Zone, and especially, a small zone with lower temperature of about 600K forms in the tip of the crack. With the evolution of adiabatic shear fracture and the spread of the high-temperature zone, temperature in the whole Shear Zone rises rapidly. At this time, the distribution of temperature in the Shear Zone behaves as higher in the two ends and lower in the inner. Moreover, the width of high-temperature zone almost doesn't change in the whole process of deformation.

Within the loading rate range of V_0 <85m/s, the deformation of specimen is similar to that of the case V_0=20.5m/s. Corresponding to different loading rates, the fracture in specimen always extends from the two ends of Shear Zone to its centre. Besides, material temperature always rises to about 700K before it fractures. However, with increasing the loading rate, the initiation time of temperature rising in specimen moves up quickly, and the initiation of shear fracture in the Shear Zone becomes earlier. Also the duration of shear fracture shortens rapidly. Moreover, within this loading rate range, the width of the temperature-rising zone is almost the same as that at loading rate V_0=20.5m/s.

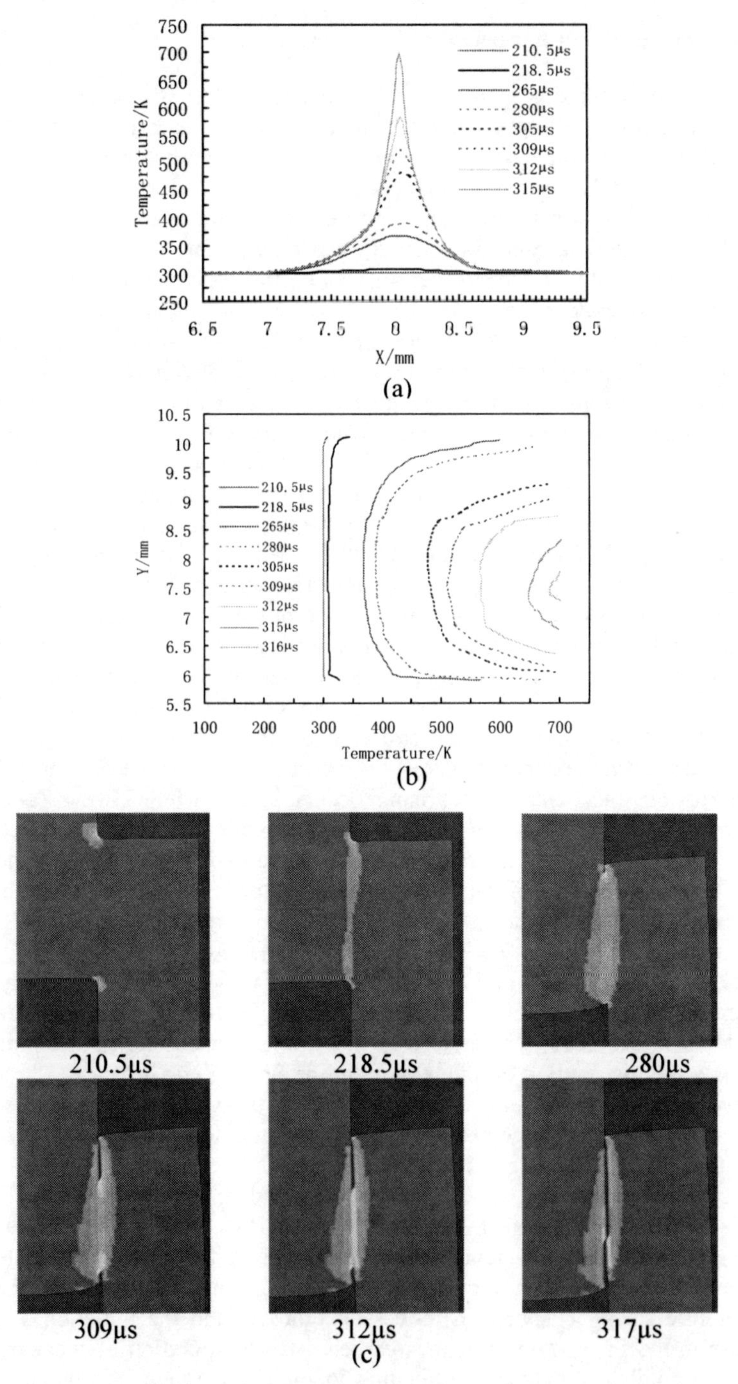

Figure 6: Temperature distribution and deformation evolution of the analytical zone under V_0 =20.5m/s.

When V_0 exceeds 85m/s, the Cap of specimen yields and becomes stub before the fracture of Shear Zone (seen in [4]), also the stub Cap compresses the Lap transversely. Before the initiation of ASB, the temperature rising in the Shear Zone and its surrounding regions is higher than that under loading rates of V_0 <85m/s, and temperature of the Cap also rises rapidly because of its intense plastic deformation. Therein the zone with high temperature is a little wider than that under loading rates of V_0<85m/s.

3.3 Distributions of stress, strain and temperature in the tip of ASB

Kuriyama and Meyers [7] predicted that there is an unstable zone with a width of about 5μm, high strain and comparatively low stress in the tip of ASB, and ASB extends following it. Figures 6(b, c) show that ASB in the specimen is formed due to the spread of a high-temperature zone from the ends of Shear Zone towards its centre. Thereinafter the distributions and evolutions of stress, strain and temperature in the tip of shear crack will be analyzed in detail.

The case of V_0=60m/s is selected as an example, and the contours of effective stress σ_{eff}, effective plastic strain ε_p and temperature T around the crack tip of Shear Zone at t=225.5μs are shown in Figure 7. From the figure it can be seen that the effective stress, effective plastic strain and temperature in the Shear Zone are all abruptly higher than that in other regions of the specimen. In the tip of the shear crack there assuredly exists a small zone with higher effective plastic strain and higher temperature than that in other parts of the Shear Zone, and the characteristic scale of this small zone is almost the same as the size of meshes (~30μm). However, there is no obvious difference of the effective stress in the Shear Zone. Moreover, simulations show that before the initiation of fracture in the specimen, effective plastic strain in the two ends of Shear Zone is already obviously higher than that in other regions. After the shear fracture initiates, a small zone with higher temperature and higher effective plastic strain always exists in the front of the shear crack until the Shear Zone fractures thoroughly. The smallest mesh in the simulations is about 30μm, which is far larger than 5μm, and thus it may be too large to show the lower stress zone in the tip of the shear crack. However, ASB forms following the spread of the unstable zone with higher temperature and higher strain in the front of the shear crack, which are absolutely consistent with Kuriyama and Meyers [7].

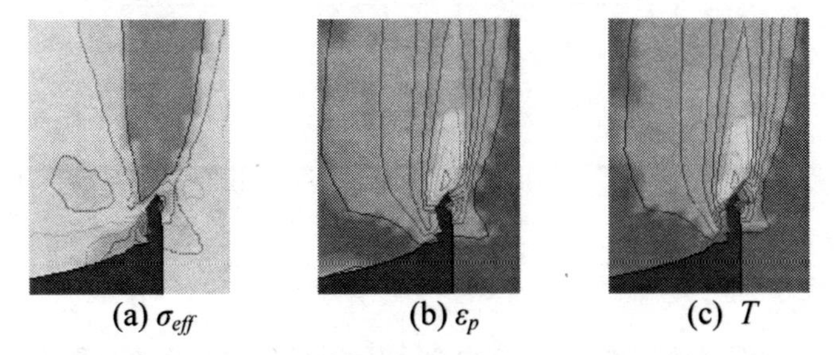

(a) σ_{eff} (b) ε_p (c) T

Figure 7: Contours of effective stress, effective plastic strain and temperature in the Shear Zone in case of V_0=60m/s and t=225.5μs.

3.4 Spread speeds of ASB and fracture

According to the analysis in Section 3.1, the duration of fracture in the Shear Zone shortens with increasing the loading rate, and also the element erodes rapidly after the initiation of ASB as shown in Figure 5. Therein the spread speed of ASB in the Shear Zone increases with increasing the loading rate. Supposing the time of ε_p =0.5 in elements 22832 and 22300, as shown in Figure 2, as the instant initiation of ASB in the end of Shear Zone, and that the ASB runs through all the Shear Zone, respectively, and thus the interval between these two instant time is the duration of ASB spreading in the Shear Zone. Likewise, the time interval between the erosion of these two elements is the duration of spread of fracture in the Shear Zone. Thus the average spread speeds of ASB and element erosion versus the loading rates could be calculated integrated with the length of Shear Zone, respectively, as shown in Figure 8.

Figure 8 shows that both the average speeds increase with increasing the loading rate, and the increase of spread speed of element erosion is quicker than that of ASB. Corresponding to comparatively high loading rates, the average spread speeds of ASB and element erosion approaches to a saturated value of 200m/s and 235m/s, respectively. There exists two phases in the spread of ASB according to Lebouvier and Lipinski [8], i.e., the spread speed of ASB depends linearly and intensively on the loading rate when the loading rate is comparatively low, while under a higher loading rate, the spread speed of ASB increases little and approaches to a saturated value. Figure 8 consists with Lebouvier and Lipinski [8] well.

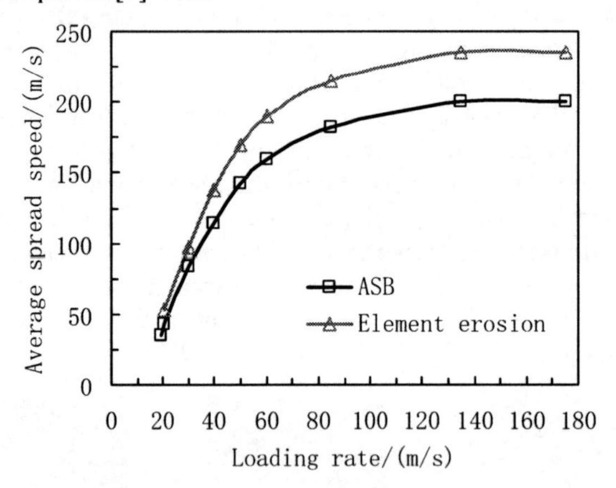

Figure 8: Average spread speeds of ASB and element erosion *vs* loading rates.

3.5 The width of ASB and its type

This section discusses the dependence of the width of ASB on the loading rate. The temperature distributions of the transverse analytical zone at the initiation of erosion of element 22300 under different loading rates are shown in Figure 9.

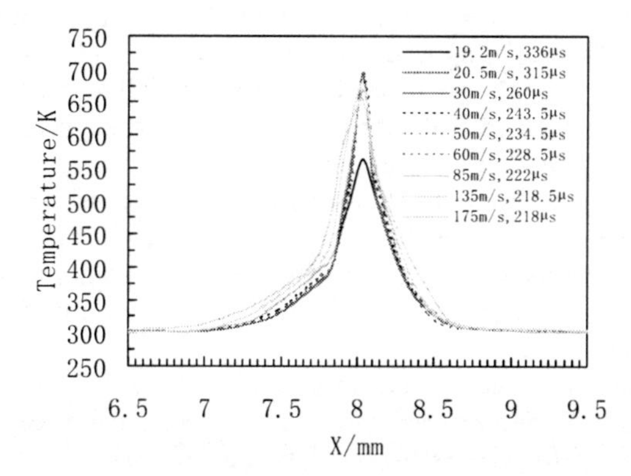

Figure 9: Temperature distributions of the transverse analytical zone at time of erosion initiation of element 22300 under different loading rates.

When V_0=19.2m/s, because the bars stop to compress the specimen after the shear fracture occurs in the ends of Shear Zone, the temperature of element 22300 only reaches about 550K and no longer rises up; while if V_0 ≥20.5m/s, temperature of the element always rises rapidly to about 700K before it erodes. Besides, in the loading rate range of 20.5m/s< V_0 <85m/s, the width of the high temperature zone is almost the same. Integrated with the temperature distributions and the specimen deformations, it can be found that the width of ASB is about 70μm and it approximates to the width of Shear Zone.

However, while if V_0≥85m/s, the Cap of specimen becomes stub (seen in [4]), and severe plastic deformation of the Cap leads the obvious temperature rising in specimen rather than the Shear Zone, and thus the zone with high temperature becomes wider. Essentially in that case, the deformation of specimen no longer satisfies the condition of pure shear, so the enlargement of high-temperature zone doesn't mean that the ASB becomes wider. Therein the width of ASB in the specimen is almost independent of the loading rate.

In general, it is supposed that when temperature achieves a value of T=0.4T_m, where T_m is the melting temperature of metal material, a second crystallization will occur in the material and the ASB will transform to a transformed band from deformed band. In other words, if T<0.4T_m, the corresponding ASB is a deformed band. The melting temperature of 921A steel is 1765K, and 0.4T_m =706K correspondingly. Regarding all of the loading rates in the simulations, the maximum temperature in ASB is always less than 706K, thus it can be concluded that transformed band never occurs in the hat-shaped specimen during its deformation process, i.e. corresponding ASBs are all deformed bands.

4 Conclusions

Integrated with relative tests, the adiabatic shear behaviour of 921A steel pure shear hat-shaped specimen is analyzed by means of the numerical SHPB tests.

When the specimen is compressed, the temperature rising mainly centralizes in the Shear Zone. If the loading rate is comparatively low (V_0 <18 m/s), adiabatic shear deformation doesn't occur in the specimen; while if $V_0 \geq 18$ m/s, ASB begins to initiate in the two ends of Shear Zone. When V_0 increases to 19.2 m/s, adiabatic shear deformation has run through all the Shear Zone; especially if $V_0 \geq 20.5$ m/s, the Shear Zone fractures thoroughly because of adiabatic shear deformation. ASB always develops from the two ends of Shear Zone to its centre. There is a small unstable zone with higher temperature and higher strain in the tip of shear crack, and it extends towards the centre of Shear Zone along with the evolution of fracture.

Moreover, both average spread speeds of ASB and fracture increase with increasing the loading rate, and approach to a respective saturated value. Besides, the spread speed of fracture is higher than that of ASB. The width of ASB is little dependent on the loading rate and equals to about 70μm, which approximates to the width of Shear Zone. All ASBs corresponding to various loading rates in this paper are deformed bands.

References

[1] Meyers, M.A., Meyer, L.W., Beatty, J., et al. High strain, high strain-rate deformation of copper. *Proc. of "Shock waves and high-strain-rate phenomena in materials"*, eds. M.A. Meyers, L.E. Murr, K.P. Staudhammer and M. Dekker. USA: New York, pp. 529-542, 1992.

[2] Chen, G., Chen, X.W., Pan, X.X., et al. Numerical simulation and experiments on the dynamic deformation of 921A steel hat-specimen. *Transactions of Beijing Institute of Technology*, 29(Suppl. 1), pp. 106-110, 2009. (in Chinese)

[3] Johnson, G.R., Cook, W.H. A constitutive model and data for metals subjected to large strains, high strain rates and high temperatures. *Proc. of the 7th International Symposium on Ballistics*. Netherlands: The Hague, pp. 541-552, 1983.

[4] Li, J.C., Chen, X.W., Chen, G. Numerical simulations of SHPB test using 921A steel pure shear hat-shaped specimens. *Proc. of 9th International Conference on the Mechanical and Physical Behaviour of Materials under Dynamic Loading (DYMAT 2009)*. Brussels, Belgium, pp.1685-1692, Sept. 7-11, 2009.

[5] Wright, T.W., Batra, R.C. The initiation and growth of adiabatic shear bands. *Int J Plasticity*, 1, pp. 205-212, 1985.

[6] Wright, T.W., Ockendon, H. A model for fully formed shear bands. *J Mech Phys Solids*, 40(6), pp. 1217-1226, 1992.

[7] Kuriyama, S., Meyers, M.A. Numerical modeling of the propagation of an adiabatic shear band. *Metallurgical Trans A*, 17A, pp. 443-449, 1986.

[8] Lebouvier, A.S., Lipinski, P. Numerical study of the propagation of an adiabatic band. *J Phys IV France*, 10, pp. 403-408, 2000.

Author Index

Tribology and Design

Edited by: ***M. HADFIEL,*** *Bournemouth University, UK,* ***C.A. BREBBIA****, Wessex Institute of Technology, UK and* ***J. SEABRA,*** *University of Porto, Portugal*

Tribology and Design 2010 is the 3rd International Conference in a series that originated with two meetings held at Bournemouth University, UK in 2005 and 2007. The Tribology and Design Conference explores the role of technology and design in the broader sense. It brings together colleagues from different disciplines interested in problems of surface interaction and design. The applications covered range from geomechanics to nano problems and from sustainability issues to advanced materials. It has never been so important for the designer to consider product and system durability in relation to reliability and sustainability issues. The topics for discussion also cover studies of tribology in nature and how the resulting lessons can be applied by the designers. Another important theme is the application of tribology in biomechanics, a field in which surface mechanics in general is of fundamental importance.

This book contains the papers presented at the Third International Conference, arranged into the following subject areas: Tribology in Space Applications; Reliability in Product Design; Nano-Tribology and Design; Tribology Under Extreme Conditions; Tribology in Geo-Mechanics; Energy Applications; Surface Measurements; Tribology in Biomechanics; Life-Oriented Design; Tribology and Nature; Design Tools; Surface Engineering; Lubricant Design; Test Methods; Advanced Materials; Analytical Studies; Sustainability and Tribology; Product Reliability; Corrosion Problems

WIT Transactions on Engineering Sciences, Vol 66
ISBN: 978-1-84564-440-6 eISBN: 978-1-84564-441-3
2010 304pp apx £115.00

WIT*Press*
Ashurst Lodge, Ashurst, Southampton,
SO40 7AA, UK.
Tel: 44 (0) 238 029 3223
Fax: 44 (0) 238 029 2853
E-Mail: witpress@witpress.com

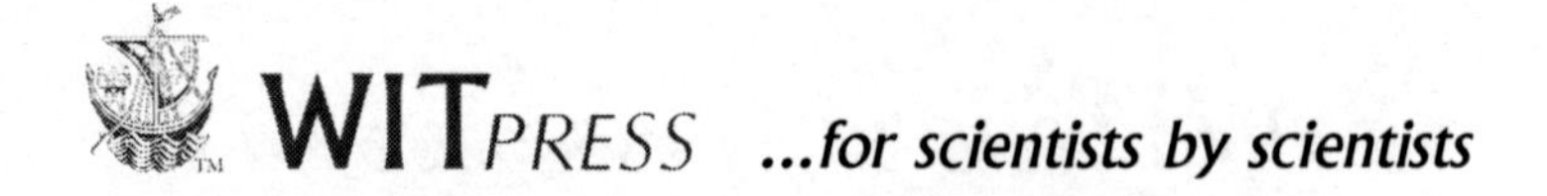

High Performance Structures and Materials V

Edited by: ***W.P. De WILDE****, Vrije Universiteit Brussel, Belgium and* ***C.A. BREBBIA****, Wessex Institute of Technology, UK,* ***Ü. MANDER****, University of Tartu, Estonia*

This book contains the edited papers presented at the Fifth International Conference on High Performance Structures and Materials. The Conference addresses issues involving advanced types of structures, particularly those based on new concepts or new types of materials. This responds to the need to develop a generation of new materials that are suitable for high performance structures which can easily resist a wide range of external stimuli and react in a non-conventional manner.

The papers presented are arranged into the following subject areas: Material Characterisation; High Performance Concretes; Composite Materials and Structures; Natural Fibre Composites; Adhesion and Adhesives; Damage and Fracture Mechanics; Structural Dynamics and Impact Behaviour; Optimal Design; Behaviour of FRP Structures; Structural Characterisation; Durability and Sustainability of Materials and Structures; Timber Structures; Lightweight Structures; Biomass Composites.

WIT Transactions on The Built Environment, Vol 112
ISBN: 978-1-84564-464-2 eISBN: 978-1-84564-465-9
Forthcoming apx 500pp apx £190.00

We are now able to supply you with details of new WIT Press titles via E-Mail. To subscribe to this free service, or for information on any of our titles, please contact the Marketing Department, WIT Press, Ashurst Lodge, Ashurst, Southampton, SO40 7AA, UK
Tel: +44 (0) 238 029 3223
Fax: +44 (0) 238 029 2853
E-mail: marketing@witpress.com

WIT*PRESS* ...*for scientists by scientists*

Elements of Plasticity

Theory and Computation

***I. DOLTSINIS**, University of Stuttgart, Germany*

"...clearly written, and it is easy to follow the text...[the book] gives a good understanding of the inelastic behaviour of materials. The examples chosen are good for illustration... Elements of Plasticity will certainly benefit graduate students, engineers, and consultants working on the numerical aspects of the inelastic behaviour of solids and structures."

APPLIED MECHANICS REVIEWS

In this revised second edition, Elements of Plasticity continues to bridge the gap between classical theory and modern computational techniques and does so by focussing on the most important elements of theory and computation using matrix notation.

Some complementary aspects of creep and viscoplasticity are considered and a number of applications from engineering practice are used to demonstrate the use of computational techniques. Practicing engineers, consultants and graduate students from civil, mechanical, automotive and aerospace engineering will find this revised edition indispensable.

Series: High Performance Structures and Materials, Vol 1
ISBN: 978-1-84564-428-4 eISBN: 978-1-84564-429-1
2nd Edition 2010 328pp £125.00

WIT eLibrary

Home of the Transactions of the Wessex Institute, the WIT electronic-library provides the international scientific community with immediate and permanent access to individual papers presented at WIT conferences. Visitors to the WIT eLibrary can freely browse and search abstracts of all papers in the collection before progressing to download their full text.

Visit the WIT eLibrary at
http://library.witpress.com

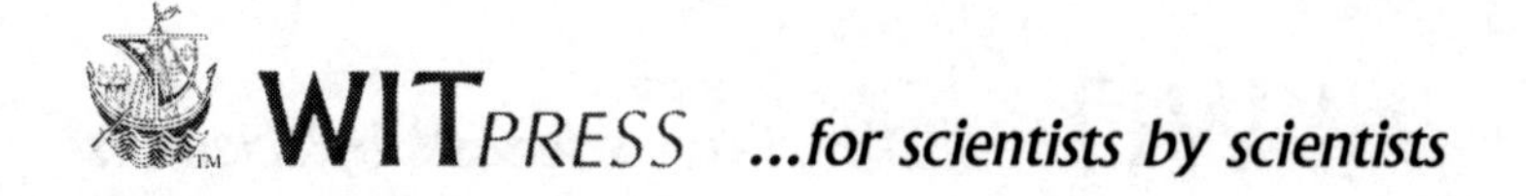

Influence Function Approach

Selected Topics of Structural Mechanics

Edited by: ***Y. MELNIKOV**, Middle Tennessee State University, USA*

Structural mechanics is the study of the effects that forces of different physical origin (mechanical, thermal, magnetic and so on) produce on elements of structures such as cables, pillars, beams, plates and shells.

This text represents the first ever attempt to include in book format a number of standard problems from structural mechanics. It is innovative in treating each problem by means of a single mathematical approach (the influence function method).

The book covers only a limited number of topics from the undergraduate/ graduate course on structural mechanics and as such is intended as a supplementary, rather than a primary, text. It can also be used in other core courses in the mechanical/civil engineering curriculum, as well as in the applied or industrial mathematics curriculum. It can even be adapted as a graduate text for a course on computational mechanics, where a student could use a strong mathematical background in modelling and solving actual problems from mechanics.

ISBN: 978-1-84564-129-0 eISBN: 978-1-84564-301-0
2008 400pp £132.00

All prices correct at time of going to press but subject to change.
WIT Press books are available through your bookseller or direct from the publisher.